Springer Transactions in Civil and Environmental Engineering

Editor-in-Chief

Springer Transactions in Civil and Environmental Engineering (STICEE) publishes the latest developments in Civil and Environmental Engineering. The intent is to cover all the main branches of Civil and Environmental Engineering, both theoretical and applied, including, but not limited to: Structural Mechanics Steel Structures Concrete Structures Reinforced Cement Concrete Civil Engineering Materials Soil Mechanics Ground Improvement Geotechnical Engineering Foundation Engineering Earthquake Engineering Structural Health and Monitoring Water Resources Engineering Engineering Hydrology Solid Waste Engineering Environmental Engineering Wastewater Management Transportation Engineering Sustainable Civil Infrastructure Fluid mechanics Pavement engineering Soil dynamics Rock mechanics Timber engineering Hazardous waste disposal Instrumentation and monitoring Construction management Civil engineering construction Surveying and GIS Strength of Materials (Mechanics of Materials) Environmental geotechnics Concrete engineering timber structures Within the scopes of the series are monographs, professional books or graduate textbooks, edited volumes as well as outstanding PhD theses and books purposely devoted to support education in mechanical engineering at undergraduate and graduate levels.

More information about this series at http://www.springer.com/series/13593

B. V. Venkatarama Reddy ·
Monto Mani · Pete Walker
Editors

Earthen Dwellings and Structures

Current Status in their Adoption

Springer

Editors
B. V. Venkatarama Reddy
Department of Civil Engineering
Indian Institute of Science
Bangalore, Karnataka, India

Monto Mani
Centre for Sustainable Technologies
Indian Institute of Science
Bangalore, Karnataka, India

Pete Walker
Department of Architecture and Civil Engineering
University of Bath
Bath, UK

ISSN 2363-7633 ISSN 2363-7641 (electronic)
Springer Transactions in Civil and Environmental Engineering
ISBN 978-981-13-5882-1 ISBN 978-981-13-5883-8 (eBook)
https://doi.org/10.1007/978-981-13-5883-8

Library of Congress Control Number: 2018966400

This Springer imprint is published by the registered company Springer Nature Singapore Pte Ltd.
The registered company address is: 152 Beach Road, #21-01/04 Gateway East, Singapore 189721, Singapore

Technical Review Committee

Preface

The word Earth has many interpretations; as a planet that sustains all life and ecosystems, terrestrially it is the land mass we live on; as a material resource, it represents the complex mixture of clay, sand, minerals and nutrients that sustains agriculture, subterranean biodiversity and human civilisations; and spiritually, it goes full circle to establish our primal oneness with the planet. To all native civilisations, the earth is all encompassing and revered spiritually; to all modern civilisations, the earth supports all the ecosystem services that drives development, industrialisation and economic growth. Through millennia, the earth has served to house civilisations across the world, in nearly all climates and terrains, and has sustained the ravages of time as a durable and sustainable building material. In modern interpretations, the earth carries a nearly zero-carbon footprint, negligible life-cycle impact and complete recyclability with no end of life (disposal). It is accessible in its diversity to nearly all civilisations across the world, where unique construction techniques have evolved from cob wall, wattle and daub, to modern rammed earth building technologies.

However, never in recorded history have we faced with the challenges to sustainability such as now, with buildings driving more than half of global energy and resource consumption and CO_2 emissions. Modern pursuits, driven by high energy efficiency in buildings, are also proving to be counterintuitive with rebound effects yielding an exponentially higher net-energy consumption, rather than energy saving. Recent reports on measures to mitigate climate change have revealed the need to reduce building energy and resource footprint by half in the coming decades, if global warming temperatures are to be kept below 2 °C.

It is but timely that earthen constructions are revisited for their potential to meet the growing demand for modern housing, relieve the increasing burden of urbanisation, and as an alternative material which is environmentally benign, renewable, globally accessible and affordable. Scientific research needs to step in to reinforce modern faith on the durability, structural performance, climate responsiveness and best building practices in the adoption of earthen construction to suit modern lifestyles. Researchers working on various facets of earthen construction, ranging from its cultural heritage to climatic and structural performance, are few and scattered. The current volume is a compilation of well-written diverse articles, with earth being the common connecting theme, from researchers worldwide exploring the earth for sustainable construction.

Bangalore, India B. V. Venkatarama Reddy
Bangalore, India Monto Mani
Bath, UK Pete Walker

Acknowledgements

We gratefully acknowledge the authors for their contribution and also extend our sincere appreciation for the support extended by a large panel of reviewers for critically reviewing all the papers and enhancing the quality of the book.

This book is an important outcome of the International Symposium of Earthen Structures 2018 (ISES 2018). The prime mover for organising ISES 2018 has been the UK-India collaborative (UKIERI) project (UKIERI 2016-17-063) on developing earth-based building products utilising solid wastes. On behalf of the organisers, we take immense pride in expressing our gratitude for the generous financial assistance from the UKIERI scheme.

Special thanks also to Springer for taking this up as an edited publication and making it accessible worldwide.

B. V. Venkatarama Reddy
Monto Mani
Pete Walker

Contents

Part VII Codes and Design Guidelines

Editors and Contributors

About the Editors

B. V. Venkatarama Reddy is a Professor in the Department of Civil Engineering and the Chairman of the Centre for Sustainable Technologies in Indian Institute of Science (IISc), Bangalore. He holds a doctorate in structural engineering from IISC, where he has been working since 1979. He has a strong interest in structural masonry, mechanics of materials, energy in buildings, green buildings, low carbon construction materials, and recycling of solid wastes into construction materials, and has jointly authored a book on alternative building materials and technologies, edited a couple of books on masonry, and developed a large number of low embodied carbon construction materials and techniques. Prof Reddy has served as consultant for several innovative projects on alternative building technologies and as a member of several technical committees in Bureau of Indian Standards and other government agencies. He was the DAAD Visiting professor at Bauhaus University, Germany and a Visiting Professor at University of Bath, UK. He is a member of many professional bodies including the Indian Concrete Institute, ISCEAH, IC-NOCMAT, British Masonry Society and RILEM.

Monto Mani is an Associate Professor in the Centre for Sustainable Technologies and Centre for Product Design & Manufacturing in IISc Bangalore. He is an Architect, with a master's degree in Civil Engineering. As part of his PhD from Indian Institute of Technology Madras, he developed a systems-framework identifying societal attitude as a critical determinant of sustainability. Dr Monto's research deals with the interdisciplinary domain of sustainability science, specifically focusing on its theoretical basis, and application in the architecture (buildings) and design. His Sustainability and Design lab (SuDesi) comprises multidisciplinary researchers working in diverse areas of sustainability. He is best known for his contributions in functional performance of buildings, PV performance and building integration (BIPV) and sustainability evaluation in design. He has extensive interdisciplinary publications pertaining to sustainability.

Prof. Pete Walker is a Fellow of the Institution of Structural Engineers (UK), and has been Director of the BRE Centre for Innovative Construction Materials at the University of Bath since 2006. He has published over 250 articles on his research work on natural building materials, covering earthen construction, crop-based building materials and timber engineering. He teaches materials and structural engineering to undergraduate and postgraduate civil engineering and architecture students at the University of Bath.

Contributors

Holur Narayanaswamy Abhilash Indian Institute of Science, Bangalore, India

Rodrigo Amaral Rocha Olnee Constructions and Earth House Australia, Melbourne, Australia

Federico C. Antico Universidad Adolfo Ibáñez, Viña del Mar, Chile

Gerardo Araya-Letelier Pontificia Universidad Católica de Chile, Santiago, Chile

K. P. I. E. Ariyaratne Department of Civil Engineering, University of Moratuwa, Moratuwa, Sri Lanka

K. Arnold Biosystems Engineering, University of Manitoba, Winnipeg, Canada

K. Arpitha ACS College of Engineering, Bangalore, India

Charles Augarde Department of Engineering, Durham University, Durham, UK

N. C. Balaji Department of Civil Engineering, The National Institute of Engineering, Mysore, India

Danijela Bojic ZRS Ingenieure, Berlin, Germany

Renee M. Borges Centre for Ecological Sciences, Indian Institute of Science, Bangalore, India

Mohamed Boutouil ESITC Caen, Epron, France

Agostino Walter Bruno Laboratoire SIAME, Fédération IPRA, Université de Pau et des Pays de l'Adour, Anglet, France

Valeria Cascione Department of Architecture and Civil Engineering, University of Bath, Bath, UK

Sridevi Changali Masons Ink, Bangalore, India

Sujoy Chaudhury Center for Sustainable Solutions, Kolkata, India

T. Colinart Université Bretagne Sud, Lorient, France

Jose Concha-Riedel Universidad Adolfo Ibáñez, Santiago, Chile

Alessia Cuccurullo Department of Engineering, Durham University, Durham, UK
Laboratoire SIAME, Fédération IPRA, Université de Pau et des Pays de l'Adour, Anglet, France

Aritra Das Centre for Neuroscience, Indian Institute of Science, Bangalore, India

Sérgio Ricardo Honório de Assis Federal University of Paraiba, PPGECAM, João Pessoa, Brazil

Aline Figueirêdo Nóbrega de Azerêdo Federal Institute of Education, Science Technology of Paraiba, João Pessoa, Brazil

Givanildo Alves de Azerêdo Department of Civil and Environmental Engineering, Federal University of Paraiba, João Pessoa, Brazil

A. Hellouin de Ménibus Eco-Pertica, Hôtel Buissonnet, Perche-En-Nocé, France
Association Nationale Des Chanvriers en Circuits Courts, Bouquet, France

Kristopher J. Dick Biosystems Engineering, University of Manitoba, Winnipeg, Canada

Mark Evernden Department of Architecture & Civil Engineering, University of Bath, Bath, UK

Antonin Fabbri LTDS, UMR5513 CNRS, ENTPE, Université de Lyon, Vaulx-en-Velin, France

Paulina Faria CERIS and Civil Engineering Department, NOVA University of Lisbon (FCT NOVA), Caparica, Portugal

Patrick Fontana RISE Research Institutes of Sweden, Göteborg, Sweden

Domenico Gallipoli Laboratoire SIAME, Fédération IPRA, Université de Pau et des Pays de l'Adour, Anglet, France

Andres Glade Universidad Adolfo Ibáñez, Viña del Mar, Chile

João Gomes Chemical Engineering Department, Lisbon Engineering Superior Institute (ISEL), Lisbon Polytechnic Institute (IPL), Lisbon, Portugal

Maria Idália Gomes Civil Engineering Department, Lisbon Engineering Superior Institute (ISEL), Lisbon Polytechnic Institute (IPL), Lisbon, Portugal

K. Gourav Department of Civil Engineering, The National Institute of Engineering, Mysore, India

E. Hamard IFSTTAR, MAST, Bouguenais, France

Andrew Heath Department of Architecture & Civil Engineering, University of Bath, Bath, UK

Johannes Hoppe Bundesanstalt für Materialforschung und -prüfung (BAM), Berlin, Germany

Paul Hughes Department of Engineering, Durham University, Durham, UK

K. S. Jagadish Department of Civil Engineering, Indian Institute of Science, Bangalore, India

Chintha Jayasinghe Department of Civil Engineering, University of Moratuwa, Moratuwa, Sri Lanka

M. T. R. Jayasinghe Department of Civil Engineering, University of Moratuwa, Moratuwa, Sri Lanka

Shiva Ji Indian Institute of Technology Hyderabad, Hyderabad, India

P. T. Jitha Department of Civil Engineering, BMS College of Engineering, Bangalore, India

H. M. Jonkers Faculty of Civil Engineering and Geosciences, Delft University of Technology, Delft, The Netherlands

T. K. Jyothi Department of Civil Engineering, Government Engineering College, Ramanagaram, India

Caroline Kaiser ZRS Ingenieure, Berlin, Germany

Ramesh K. Kandasami Schofield Centre, West Cambridge Site, University of Cambridge, Cambridge, UK

Inayath Kharoti Department of Architecture and Civil Engineering, University of Bath, Bath, UK

Amit C. Kinjawadekar Faculty of Architecture, MIT, MAHE Manipal, Manipal, India

Andrea Klinge ZRS Architekten, Berlin, Germany

Tejas Kotak Hunnarshala, Foundation for Building Technology and Innovation, Bhuj District, Gujarat, India

Timothy J. Krahn Building Alternatives Inc., Codrington, ON, Canada

Y. Kulshreshtha Faculty of Civil Engineering and Geosciences, Delft University of Technology, Delft, The Netherlands

B. Sunil Kumar Cushman & Wakefield India Private Limited, Bangalore, India

Malo Le Guern ESITC Caen, Epron, France

T. Lecompte Université Bretagne Sud, Lorient, France

H. Lenormand UniLaSalle, Mont-Saint-Aignan, France

R. Lepakshi Department of Civil Engineering, Indian Institute of Science, Bangalore, India

P. Logan Biosystems Engineering, University of Manitoba, Winnipeg, Canada

Hasna Louahlia LUSAC—Caen Normandy University, Saint-Lô, France

George Lucas Department of Engineering, Durham University, Durham, UK

Monto Mani Centre for Sustainable Technologies, Indian Institute of Science, Bangalore, India

Alastair Marsh Department of Civil Engineering, University of Bath, Bath, UK

Daniel Maskell Department of Architecture and Civil Engineering, University of Bath, Bath, UK

Fionn McGregor LTDS, UMR5513 CNRS, ENTPE, Université de Lyon, Vaulx-en-Velin, France

Pedro Henryque Melo de Oliveira UFU, Uberlândia, Minas Gerais, Brazil

Jean-Claude Morel Coventry University, Coventry, UK

N. J. A. Mota Faculty of Architecture and the Built Environment, Delft University of Technology, Delft, The Netherlands

Sravan Muguda Department of Engineering, Durham University, Durham, UK
Laboratoire SIAME, Fédération IPRA, Université de Pau et des Pays de l'Adour, Anglet, France

Tejas G. Murthy Department of Civil Engineering, Indian Institute of Science, Bangalore, India

K. S. Nanjunda Rao Department of Civil Engineering, Indian Institute of Science, Bangalore, India

V. Nikhil Centre for Sustainable Technologies (CST), Indian Institute of Science, Bangalore, India

M. Nikhilash Centre for Sustainable Technologies (CST), Indian Institute of Science, Bangalore, India

Pascaline Patureau Department of Chemistry, University of Bath, Bath, UK

Rosie Paul CRAterre-ENSAG, Grenoble, France

G. S. Pavan Department of Civil Engineering, Indian Institute of Science, Bangalore, India

Jerome Payet Cycleco, Ambérieu-En-Bugey, France

Celine Perlot Laboratoire SIAME, Fédération IPRA, Université de Pau et des Pays de l'Adour, Anglet, France

Tuan Anh Phung ESITC Caen, Epron, France

J. Pieniuta Biosystems Engineering, University of Manitoba, Winnipeg, Canada

R. K. Preethi Department of Civil Engineering, Indian Institute of Science, Bangalore, India

Ravi Mokashi Punekar Indian Institute of Technology Guwahati, Guwahati, India

S. Raghunath Department of Civil Engineering, BMS College of Engineering, Bangalore, India

R. V. Ranganath Department of Civil Engineering, BMS College of Engineering, Bangalore, India

Ursula Reidel Sika S.A. Chile, Santiago, Chile

Matthias Richter Bundesanstalt für Materialforschung und -prüfung (BAM), Berlin, Germany

Eike Roswag-Klinge ZRS Architekten, Berlin, Germany

Lucas Miranda Araújo Santos Federal University of Paraiba, PPGECAM, João Pessoa, Brazil

Andy Shea Department of Architecture and Civil Engineering, University of Bath, Bath, UK

V. Suresh Department of Civil Engineering, Indian Institute of Science, Bangalore, India

S. N. Ullas Centre for Sustainable Technologies, Indian Institute of Science, Bangalore, India

M. C. M. van Loosdrecht Faculty of Applied Sciences, Delft University of Technology, Delft, The Netherlands

Kiran Vaghela Hunnarshala, Foundation for Building Technology and Innovation, Bhuj, Gujarat, India

P. J. Vardon Faculty of Civil Engineering and Geosciences, Delft University of Technology, Delft, The Netherlands

B. V. Venkatarama Reddy Centre for Sustainable Technologies (CST), Indian Institute of Science, Bangalore, India
Department of Civil Engineering, Indian Institute of Science, Bangalore, India

T. Vinceslas Université Bretagne Sud, Lorient, France

H. G. Vivek Prasad Department of Construction Technology and Management, Sri Jayachamarajendra College of Engineering, Mysore, India

Pete Walker BRE Centre for Innovative Construction Materials, University of Bath, Bath, UK
Department of Architecture & Civil Engineering, University of Bath, Bath, UK

María J. Wiener Purdue University, West Lafayette, USA

Henry Wong LTDS, UMR5513 CNRS, ENTPE, Université de Lyon, Vaulx-en-Velin, France

Longfei Xu LTDS, UMR5513 CNRS, ENTPE, Université de Lyon, Vaulx-en-Velin, France

Nikita Zachariah Centre for Ecological Sciences, Indian Institute of Science, Bangalore, India

Mohammad Sharif Zami King Fahd University of Petroleum and Minerals, Dhahran, Saudi Arabia

Christof Ziegert ZRS Ingenieure, Berlin, Germany

Introduction

Earthen Structures

The earliest attempts by humans to build shelters included the use of earth/soil along with twigs/leaves and branches as basic building materials. More than a third of humanity dwells in earthen buildings even in today's twenty-first-century super modern world. Earthen structures are more climatically suited, environmentally benign, accessible and affordable to all and provide a very primal cultural connection with nature. This is valid, despite the diversity in culture, soil and natural conditions and climatic conditions. Civilisations have thrived in earthen construction even in extremes of climatic conditions. The earth that is accessible to diverse geographical locations is in itself diverse in its characteristics and represents a material that is inherently durable given their availability and occurrences despite millennia of climatic exposure and weathering. In our current pursuit of sustainable development, earthen structures hold enormous relevance and potential in providing solutions for environmentally friendly buildings that are energy efficient, comfortable, durable and recoverable/recyclable. The earth in the native cultures is associated with poverty, deprivation and underdevelopment, which in modern civilisations is associated with abundance, choice and wealth. Given the incessant demand for housing, earth holds immense potential as a sustainable material for the larger share of human society. Native cultures have always found a spiritual connection with the earth as a supporter of life (and dwellings). They have also developed a natural physiological resilience to withstand wider climatic variabilities moderated within earthen dwellings. The indoor air quality in naturally ventilated earthen dwellings is generally healthier than that found in conditioned buildings. Their acoustic performance is also superior to that of modern building materials. The design of earthen dwellings is generally organic with spatial inclusiveness that fosters greater social interaction in comparison with modern conditioned dwellings that tend to adopt a compartmentalised layout that inhibits social interaction. Faith in the appropriateness of earth as a climate responsive, durable and environmentally friendly building material needs to be restored, by revisiting them based on modern scientific scrutiny and the subsequent generation of codes and best practices.

Lack of tested earth-based construction practices, rapid urbanisation, changing lifestyles and increased adoption of (energy-intensive) modern construction materials have led to a steep decline in the adoption of traditional/vernacular earthen structures. Modern architecture is characterised by dwellings that are not climatically responsive, adopt exotic energy-intensive materials and rely heavily on (fossil fuel) energy for operation and maintenance. The inhabitants exhibit much lower physiological resilience to climatic variabilities, being completely habituated (and vulnerable) to artificially conditioned indoor environments.

Vernacular dwellings typically carry a low embodied energy less than 2 GJ/m^2, while modern dwelling can exceed 8 GJ/m^2. Consequently, operational energy (for comfortable indoor environments) is barely 1 GJ/household/year for vernacular dwellings and can exceed 30 GJ/household/year for modern dwellings. Accounting for modern transitions in vernacular dwellings (that house nearly 2 billion in China and India), the implication on resource and energy demand for housing is worrisome. These transitions, if not regulated appropriately, can further exasperate sustainability at the global scale, which is already threatened by energy- and resource-intensive lifestyles of industrialised regions. Materials derived from the earth, that are easily accessible, have the potential to provide for sustainable dwelling alternatives and dampen the otherwise insatiable reliance on energy-intensive materials (and lifestyle).

In the past, a variety of earth-based technologies and techniques have been adopted for building construction. Adobe masonry, cob walls, rammed earth, natural fibre-reinforced earth, wattle and daub, etc., are few of the traditional methods used for earthen construction. Limitations in the widespread adoption of earth-based techniques for buildings include lack of standardised engineering methodologies, loss of traditional (undocumented) skill and wisdom, poor seismic resistance, lack of strength upon saturation, poor resistance against rain impact, uncertified products, lack of sustained R&D efforts, insufficient education and training, poor regulatory mechanisms and the perceived stigma of poverty associated with earthen construction. Fortunately, interest in traditional and modern methods of earthen structures has been steadily growing as more sustainable and healthier buildings are sought globally. There is considerable interest in the adoption of earth-based materials such as stabilised earth blocks, rammed earth, fibre-reinforced earthen materials, cob walls and earthen mortars. Currently, there are focused R&D efforts in the areas of earthen materials, thermal performance of such materials and buildings, durability studies, standardisation of earthen building products, seismic response of earthen structural systems, knowledge dissemination, education and teaching, across the world.

The edited book is an amalgamation of diverse and interconnected topics on earthen structures, derived from peer-reviewed papers submitted to the International Symposium on Earthen Structures (ISES 2018). The book provides an in-depth analysis on various aspects of earthen structures, with science-based technical content on materials and technologies, structural design and seismic performance, durability, seismic response, climatic response, hygrothermal performance and durability, design and codes, architecture, heritage and conservation and technology dissemination. The book will be useful to architects, engineers, scientists, teaching professionals, construction professionals and students, providing a useful document on the current status and knowledge of earthen structures.

Part I
Earthen Materials and Technology

Chapter 1
Studies on Geopolymer-Based Earthen Compacts

R. K. Preethi and B. V. Venkatarama Reddy

1.1 Introduction

Geopolymer mechanism involves the silicates and aluminates in the presence of alkali to undergo the process of geopolymerisation. Geopolymer products are originated by poly-condensation of aluminosilicates with alkali-activating metals yielding polymeric Si–O–Al bonds (Davidovits 1999; Duxson et al. 2007; Provis 2014). Earlier geopolymer was named as "Gruntosilikat" and "Gruntocement-geocement" (Gluchovskij 1959). Sodium hydroxide (NaOH) or potassium hydroxide (KOH) along with sodium silicate solution is used as an alkali activator solution in preparing geopolymer products (Davidovits 1988, 1994). Hardening process of the geopolymers in the presence of alkali metals takes place at the temperatures between 25 and 90 °C. Curing the geopolymer specimens beyond 90 °C results in the dehydration which will lead to the formation of cracks in the specimens (Hardjito et al. 2003; Khale 2007; Rovnanik 2010; Heah and Kamarudin 2011; Slaty et al. 2013).

Cement is the most commonly and widely used binder material in the construction industry. To reduce the consumption of cement in the building industry, alkali-activated products (Geopolymers) are emerging as alternative binder materials. Replacing Portland cement with geopolymer binder as an alternative in the conventional concrete has been attempted (Rangan 2008a, b, 2009; Hardjito 2004; Kunal Kupawade Patil and Allouche 2013). Geopolymer binders are energy efficient as they result in reduced carbon emission (Mclellan et al. 2011).

In the manufacturing process of clay bricks, clay is subjected to high temperature (1000–1400 °C) where the clay mineral changes from its natural form to a stable form

R. K. Preethi (✉) · B. V. Venkatarama Reddy
Department of Civil Engineering, Indian Institute of Science, Bangalore, India
e-mail: preethirk@iisc.ac.in

B. V. Venkatarama Reddy
e-mail: venkat@iisc.ac.in

B. V. V. Reddy et al. (eds.), *Earthen Dwellings and Structures*,
Springer Transactions in Civil and Environmental Engineering,
https://doi.org/10.1007/978-981-13-5883-8_1

called mullite (Grim and Bradley 1940). Burnt clay bricks possess high embodied energy (Reddy and Jagadish 2003; Praseeda et al. 2015). Production of masonry units using Portland cement, autoclaving or firing at higher temperature results in higher amount of energy consumption and carbon emissions. Alkali activation of natural clays and natural soils is an alternative method in the manufacturing process of masonry units (Munoz et al. 2015; Maskell et al. 2014). The current study is focused on exploring geopolymer binder using natural soil and clay minerals for the manufacture of masonry units.

1.2 Scope of the Study and Experimental Programme

The scope of the present study included the utilisation of geopolymer binders in the manufacturing process of masonry units. The earlier studies have indicated the benefit of using the geopolymer binders in manufacturing the masonry units. An attempt was made to examine the wet compressive strength of the alkali-activated earthen compacts in the presence of ground granulated blast-furnace slag (GGBS) and fly ash materials, with various molar concentrations of NaOH solution.

Different mix proportions were considered for casting the specimens. One set of specimens were cast by varying the clay content in the mix. Additional source of silica and alumina materials such as GGBS and fly ash was also used in casting the specimens. Second set of specimens were cast using GGBS and fly ash with fixed

Table 1.1 Details of the experimental programme

Materials	Clay (%)	GGBS or fly ash (%)	NaOH		
			8 M	10 M	12 M
Kaolinite/ Montmorillonite mineral	10	0	√	√	√
	15	0	√	√	√
	20	0	√	√	√
		4	√	√	√
		8	√	√	√
		12	√	√	√
		15	√	√	√
Red soil	20	0	√	√	√
	30	0	√	√	√
		5	√	√	√
		10	√	√	√
		15	√	√	√
		30	√	√	√
	41	0	√	√	√

clay content in the mix. The experimental programme considered in the study is given in Table 1.1.

1.3 Materials Used in the Study

The materials used in the investigations include locally available soil, river sand, natural clay minerals (kaolinite and montmorillonite), ground granulated blast-furnace slag (GGBS) and fly ash. Laboratory grade sodium hydroxide (NaOH) with 99% purity was used in the study.

The lime reactivity of GGBS and fly ash was tested as per IS: 1727–2004 code guidelines; the results were 9.74 and 2.99 MPa, respectively. Figure 1.1 gives the grain size distribution curves of kaolinite, soil, sand, soil with 20% clay fraction and 20% kaolinite in the mix used. It was difficult to obtain grain size distribution curve for montmorillonite clay mineral using hydrometer analysis. The natural soil has 41% clay fraction (<2 μm) containing predominantly kaolinite clay mineral. Kaolinite clay mineral possesses clay size fraction of 54.69%. The clay fraction (<2 μm) of the soil mix with 20% clay content and that of mix with 20% kaolinite are 18.73 and 10.96%, respectively.

The chemical composition and physical properties of some of the materials used in the study are given in Table 1.2. The elemental composition was determined by energy-dispersive X-ray (EDX) spectroscopy. Silica (Si) and alumina (Al) are the major components present in the materials.

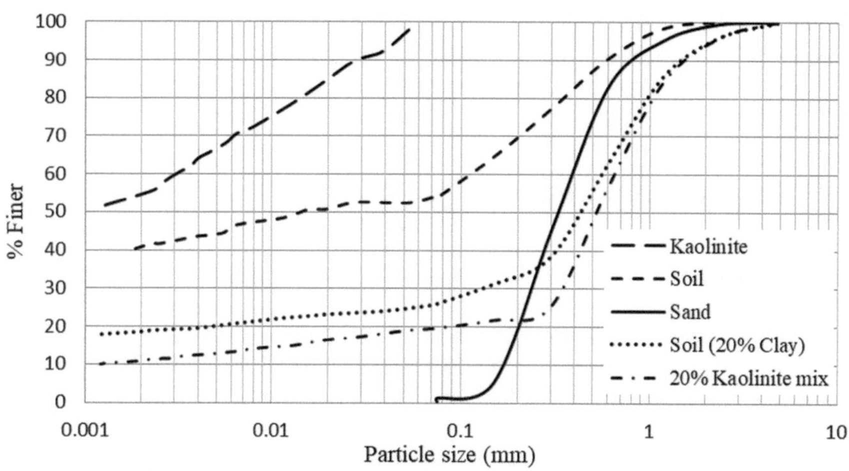

Fig. 1.1 Particle size distribution curve for river sand, natural soil, kaolinite, soil (20% clay fraction) and 20% kaolinite mix

Table 1.2 Chemical composition and physical properties of the materials used in the study

Element	Composition (% by weight)				
	Red soil	Kaolinite	Montmorillonite	GGBS	Fly ash
Al	15.68	22.13	10.79	10.87	20.85
Si	24.56	29.82	20.31	18.31	26.77
Ca	0.32	0.61	0.26	21.61	1.27
Fe	9.28	1.41	2.83	0.48	5.08
Ti	0.8	0.81	2.02	0.41	1.93
K	1.17	0.78	–	0.4	2.08
Mg	0.19	–	1.36	4.41	–
Na	–	–	2.11	–	–
S	0.3	–	–	0.65	–
Physical properties					
Specific gravity	2.68	2.63	2.39	2.91	2.28
Liquid limit	31	37.1	264.0	–	–
Plastic limit	19.48	18.87	158.0	–	–
Shrinkage limit	15.99	15.98	–	–	–

1.4 Casting and Testing Procedure

The effectiveness of geopolymer binders was evaluated through the determination of compressive strength using the cylindrical specimens of size 38 mm diameter and 76 mm height. Sodium hydroxide pellets were dissolved in the distilled water to prepare three different molar concentrations of 8, 10 and 12 M solution. The alkali solution was used after 24 h of its preparation.

1.4.1 Mixing and Casting

The materials were mixed in the dry state to achieve a homogenous mixture; later, the alkali activator solution was added to the dry mix. The moulding moisture content (MMC) (containing alkali and silica) was in the range of 10–15% of the dry mix. MMC depends upon the quantity of clay minerals in the mix. Higher percentage of clay demanded higher MMC to achieve a consistency needed for compaction. Mortar mixer was used in mixing the ingredients for 7 min to obtain the uniform mixture. The dry density of the specimens was controlled and kept at 1.8 g/cc. The cylindrical specimens were cast by compacting the partially saturated mix in a screw press.

1.4.2 Curing and Testing

Specimens after 24 h of casting were cured in an oven at 80 °C for 72 h. Cured specimens were dried in air for 24 h before testing. The specimens were tested for the wet compressive strength by soaking them in water for 48 h prior to the testing.

1.5 Results and Discussions

1.5.1 Alkali-Activated Earthen Compacts

The wet compressive strength of the alkali-activated natural soil (containing kaolinite clay) compacts was determined. The strength results are shown in Fig. 1.2. The figure shows the relationships between strength and clay content of the natural soil with varying molar concentrations of the alkali solution. The relationships show that the wet strength of the specimens increases with the increase in clay content in the mix, irrespective of the molar concentration. The strength and clay content are linearly related. There is about 50% increase in strength as the clay content was increased from 20 to 41%. Higher the clay content in the mix, more amount of reactive silica and alumina available, which resulted in the higher strength. Also, the increase in molarity of the activator solution increased the wet strength of the specimens. High alkali content (>12 M) and higher clay content in the mix result in maximum compressive strength for the soil compacts. The maximum strength obtained was 1.72 MPa with 12 M NaOH solution and with 41% clay fraction in the mix.

An attempt was made to examine the strength of alkali-activated compacts using pure clay minerals. The compacts were prepared using pure clay minerals (kaolinite and montmorillonite) and sand. The percentage of pure clay minerals in the mix was varied between 10 and 20%. Figure 1.3 shows the variation in wet compressive

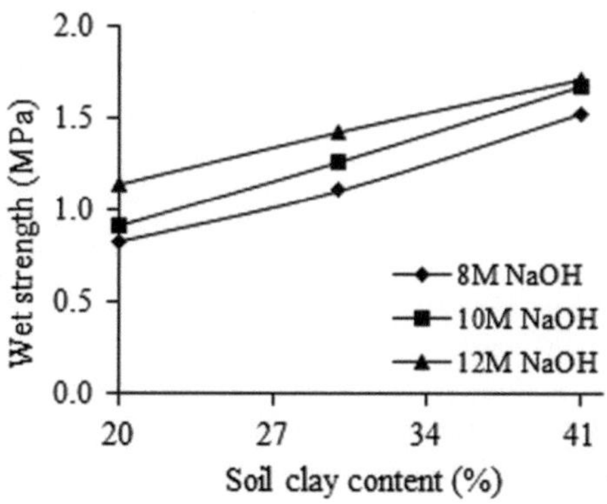

Fig. 1.2 Wet compressive strength of alkali-activated natural soil compacts

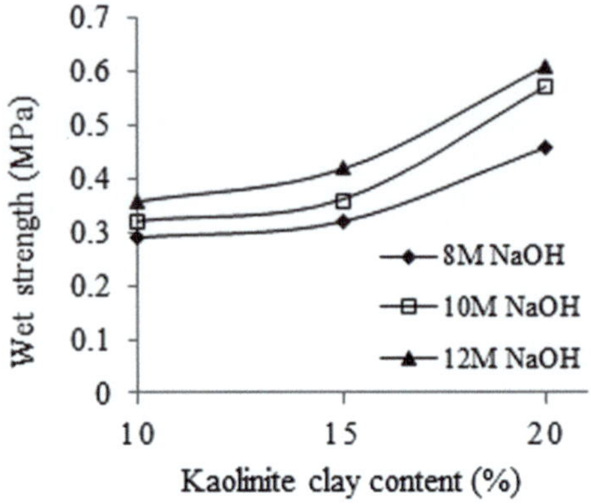

Fig. 1.3 Variation in wet compressive strength of clay–sand compacts

Fig. 1.4 Disintegrated montmorillonite compacts when soaked in water for 48 h

strength with kaolinite mineral as well as molarity of the alkali activator solution. Strength increases with increase in clay content. There is about 70% increase in wet strength as the clay percentage was increased from 10 to 20%. The strength increases marginally (10–12%) as the molarity was increased from 8 to 12 M. The maximum wet strength of 0.61 MPa was obtained with 20% kaolinite and 12 M molarity. This strength is nearly half of that obtained using natural soil using similar clay content and 12 M solution.

The compacts using montmorillonite clay mineral disintegrated upon soaking in water for 48 h prior to the testing. Figure 1.4 shows the condition of compacts using montmorillonite upon soaking in the water. The compacts using natural soil and the pure clay minerals form lumps upon mixing with higher molarity alkali solution (>12 M) as shown in Fig. 1.5. It becomes very difficult to prepare cylindrical compacts using such a lumpy mass.

Fig. 1.5 Increase in alkalinity and clay content resulted in the lumpy mix

1.5.2 Effect of Using GGBS and Fly Ash as Additional Source of Silica and Alumina

The wet strength achieved using soil and pure clay minerals was low (<1.75 MPa). Hence, addition of GGBS and fly ash was explored to improve the strength of the compacts.

1.5.2.1 Effect of Using GGBS

Natural soil with 30% clay fraction and the mix using natural clay and sand containing 20% kaolinite clay were used in preparing the compacted specimens. For natural soil compacts, GGBS was varied between 5 and 30%, and for kaolinite clay compacts, it was varied between 4 and 15%. The strength results of natural soil compacts and compacts with 20% kaolinite clay using GGBS are shown in Figs. 1.6 and 1.7, respectively. The figures show the strength variation in the compacts with GGBS and varying molarity of the activator solution. The following points emerge from the strength results shown in the figures;

- The strength varies linearly with GGBS content. As the GGBS content increased, the wet strength increased irrespective of the molarity of the alkali activator.
- The maximum wet compressive strength achieved was 7.68 MPa with the soil compacts, where soil with 30% clay content, 30% GGBS and 12 M NaOH solution were used. Increase in GGBS content from 5 to 30% results in about two-times increase in wet strength.
- Molarity of alkali solution has strong influence on the strength at lower dosages of GGBS. Its effect reduces at higher GGBS contents. For example, at 5% GGBS the strength of the soil compacts with 30% clay increased by about 50% as molarity

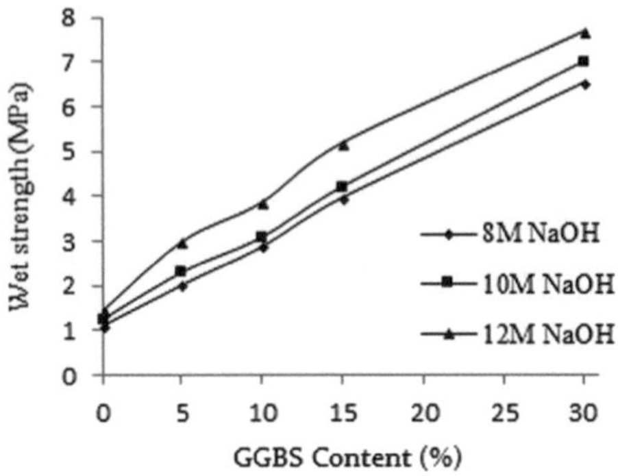

Fig. 1.6 Variation in wet strength with GGBS content for soil compacts with 30% clay

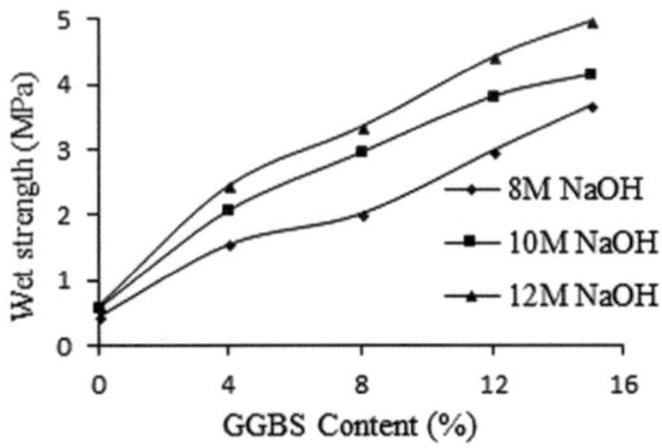

Fig. 1.7 Variation in wet strength of specimens with 20% kaolinite and with GGBS content

was increased from 8 to 12 M, whereas at 30% GGBS content, there is only 25% increase in strength.

- The maximum wet compressive strength of the compacts with 20% kaolinite was 4.96 MPa with 15% GGBS and 12 M NaOH. Strength increased by about 100% with the increase in GGBS content from 4 to 15%. Increase in molarity of the activator solution from 8 to 12 M increased the strength of the 20% kaolinite specimens. The strength increases by about 60% and 35% at 4% and 15% GGBS contents, respectively.

1.5.2.2 Effect of Using Fly Ash

The strength of alkali-activated soil compacts with 30% clay fraction as well as with 20% kaolinite clay and using fly ash was examined. The fly ash content was varied between 5 and 30% with soil compacts and between 4 and 15% with kaolinite compacts. The variation in compressive strength of soil compacts and pure clay compacts using fly ash is shown in Figs. 1.8 and 1.9, respectively. The figures show the relationships between strength and fly ash content with varying molar concentrations. From these results, the following observations can be made:

- Increase in fly ash content increased the wet strength of the compacts. There is a linear relationship between the strength and the fly ash content irrespective of the molar concentrations of the alkali activator.
- Strength increases by about 1.6 times as the fly ash content was increased from 5 to 30% in natural soil compacts. The maximum strength obtained was about 6 MPa with 15% fly ash and 12 M molarity with 30% clay fraction. Compressive strength of the compacts using natural soil increased as the molarity increased from 8 to 12 M. Strength increased by about 60% and 46% at 5% and 30% fly ash content with increase in molarity from 8 to 12 M.
- Increase in fly ash content from 4 to 15% resulted in 100% increase in wet strength of the compacts with 20% kaolinite, irrespective of the molarity of the solution. The maximum strength obtained was about 3 MPa with 15% fly ash. It was observed that the strength increased by about 70% with the increase in molarity from 8 to 12 M at 4% fly ash content and that about 60% at 15% fly ash content.

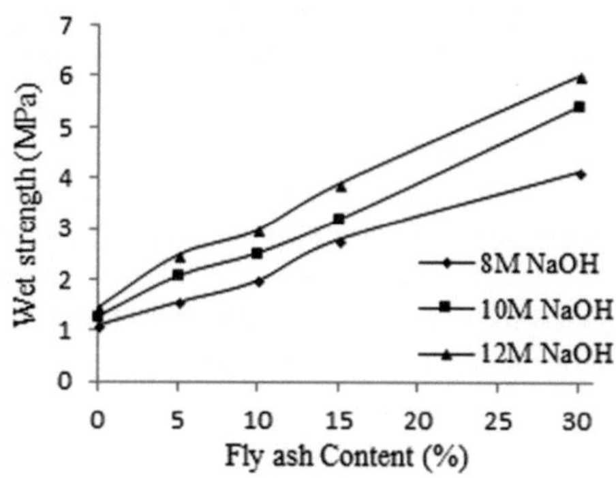

Fig. 1.8 Wet strength variation of soil compacts with 30% clay and with fly ash content

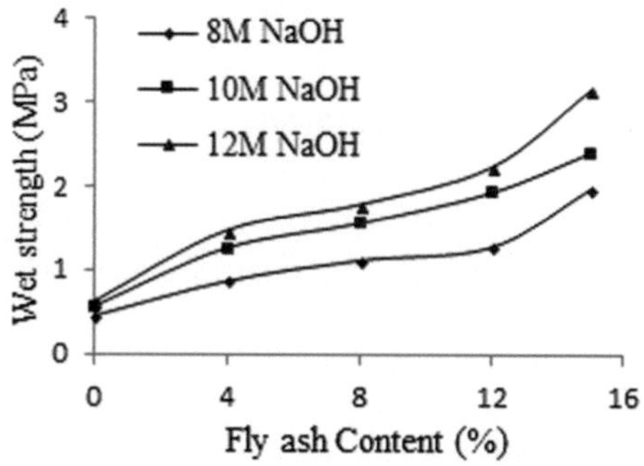

Fig. 1.9 Wet strength variation of kaolinite compacts with 20% clay and fly ash content

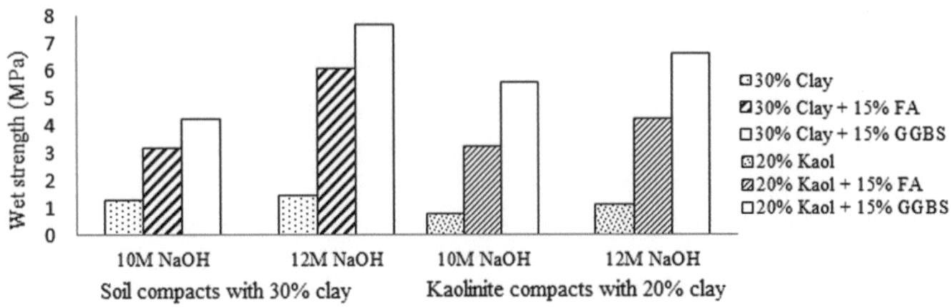

Fig. 1.10 Variation in wet strength of geopolymer compacts with natural soil (30% clay) and 20% kaolinite with 10 and 12 M activator solution using GGBS and fly ash (FA)

1.5.2.3 Comparison of Results Using GGBS and Fly Ash

Addition of GGBS and fly ash as an additional source of silica and alumina plays a key role in improving the wet compressive strength of the specimens using natural soil and kaolinite mineral. The strengths increased with the addition of GGBS and fly ash. The comparison of compressive strength results of soil compacts with 30% clay fraction and compacts with 20% kaolinite using GGBS and fly ash with 10 and 12 M alkali activator are shown in Fig. 1.10. Addition of GGBS resulted in higher strength when compared with the compacts using fly ash. The increase in strength using GGBS is associated with the high lime reactivity of GGBS than that of fly ash and with the high calcium content present in the GGBS, the presence of calcium ensures the lime pozzolana reaction along with geopolymerisation. The GGBS shows higher lime reactivity than fly ash. This is attributed to the fact that the higher lime reactivity is due to the presence of higher quantity of reactive alumina and silica. Hence, the material has more tendencies to react with the alkali activator used. The specific surface area of GGBS is >500 m^2/kg and that of low-calcium fly ash ranges between 200 and 300 m^2/kg (Mehta and Monterio 2014). Finer material possesses higher pozzolanic activity and results in formation of more cementitious products.

1.5.3 Efflorescence on the Specimens

Alkali-activated cylindrical specimens showed the efflorescence on the surface prior to curing. Deposition of salts on the cylindrical specimens with kaolinite clay and natural soil before curing is shown in Fig. 1.11a, b respectively. The sodium-rich solution evaporates from exposed surface of the cylinders, leaving the salts as white deposits on the surface. Leaching of salts was also observed when the specimens were soaked in water for 48 h prior to the testing. Excessive leaching of salts is a serious concern. There is a need for understanding the effect of leaching on the strength and durability of alkali-activated compacts.

Fig. 1.11 Deposition of salts, **a** kaolinite compacts and **b** soil compacts

1.6 Conclusions

Tests on various mix proportions indicate a linear relationship between strength and clay content. Strength increased with increase in clay fraction in the mix and with increase in molarity. Alkali-activated natural soil compacts resulted in wet strength <2 MPa. Pure clay specimens showed very low strength and are practically insufficient. The specimens using kaolinite clay resulted in very low strength of 0.61 MPa with 20% clay fraction and 12 M solution in wet condition. Cylindrical specimens with montmorillonite disintegrated when soaked in water prior to the testing.

The wet strength of alkali-activated earthen compacts improved with the addition of fly ash and GGBS as an additional source of silica and alumina. The maximum wet strength of cylindrical specimens achieved was 7.68 MPa with the soil compacts, where the red soil with 30% clay content, 30% GGBS and 12 M sodium hydroxide solution. The specimens with kaolinite showed up to 5 MPa wet compressive strength with 15% GGBS and 20% clay fraction. Test results reveal that strength increases with increase in GGBS or fly ash content. High strengths were obtained with the use of GGBS rather than fly ash due to high lime reactivity of GGBS.

MMC to achieve a homogenous mixture was controlled by the clay fraction in the mix. Increase in clay content and increase in alkalinity (>12 M) of the solution resulted in the formation of lumpy mix, which is difficult to compact.

Alkali-activated specimens showed salt deposition on the surface prior to curing. Leaching of salts was observed when the specimens were soaked in water for 48 h prior to the testing. Addressing the leaching in an earth-based geopolymer products is a serious concern.

References

Davidovits J (1988) Geopolymer chemistry and properties. In: Geopolymer'88, first European conference of soft minerology, Compiegne, France

Davidovits J (1994) Properties of geopolymer cements. In: First international conference of alkaline cements and concretes, Kiev, pp 131–149

Davidovits J (1999) Chemistry of geopolymeric systems, terminology. In: Proceedings of the 2nd international conference on Geopolymer'99, Saint Qunentin, pp 9–39

Duxson P, Fernańdez-Jimeńez A, Provis JL, Lukey GC, Palomo A, Van Deventer JSJ (2007) Geopolymer technology: the current state of the art. J Mater Sci 42(9):2917–2933

Gluchovskij VD (1959) "Gruntosilikaty" Gosstrojizdat Kiev, Patent USSR 245 627 (1967), Patent USSR 449894 (Patent appl. 1958, filled 1974)

Grim RE, Bradley WF (1940) Investigation of the effect of heat on the clay minerals Illite and Montmorillonite. J Am Ceram Soc 23(8):242–248

Hardjito (2004) On the development of fly ash-based geopolymer concrete. ACI Mat J 101(6):467–472

Hardjito D, Wallah SE, Sumajouw DMJ, Rangan BV (2003) Geopolymer concrete: turn waste into environmentally friendly concrete. In: Krishnamoorthy R (ed) International conference on recent trends in concrete technology and structures, 10–11, September, Kumaraguru College of Technology, Coimbatore, India

Heah CY, Kamarudin H (2011) Effect of curing profile on Kaolin-based geopolymers. In: International conference on physics science and technology (ICPST), Physics Procedia, vol 22, pp 305–311

IS: 1727–2004 Methods of test for pozzolanic materials. Bureau of Indian Standards, New Delhi

Khale D, Chaudhary R (2007) Mechanism of geopolymerization and factors influencing its development: a review. J Mat Sci 42(3):729–746

Kunal Kupawade Patil, Allouche EN (2013) Impact of Alkali Silica reaction on fly ash-based geopolymer concrete. In: 2013 American society of civil engineers

Maskell D, Heath A, Walker PJ (2014) Geopolymer stabilisation of unfired earth masonry units. Key Eng Mat 600:175–185. ISSN 1662-9795

McLellan BC, Williams RP, Lay J, van Riessen A, Corder GD (2011) Costs and carbon emissions for geopolymer pastes in comparison to ordinary Portland cement. J Clean Prod 19:1080–1090

Mehta PK, Monterio PJM (2014) CONCRETE, microstructure, properties and materials. Mc Graw Hill Education, India (Edition 2014)

Munoz JF, Easton T, Dahmen J (2015) Using alkali-activated natural aluminosilicate minerals to produce compressed masonry construction materials. Constr Build Mater 95:86–95

Praseeda KI, Reddy BVV, Mani Monto (2015) Embodied energy assessment of building materials in India using process and input-output analysis. Energy Build 85:677–686

Provis LJ (2014) Geopolymers and other alkali activated materials: why, how, and what? Mater Struct 47:11–25

Rangan BV (2008a) Low-calcium fly ash-based geopolymer concrete. In: Construction engineering handbook (Chapter 11), 2nd edn. CRC Press, New York

Rangan BV (2008b) Mix design and production of fly ash based geopolymer concrete. Ind Concr J 82(5):7–15

Rangan BV (2009) Engineering properties of geopolymer Concrete. In: Geopolymers: processing properties and applications (Chapter 11). Woodhead publishing Limited, London

Reddy BVV, Jagadish KS (2003) Embodied energy of common and alternative building materials and technologies. Energy Build 35:129–137

Rovnanik P (2010) Effect of curing temperature on the development of hard structure of metakaolin-based geopolymer. Constr Build Mater 24:1176–1183

Slaty F, Khoury H, Wastiels J, Rahier H (2013) Characterization of alkali activated kaolinitic clay. Appl Clay Sci 75–76:120–125

Chapter 2
Stabilisation of Clay Mixtures and Soils by Alkali Activation

Alastair Marsh, Andrew Heath, Pascaline Patureau, Mark Evernden and Pete Walker

2.1 Introduction

Alkali-activated materials (AAM) have emerged in recent decades as novel materials for several applications, including construction materials (Davidovits 2011; Provis 2014). One of their main selling points is their prospect as a low-carbon building material. This is due to their typically low curing temperature of <100 °C and also that their precursor preparation does not chemically require the release of carbon, unlike Portland cement (Heath et al. 2014; Khale and Chaudhary 2007). A geopolymer is an amorphous, alkali aluminosilicate phase that is typically produced by a dissolution–condensation reaction between an aluminosilicate precursor and an alkaline activating solution, such as a concentrated aqueous sodium hydroxide solution (Duxson et al. 2007a). The formation of a geopolymer depends on several compositional and processing conditions, including the extent of dissolution and Si:Al molar ratio of the system (Duxson et al. 2007b). Alkali activation can also produce zeolitic reaction products (Criado et al. 2007).

A. Marsh (✉)
School of Civil Engineering, University of Leeds, Leeds LS2 9JT, UK
e-mail: a.marsh@leeds.ac.uk

A. Heath · M. Evernden · P. Walker
Department of Architecture & Civil Engineering, University of Bath, Bath BA2 7AY, UK
e-mail: a.heath@bath.ac.uk

M. Evernden
e-mail: m.evernden@bath.ac.uk

P. Walker
e-mail: p.walker@bath.ac.uk

P. Patureau
Department of Chemistry, University of Bath, Bath BA2 7AY, UK
e-mail: p.m.f.patureau@bath.ac.uk

B. V. V. Reddy et al. (eds.), *Earthen Dwellings and Structures*,
Springer Transactions in Civil and Environmental Engineering,
https://doi.org/10.1007/978-981-13-5883-8_2

The aluminosilicate precursors used in AAM are commonly fly ash, blast furnace slag, other industrial by-products, metakaolin, or mixtures thereof (Pacheco-Torgal et al. 2008). Subsoil has the benefit of being a precursor that is widely available and available at very low environmental cost (Diop and Grutzeck 2008). In these systems, the clay minerals in the soil are the aluminosilicate reactant, with the other less reactive phases acting as an aggregate. Thus, the alkali aluminosilicate product phase performs the role of stabiliser for the remnant components in the soil. In effect, the clay minerals become a water-resistant binder, replacing the role of cement in a concrete block. Some researchers have investigated adding industrial by-products (e.g. fly ash) to soils before alkali activation, but this research is focussed on natural soils without any other precursors.

Soil materials stabilised by alkali activation have good environmental prospects by virtue of their low curing temperature, avoidance of chemical production of carbon during preparation and availability of subsoil (Diop and Grutzeck 2008; Murmu and Patel 2018). However, there is still a significant knowledge gap around how soil composition influences the alkali activation reaction, and in particular the reaction products formed. The mix of clay minerals in soil is of most interest, as out of the minerals typically found in soil, the clay minerals are the largest reactive components in the activation reaction (Xu and van Deventer 2000; Autef et al. 2012). The clay minerals most commonly found in soils are kaolinite, montmorillonite and illite (Nickovic et al. 2012). In this study, alkali activation was done on samples of these individual clays and also on a mixture of all three. The aim was to determine whether phase formation behaviour for the mixture differed from that expected by the behaviour of the individual clay minerals.

2.2 Materials and Methods

Imerys Speswhite kaolin (abbreviation = Kao) (mined from Cornwall, U.K.), K10 montmorillonite (abbreviation = Mont) (Sigma-Aldrich, product no. 69866-1KG) and Clay Minerals Society IMt-2 (Silver Hill) illite (abbreviation = ILL) were used as the precursor clays. The clays were activated by adding sodium hydroxide solution and mixing. The concentrations and quantities of sodium hydroxide solution were selected to give an Na:Al molar ratio of 1 for each system, whilst keeping the wet mix workability at approximately the plastic limit (Marsh et al. 2018b). The first constraint was used as a molar ratio of Na:Al = 1 is the stoichiometric balance theoretically required for geopolymer formation (Barbosa et al. 2000), the second constraint used to be compatible with extrusion processing (Maskell et al. 2013). This was achieved for all systems except the activated illite, which due to its lower plastic limit had a maximum ratio of Na:Al = 0.75.

Solutions of different concentrations were prepared by adding sodium hydroxide pellets to distilled water, mixed with a magnetic stirrer (Stuart UC152 heat-stir) for a minimum of 2 h until fully dissolved and then allowed to cool. The clays were pre-dried in a 105 °C oven and left to cool. For the mixture, the constituent clays

Table 2.1 Clay contents of the samples given in wt%

Sample	Kao content	Mont content	ILL content	[NaOH] molarity	NaOH solution: clay mass ratio
Kao-activated	100%	n/a	n/a	16.1	0.73
Mont-activated	0%	100%	n/a	6.4	0.77
ILL-activated	0%	n/a	100%	19.7	0.39
Kao-Mont-ILL-activated	33%	33%	33%	13.6	0.65

were then dry-mixed together using a magnetic stirrer for 5 min. Varying quantities of activating solutions were added to each clay or clay mixture, as given in Table 2.1. Each wet mix of activating solution and clay was mixed by hand for 3 min, providing a consistent and well-distributed mixture. The high viscosity of the samples allowed them to be compacted by hand into 18 mm × 36 mm cylindrical Teflon moulds by tamping with a glass rod in three layers for each sample, using 25 blows for each layer. Samples were cured in an air atmosphere in a 80 °C oven for 24 h in their moulds. A control sample was made for each composition, by adding distilled water and then mixing and curing in the same manner.

Activated samples of illite and 33Kao–33Mont–33ILL did not fully dry with curing, so were forcibly dried in a vacuum desiccator for 72 h.

The set of characterisations were done at 28 ± 2 days ageing time and (with the exception of SEM imaging) were done using powders prepared from the cured samples. Powders were prepared by grinding by hand, having been wetted with isopropanol to avoid damaging the clay minerals' crystal structures (Moore and Reynolds 1997). Powder X-ray diffraction (PXRD) analysis was done to identify phases with a Bruker D8 Advance instrument using monochromatic CuKα1 L3 (λ = 1.540598 Å) X-radiation and a Vantec superspeed detector. A step size of 0.016°(2θ) and step duration of 0.3 s were used. Phase identification was done using Bruker EVA software. Scanning electron microscope (SEM) imaging was used on a fracture surface of the bulk samples sputter coated with gold for 3 min, to characterise phase size and morphology. A JEOL SEM6480LV was used in secondary electron mode with an accelerating voltage (AV) of 10 kV.

2.3 Results

2.3.1 Precursors

The PXRD patterns of the precursors are given in Fig. 2.1. The kaolinite precursor contained kaolinite clay mineral as the major phase, with muscovite and quartz present as minor phases, as expected from a Cornish residual deposit. The montmorillonite precursor contained Ca-montmorillonite clay mineral as the major phase, as well as muscovite, quartz and minor amounts of kaolinite. The illite precursor contained illite clay mineral as the major phase, with quartz, microcline and kaolinite present as minor phases. Previous studies on this source clay identified the illite clay mineral to be mostly of the 1 M/1 Md polytype (Haines and van der Pluijm 2008). Their chemical compositions are given in Table 2.2.

Table 2.2 Chemical composition of clay precursors in oxide wt%

Oxide	Al_2O_3	CaO	Fe_2O_3	K_2O	MgO	Na_2O	SiO_2	SO_3	TiO_2	Total
Kaolinite (std error)	40.11 (0.15)	0.00	0.95 (0.06)	2.06 (0.09)	0.04 (0.04)	0.00	56.83 (0.15)	0.00	0.00	100
K10 montmorillonite (std error)	13.53 (0.66)	0.47 (0.14)	4.53 (1.05)	1.56 (0.22)	1.67 (0.11)	0.03 (0.03)	77.60 (2.12)	0.12 (0.07)	0.49 (0.02)	100
Illite (std error)	20.80 (0.34)	0.00	8.32 (0.38)	8.67 (0.18)	2.28 (0.06)	0.00	59.14 (0.26)	0.00	0.78 (0.06)	100

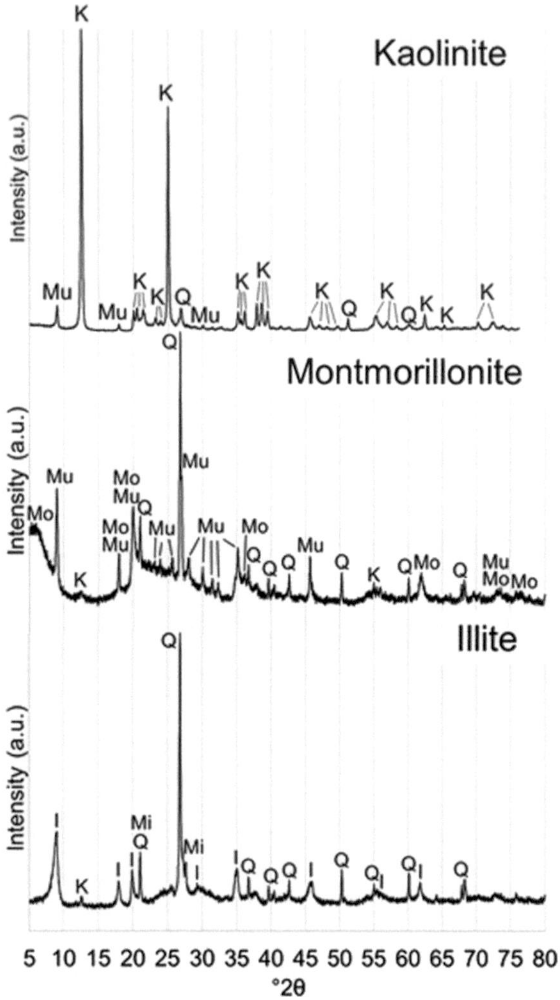

Fig. 2.1 PXRD patterns of **a** kaolinite precursor; **b** K10 montmorillonite precursor; **c** IMt-2 illite precursor. Indexed as: I = illite; K = kaolinite; Mi = microcline; Mo = montmorillonite; Mu = muscovite; Q = quartz

2.3.2 *Kaolinite*

Alkali activation of kaolinite produced the hydrosodalite $Na_8[AlSiO_6]_4.(OH)_2 \cdot 2H_2O$ (abbreviated as 8:2:2) as the product phase, a member of the zeolite family (Marsh et al. 2018b). This was clearly evident in the strong crystalline peaks in the PXRD pattern (Fig. 2.2), as well as the 0.5–1 μm crystallites in the SEM image (Fig. 2.3). As seen in both the XRD and SEM, a significant amount of kaolinite was consumed in the reaction, but some remained unreacted. No shrinkage was observed in the cured sample, and no unusual morphological or colour changes were observed either (Fig. 2.4).

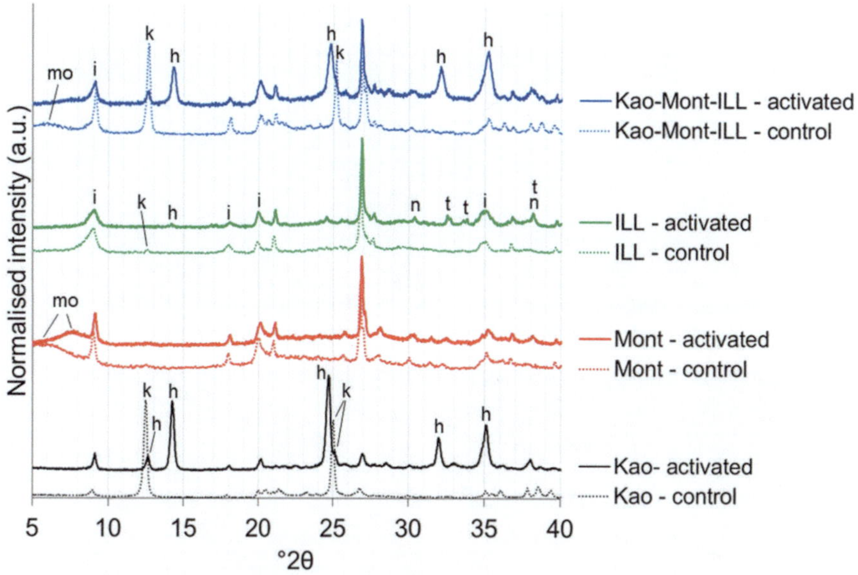

Fig. 2.2 XRD patterns of control and activated samples of kaolinite, montmorillonite, illite and a mixture of all three. For simplicity, only the clay minerals and product phases have been indexed

2.3.3 Montmorillonite

Activation of montmorillonite produced a geopolymer as the product phase. This can be seen in the PXRD pattern from the characteristic shift in the background in the region of 22–35°2θ (Duxson et al. 2007a) (Fig. 2.2). No new crystalline peaks were observed. Some of the montmorillonite clay mineral was consumed, but some remained unreacted. The 001 reflection shifted after activation, a change in d-value from 14.4 to 11.6 Å. This decrease in interlayer space was partly attributed to cation exchange of Na^+ in the sodium hydroxide activating solution for Ca^{2+} in the montmorillonite's interlayer sites (Marsh et al. 2018a). The microstructure of the activated sample was very different to that of the plate-like clay minerals in the precursor, with a semi-continuous morphology indicative of geopolymer formation (Provis et al. 2005) (Fig. 2.3). The cured sample showed very distinctive radial shrinkage cracks, aligned upwards (Fig. 2.4).

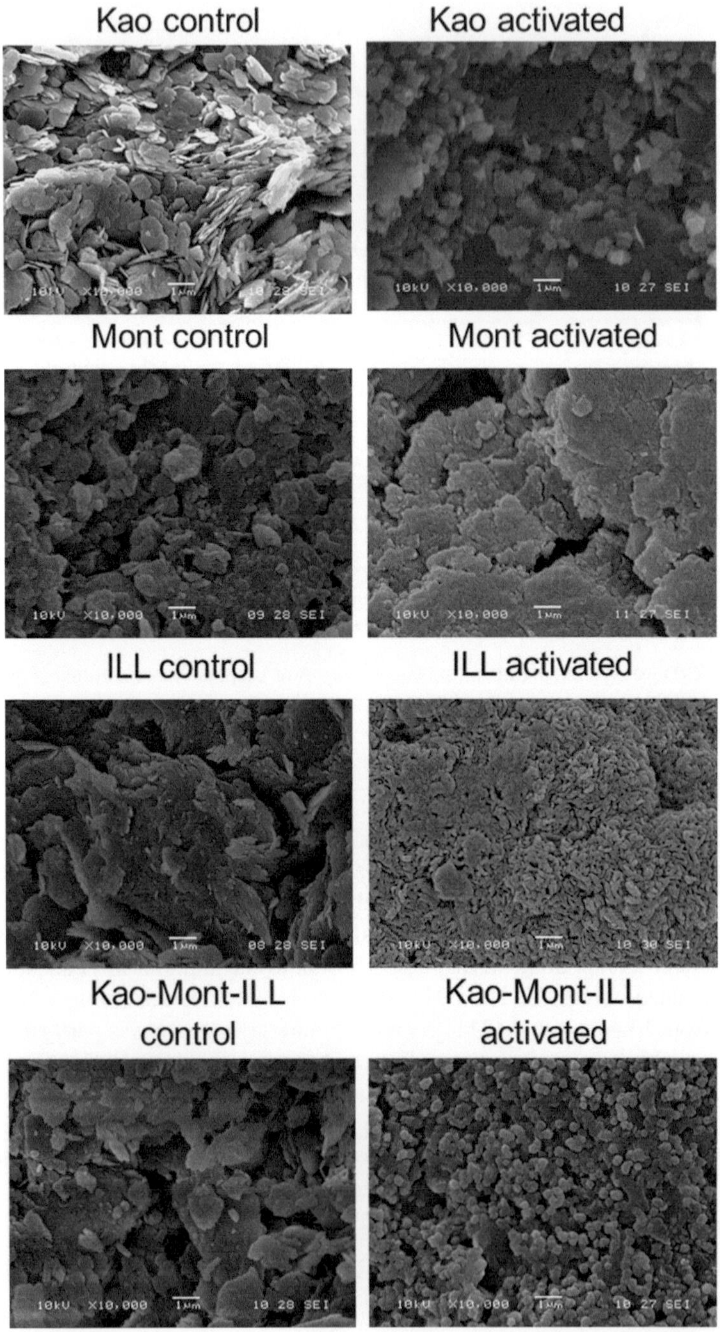

Fig. 2.3 SEM images of control and activated samples of kaolinite, montmorillonite, illite and a mixture of all three

Fig. 2.4 Photographs of activated samples of kaolinite, montmorillonite, illite and a mixture of all three

2.3.4 Illite

Activation of illite did not result in a product phase, but instead resulted in the alteration of the illite clay mineral. The XRD pattern of the activated sample contained no major new crystalline peaks (Fig. 2.2), but contained minor peaks attributed to hydrosodalite, natrite (Na_2CO_3) and thermonatrite ($Na_2CO_3 \bullet H_2O$) (Marsh et al. 2018a). The microstructure of the activated illite is significantly different to the precursor, mostly due to the emergent porosity (Fig. 2.3). This porosity on the micro-scale may have contributed to the significant expansion which was observed upon curing (Fig. 2.4).

2.3.5 Kaolinite–Montmorillonite–Illite Mixture

The major reaction product was 8:2:2 hydrosodalite, as seen in the XRD pattern of the activated sample (Fig. 2.2). There was some evidence of a background shift in the region of 22–35°2θ, but not enough to conclusively show that a geopolymer was formed. None of the clay minerals in the precursors were fully consumed. Particles of ~300 nm size were observed in the SEM images of the activated sample (Fig. 2.3). No structural defects, shrinkage or expansion was observed in the cured sample, but there was noticeable darkening around the top of the sample at the open end of the mould (Fig. 2.4).

2.4 Discussion

Using a consistent Na:Al ratio, different clay minerals have vastly different reactions to alkaline activation. Kaolinite forms hydrosodalite, a crystalline phase; montmorillonite forms a geopolymer, an amorphous phase; illite does not form a product phase, but instead seems to undergo alteration. A difference between kaolinite (a 1:1 clay) and montmorillonite (a 2:1 clay) is expected, since geopolymers are favoured over crystalline reaction products for systems with Si:Al $\geq$ 1.5 (Duxson et al. 2007b). However, the difference between the montmorillonite and illite (both 2:1 clays) suggests that alkali activation behaviour is strongly influenced by clay mineralogy rather than Si:Al stoichiometry alone.

In the test case of an equal mix of all three clay precursors, the result is not trivial to interpret. On the evidence available, the activated clay mixture is likely to contain a mix of hydrosodalite and geopolymer. Whilst geopolymers are straightforward to detect in highly reactive, simple systems such as metakaolin or fly ash, it is much more difficult in less reactive, multi-component systems such as uncalcined clay mixtures and soils. This is especially the case in low Si:Al systems where zeolitic and geopolymer reaction products can coexist (Rahier et al. 1997; Buchwald et al. 2011).

The influence of mineralogy on curing defects also opens many questions. The defects observed here are extreme, given that the clay mineral content of these samples is far higher than would ever be used in earth construction. A greater proportion of aggregate phases such as quartz would reduce the extent of these. However, it is another indication of how much there is still to be understood about these systems.

2.5 Conclusion

There are large differences in phase formation behaviour after the alkali activation of individual clay minerals. The phases formed from alkali activation of an equal mixture of these clay minerals are roughly equivalent to those formed in the individual clay minerals. Hence, it seems that the phase formation of a given clay mixture or soil could roughly be predicted from knowing the amount and types of clay minerals present. However, it also seems there is an additional degree of complexity in the phase formation behaviour of a mixture, beyond that of the individual clay minerals. This means that the exact behaviour of a given clay mixture or soil can only be fully known by testing. The findings of this study have identified an emergent issue—a greater understanding is required in order to determine how much of a difference this additional complexity of mixtures could make to the properties of construction materials made with alkali-activated clays or soils.

References

Autef A, Joussein E, Gasgnier G, Rossignol S (2012) Role of the silica source on the geopolymerization rate. J Non-Cryst Solids 358(21):2886–2893

Barbosa VFF, MacKenzie KJD, Thaumaturgo C (2000) Synthesis and characterisation of materials based on inorganic polymers of alumina and silica: sodium polysialate polymers. Int J Inorg Mater 2:309–317

Buchwald A, Zellmann HD, Kaps C (2011) Condensation of aluminosilicate gels—model system for geopolymer binders. J Non-Cryst Solids 357(5):1376–1382

Criado M, Fernández-Jiménez A, Palomo A (2007) Alkali activation of fly ash: effect of the SiO2/Na2O ratio: Part I: FTIR study. Microporous Mesoporous Mater 106:180–191

Davidovits J (2011) Geopolymer chemistry and applications, 3rd edn. Institut Geopolymere, Saint-Quentin, Saint-Quentin

Diop MB, Grutzeck MW (2008) Low temperature process to create brick. Constr Build Mater 22:1114–1121

Duxson P, Fernández-Jiménez A, Provis JL, Lukey GC, Palomo A, van Deventer JSJ (2007a) Geopolymer technology: the current state of the art. J Mat Sci 42:2917–2933

Duxson P, Mallicoat SW, Lukey GC, Kriven WM, van Deventer JSJ (2007b) The effect of alkali and Si/Al ratio on the development of mechanical properties of metakaolin-based geopolymers. Colloids Surf, A 292:8–20

Haines SH, van der Pluijm BA (2008) Clay quantification and Ar–Ar dating of synthetic and natural gouge: application to the Miocene Sierra Mazatán detachment fault, Sonora, Mexico. J Struct Geol 30:525–538

Heath A, Paine K, McManus M (2014) Minimising the global warming potential of clay based geopolymers. J Clean Prod 78:75–83

Khale D, Chaudhary R (2007) Mechanism of geopolymerization and factors influencing its development: a review. J Mat Sci 42:729–746

Marsh A, Heath A, Patureau P, Evernden M, Walker P (2018a) Alkali activation behaviour of un-calcined montmorillonite and illite clay minerals. Applied Clay Sci 166:250–261

Marsh A, Heath A, Patureau P, Evernden M, Walker P (2018b) A mild conditions synthesis route to produce hydrosodalite from kaolinite, compatible with extrusion processing. Microporous Mesoporous Mater 264:125–132

Maskell D, Heath A, Walker P (2013) Laboratory scale testing of extruded earth masonry units. Mater Des 45:359–364

Moore DM, Reynolds RC (1997) X-ray diffraction and the identification and analysis of clay minerals, 2nd ed. Oxford, Oxford University Press

Murmu AL, Patel A (2018) Towards sustainable bricks production: an overview. Constr Build Mater 165:112–125

Nickovic S, Vukovic A, Vujadinovic M, Djurdjevic V, Pejanovic G (2012) High-resolution mineralogical database of dust-productive soils for atmospheric dust modeling. Atmos Chem Phys 12(2):845–855

Pacheco-Torgal F, Castro-Gomes J, Jalali S (2008) Alkali-activated binders: a review. Part 2. About materials and binders manufacture. Constr Build Mater 22:1315–1322

Provis JL (2014) Geopolymers and other alkali activated materials: why, how, and what? Mater Struct 47:11–25

Provis JL, Lukey GC, van Deventer JS (2005) Do geopolymers actually contain nanocrystalline zeolites? A reexamination of existing results. Chem Mater 17:3075–3085

Rahier H, Simons W, Van Mele B, Biesemans M (1997) Low-temperature synthesized aluminosilicate glasses: Part III Influence of the composition of the silicate solution on production, structure and properties. J Mat Sci 32(9):2237–2247

Xu H, Van Deventer JSJ (2000) The geopolymerisation of alumino-silicate minerals. Int J Miner Process 59(3):247–266

Chapter 3
Moisture Transport in Cement Stabilised Soil Brick-Mortar Interface and Implications on Masonry Bond Strength

B. V. Venkatarama Reddy, V. Nikhil and M. Nikhilash

3.1 Introduction

Cement-stabilised soil blocks (also known as stabilised compressed earth blocks) are used for the load-bearing masonry walls in India and many other parts of the world. Use of cement-stabilised soil block (CSSB) for building construction has gained momentum in the last few decades (Fitzmaurice 1958; Heathcote 1991; Walker and Stace 1997; Walker 1995; Reddy and Gupta 2005; Reddy et al 2007; Reddy and Latha 2014, 2018). Apart from the load-bearing walls, CSSB is used for the construction of roofs/floors such as masonry jack arch, vaults and domes. Figure 3.1 shows a three-storey load-bearing residential building.

Masonry has two different materials: masonry units and the mortar. Good bond between the masonry unit and the mortar is essential for the structural integrity of the masonry. Masonry's capacity to resist tensile and shear stresses is mainly dependent upon the bond strength between the masonry unit and the mortar. The development of bond between the masonry unit and the mortar is mainly due to the mechanical interlocking of cement hydration products into the brick pores. The cement hydration products such as silicate hydrates of calcium and aluminium and free lime from the fresh mortar bed joint are transported into the brick pores due to suction of the bricks/blocks. Development of bond at the brick-mortar interface is influenced by several parameters related to the masonry unit and the mortar. Surface texture, pore

B. V. Venkatarama Reddy · V. Nikhil · M. Nikhilash (✉)
Centre for Sustainable Technologies (CST), Indian Institute of Science, Bangalore, India
e-mail: nikhi.manju005@gmail.com

B. V. Venkatarama Reddy
e-mail: venkat@civil.iisc.ernet.in

V. Nikhil
e-mail: nikhilvenugopal2@gmail.com

B. V. V. Reddy et al. (eds.), *Earthen Dwellings and Structures*, Springer Transactions in Civil and Environmental Engineering,
https://doi.org/10.1007/978-981-13-5883-8_3

Fig. 3.1 Load-bearing CSSB masonry building

size and pore-size distribution, masonry unit's moisture content, mortar composition and water-holding capacity of the mortar are some of the prime parameters affecting the bond development. A brief review of the earlier studies on bond strength in masonry is as follows.

Significance of brick moisture content at the time of construction on bond strength was described by Sinha (1967). The studies showed that the bond strengths were low for masonry using saturated and dry bricks and maximum for masonry using partially saturated bricks (80% of saturation value). Lawrence and Cao (1987) concluded that the mortar-brick bond is due to the network of cement hydration products growing into brick pores, which is mechanical in nature. The addition of lime into the mortar improved the network of cement hydration products. The penetration of cement hydration products into the brick pores is influenced by the moisture content in the brick during construction.

Groot (1993) studied the effects of water on mortar-brick bond and showed that the moisture transport between the mortar and the brick influences the hydration of cementation products in the middle of the joint and in the interface region of mortar significantly. Venu Madhava Rao et al. (1996) examined the flexural bond strength of stack-bonded masonry using different types of blocks and mortars. The studies revealed that (1) partially saturated masonry units yield highest bond strength when compared to the dry and saturated ones and (2) the composite mortars such as cement-lime mortar and cement-soil mortar showed better bond strength when compared with strengths using cement-sand mortar. Lange et al. (1999) investigated the condition of the mortar/unit interface through examination of the microstructure. The studies showed that the water transport is a dominant mechanism between the mortar and the masonry units, and the water loss from the mortar joint results in two mechanisms: densification and dewatering. Densification means consolidating solid phases and limiting water at the interface, whereas dewatering is limiting water for hydration and creating air voids in the mortar as water is removed.

Sarangapani et al. (2002) examined the water loss from the fresh mortars and bond strength development in low-strength masonry. Three different types of burnt clay bricks in combination with four types of mortars (including two composite mortars) were used in the investigations. The investigations revealed that bulk of the moisture transport from mortar takes place in the first hour and the water/cement ratio is practically constant after two hours. The moisture transport from mortar to brick is very substantial when the bricks are dry. Thus, partially saturated bricks are necessary in retaining a water/cement ratio of greater than 0.4 in mortars for satisfactory setting of mortars and providing maximum bond strength. Surface texture and pore structure and pore size influence the mortar-brick bond strength (Kampf 1963; Groot 1993; Reddy and Gupta 2005). The mortar proportion, type and water-holding capacity influence the bond strength. Lime in mortar holds water and reduces shrinkage cracks, prolongs workability and enables continuous growth of hydration products (Fried 1996). During masonry construction, water movement from the fresh bed joint to the brick takes place (Brocken et al. 1998; Sarangapani et al. 2002).

The past literature reveals that the suction properties of the masonry units, mortar composition and the quantity of water in the fresh mortar influence the bond strength development. Several studies revealed that use of partially saturated bricks result in maximum bond strength for the masonry. There are limited studies on the moisture transport phenomenon and the bond strength development in the stabilised earth block masonry. Hence, the present study focuses on the moisture transport from the fresh mortar bed into the CSSB's and the implications on the bond strength of CSSB masonry.

The scope of the work includes (1) monitoring rate of moisture absorption in CSSB, (2) measuring moisture loss from the fresh mortar bed to CSSB considering three different types of mortars (3) determination of bond strength of CSSB masonry prisms. Cement-sand mortar, cement-lime mortar and cement-soil mortar were considered. CSSB's at different moisture conditions (dry, partially wet and wet conditions) were used in the experiments.

3.2 Materials Used in the Investigations

Cement and lime: Ordinary Portland cement (OPC) conforming to IS 8112 (1987) code was used in the preparation of mortars and masonry prism casting. The initial and final setting time was 212 and 341 min, respectively. Laboratory grade lime (with 95% assay) was used in the preparation of cement-lime mortars.

Soil and sand: A local red loamy soil and river sand were used in the preparation of mortars. The soil has kaolinite clay mineral and contains 44.3, 13.0 and 42.7% sand, silt and clay size fractions, respectively. The liquid limit and the plasticity index

Table 3.1 Mortar composition and characteristics

Mix proportions (by volume)				Mortar designation	Water–cement ratio at 85% flow (by mass)	28-day cube compressive strength (MPa)
Cement	Lime	Soil	Sand			
1	1	–	6	CLM	1.64	4.83
1	–	–	6	CM	1.60	4.86
1	–	2	5	CSM	1.71	4.23

for the soil were 32.8 and 14.8%, respectively. The fineness modulus of the river sand was 2.36.

Cement-stabilised soil bricks: Soil composition and grading, cement content, and block density are some of the major parameters having significant influence on the characteristics of CSSB. CSSB technology has been well researched in the past (Heathcote 1991; Walker and Stace 1997; Walker 2004; Reddy and Gupta 2005, 2006; Olivier and Mesbah 1987; Reddy et al. 2007; Reddy and Latha 2014 and many other publications).

The CSSB's used in this investigation were made using a local soil–sand mixture containing 10.9% clay size fraction and 7% OPC by mass. Manually operated machine was used to prepare CSSB of size 230 × 109 × 70 mm. Processed soil–cement mixture of known quantity was fed into the machine mould and then compacted. The brick density was controlled by controlling the mass of the soil–cement mix fed into the machine mould. Compacted bricks were cured for 28 days and then air dried for six months. Air-dried specimens were then oven dried at 60 °C to attain constant weight and then used for casting the masonry prisms. The wet compressive strength and water absorption of the CSSB were determined following IS 3495 (1992a, b) guidelines. CSSB had a wet compressive strength of 9.09 MPa (standard deviation 0.83 MPa) and water absorption value of 11.9% (standard deviation 0.012%). The dry density of the CSSB was maintained at 1.85 g/cc. The compressive strength and water absorption values represent the mean of 11 and 10 specimens, respectively.

Characteristics of mortars

Details of the three types of mortars used in the investigations are given in Table 3.1. Cement mortar (CM), cement-lime mortar (CLM) and cement-soil mortar (CSM) were used in the investigations. In the experiments, it is difficult to control the mortar quality through volume batching and hence volume proportions were converted to mass proportions based on the loose bulk densities of the cement, lime, soil and sand. The water–cement ratio corresponding to 85% flow was ascertained from the flow curves shown in Fig. 3.2. The flow tests were carried out following the guidelines given in ASTM C1437 code.

The mortar flow is sensitive to the water–cement ratio. The flow increases with increase in water–cement ratio (Fig. 3.2). For any given flow value, cement mortar and cement-lime mortars need lesser water–cement ratio when compared to the corresponding value for cement-soil mortar. For example, to achieve 90% mortar

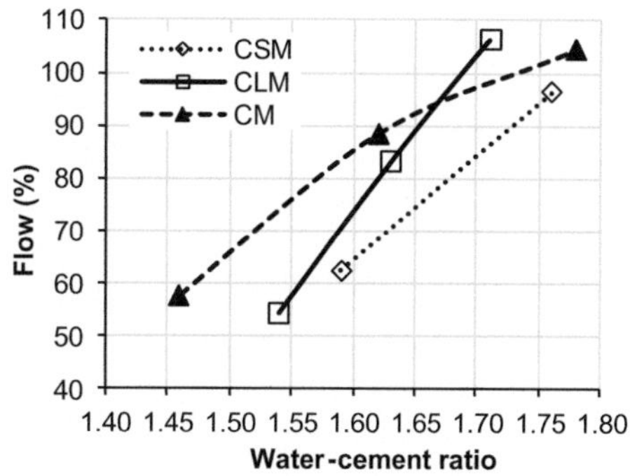

Fig. 3.2 Flow versus water–cement ratio

flow the water–cement ratio required was 1.64, 1.65 and 1.73 for CM, CLM and CSM, respectively. The 28-day compressive strengths of the three types of mortars (with 85% flow) lie in a narrow range (4. 20–4.90 MPa).

3.3 Methodology and Testing Procedures

3.3.1 Rate of Water Absorption in CSSB

Rate of water absorption in the CSSB was monitored using the following procedure.

- Cured and air-dried CSSB were oven dried at 60 °C to attain constant mass. The dry weights were ascertained.
- The oven-dried bricks were then soaked in water and the weights were recorded at the end of 2, 5, 10, 15, 30, 60, 120, 240, 360, 1440, 2880 and 4320 min of soaking.
- The moisture content of the CSSB at different soaking durations was determined based on the initial dry weight of the brick.
- A plot of moisture content of the CSSB with the soaking duration was generated (Fig. 3.3).

3.3.2 Moisture Movement from Fresh Mortar Bed Joint to CSSB

The fresh mortar is flush with water and it gets transported into brick pores during construction. The quantity of water lost by the mortar greatly depends upon the moisture content of the brick during construction. This type of moisture movement can be measured using a simple technique described as follows.

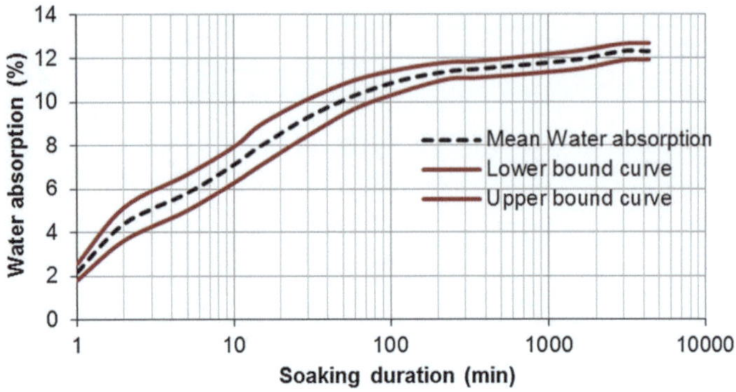

Fig. 3.3 Water absorption versus soaking duration for CSSB

- Cast a two-brick prism sandwiching a mortar bed joint.
- Lay a fresh 12-mm-thick mortar joint (using mortar with 85% flow) on the bed face of a brick and place another brick on top of the freshly laid mortar bed joint.
- Keep the two bricks on top of the freshly cast two-brick prism, as an additional mass.
- Monitor the water content of the fresh mortar bed joint at the end of 15, 30, 45, 60, 120 and 240 min after casting the prism.
- Mortar is sampled at two different places across the bed joint and in two different brick-masonry prisms. Thus, the mean of four samples is reported as moisture content of the mortar.
- The moisture content of the mortar sampled from different locations of the bed joint was determined by oven drying at 100 °C for 24 h.
- Knowing the moisture content of the mortar, water–cement ratio with respect to initial water–cement ratio of the mortar can be estimated.
- The procedure is repeated for three types of mortars (CM, CLM and CSM) using similar CSSB.

Based on the data collected following the above-mentioned procedure, a relationship between the water–cement ratio and the duration of contact with fresh mortar in the bed joint can be generated.

3.3.3 Casting and Testing Masonry Prisms for Bond Strength

The bond development between the CSSB and the mortar can be quantified through the testing of masonry prisms in flexure. The procedure adopted in ascertaining the flexure bond strength is as follows.

- Masonry prisms of size 230 × 110 × 398 mm were cast using CSSB and the three types of mortars. The prism had five bricks and four mortar joints.

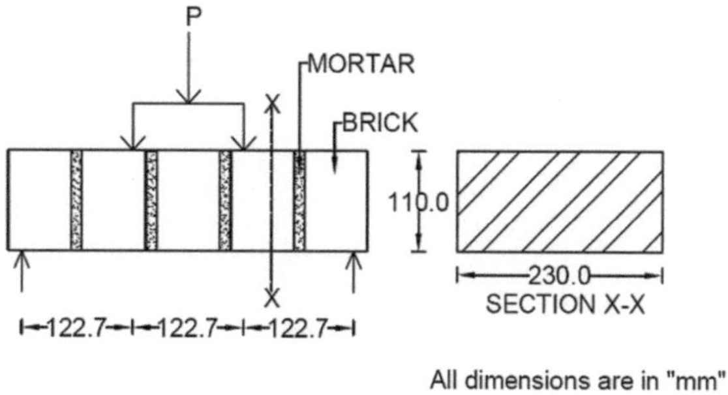

Fig. 3.4 Details of flexure strength test set-up

- The mortar flow of 85% was maintained for all the three types of mortars.
- The moisture content of the CSSBs was varied between dry and saturated condition. In each mortar type, the prisms were cast by using CSSBs having four different moisture contents during casting.
- After 28 days of curing under wet burlap, the masonry prisms were tested for flexure strength under four point bending test. Figure 3.4 shows the details of the prism and the flexure test set-up.
- The prisms soaked in water for 48 h prior to the flexure test.
- Six prisms were tested in each category and the mean values are reported.

Based on the test results generated using the above-mentioned procedure relationship between the flexure bond strength and the moisture content of CSSB during casting can be obtained.

3.4 Results and Discussion

3.4.1 Rate of Moisture Absorption

The rate at which the CSSB blocks absorb water can influence the bond strength development. This is attributed to the movement of cementitious materials from the fresh mortar bed to the block surface during masonry construction. The rate of water absorption in CSSB is dependent upon the quantity of clay, cement content, block density, porosity and pore-size distribution. Figure 3.3 shows a plot of water absorption and the soaking duration for the CSSB used in the experiments. It shows the mean curve representing average of 10 values. Also, the upper- and lower-bound-value curves are shown in the Fig. 3.3. The dry CSSB absorbs water rapidly in the first few minutes, and the rate of water absorption is highest in the initial few minutes. The CSSB absorbs 70% of 24-h water absorption value in 20 min of soaking. In one hour of soaking, the water absorption raises to 85–90% of 24-h water absorption value. Beyond 24 h of soaking hardly any change in the water absorption value. The CSSB does not get completely saturated even after 72 h of soaking in cold water in the ambient room temperature (in the range of 18–26 °C). It attains only 72% saturation even after 72 h of soaking in cold water. This can be attributed to entrapped air in the pores of the CSSBs.

3.4.2 Moisture Movement from the Fresh Mortar Bed Joint to the CSSB

The moisture from the fresh mortar bed joint is absorbed by the dry and partially saturated CSSBs and in the process the water–cement ratio of the mortar gets reduced. The plot in Fig. 3.5 shows relationships for water–cement ratio versus contact duration with the CSSBs using CSM. Similar relationships can be seen for the CM and CLM. These relationships clearly demonstrate that the water–cement ratio of the mortar decreases sharply in the initial one-hour duration of contact with the dry and partially saturated bricks. Beyond one hour of contact duration the water–cement ratio (W/C) remains nearly constant irrespective of the initial brick moisture content and the mortar type. For example, after one-hour contact with the dry brick, the W/C ratios of the mortar were 0.33, 0.34 and 0.58 for CM, CLM and CSM, respectively. The corresponding initial W/C ratios were 1.60, 1.64 and 1.71, respectively. The reduction in W/C ratio (after one-hour contact duration) is maximum for CM and CLM (~80%), and it is about 65% reduction for CSM. Use of nearly saturated bricks, there was only 15–20% reduction in W/C ratio after one hour of contact. The studies demonstrated that the use of partially saturated bricks during construction result in sufficient residual moisture in the fresh mortar for proper hydration of the cement, facilitating bond development. The minimum water–cement ratio required for proper hydration of Portland cement is 0.38.

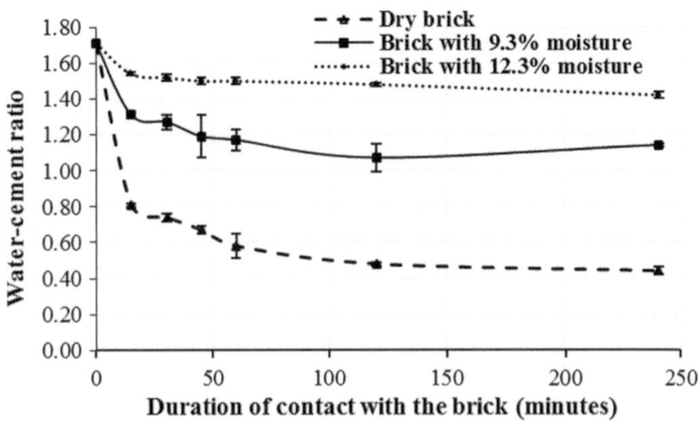

Fig. 3.5 Variation in water–cement ratio with contact duration of CSSBs for CSM

3.4.3 *Masonry Bond Strength and Moisture Content of the CSSB During Construction*

During the masonry construction, the moisture movement from the water-rich fresh mortar joint into the brick, especially when the brick is dry or partially saturated is evident from the discussions in the earlier sections. Apart from many other parameters concerning the masonry unit and the mortar, the moisture content of the masonry unit at the time of construction influences bond development. The results of the tests on flexure bond strength of CSSB masonry prisms considering three types of mortars (CM, CLM and CSM) and four different initial moisture contents for the CSSBs are shown in Fig. 3.6. The figure gives relationships for initial moisture content of the CSSB and the flexure bond strength using three types of mortars. The following observations can be made from these relationships.

1. There is an optimum initial moisture content for the CSSB yielding maximum bond strength.
2. The optimum moisture content is about 75% of 24-h water absorption value for the CSSB using CM and CSM. Whereas for CLM it is about 50%.
3. There is a considerable difference (~50%) in the flexure bond strength of CSSB masonry, between the optimum value and saturated value (water absorption at 24 h). This can be attributed to the fact that when nearly saturated brick is used during masonry construction, there will be insufficient movement of moisture to the brick. This will reduce the movement of quantity of cement hydration products getting into the brick pores, which is responsible for the mechanical interlocking of cement hydration products and the bond development.

The experimental results are in tune with similar observations on bond strength using different types of masonry units such as burnt clay bricks (Sinha 1967; Venu Madhava Rao et al. 1996; Reddy and Gupta 2006).

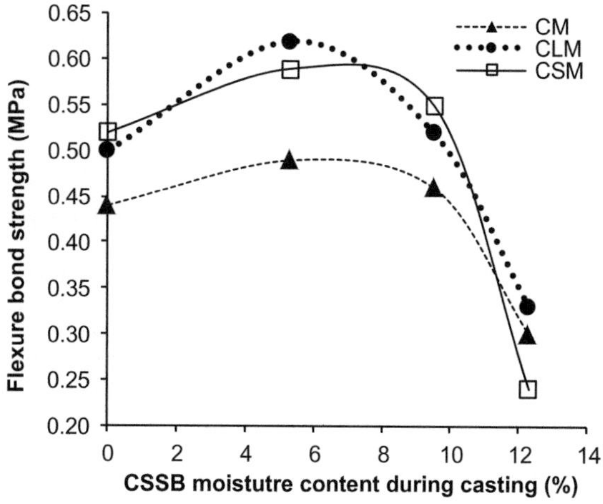

Fig. 3.6 Bond strength versus CSSB moisture content during casting

3.5 Conclusions

The investigations generated results on the rate of moisture absorption, moisture transport between the CSSB and the fresh mortar bed joint, and the influence of moisture content of CSSB on the masonry bond strength. Based on the results, the following conclusions can be drawn.

1. The dry CSSB absorbs water rapidly in the first few minutes, and the rate of water absorption is highest in the initial few minutes. The CSSB absorbs 75% of 24-h water absorption value in 20 min of soaking in water. This information will be of help in determining the soaking duration of CSSB to achieve 75% saturation during masonry construction.
2. During masonry construction using CSSBs, the water–cement ratio of the mortar decreases sharply in the initial one-hour duration of contact with the dry and partially saturated bricks. Beyond one hour of contact with the fresh mortar bed joint, the water–cement ratio remains nearly constant.
3. Use of partially saturated CSSBs (50–75% saturation) during construction result in maximum bond strength for the CSSB masonry.

Practical significance of the investigations is recommending use of partially saturated (~20 min soaking) bricks for the CSSB masonry construction.

References

ASTM C1437–07. Standard test method for flow of hydraulic cement mortar, West Conshohocken, PA, USA

Brocken HJP, Spiekman ME, Pel L, Larbi JA (1998) Water extraction out of mortar during brick laying: a NMR study. Mater Struct 31(1998):49–57

Fitzmaurice RF (1958) Manual on stabilised soil construction for housing. UN Technical Assistance Program New York

Fried AN (1996) The influence of lime and retarded mortars on masonry bond strength. In: Proceedings of 7th North American masonry conference, University of Notre Dame, South Bend, Indiana, USA, pp 217–229

Groot C (1993) Effects of water on mortar-brick bond. PhD thesis, Faculty of Civil Engineering, Delft University of Technology, The Netherlands

Heathcote K (1991) Compressive strength of cement stabilized earth blocks. Build Res Inf 19(2):101–105

IS: 8112–1989. Specification for 43 grade ordinary Portland cement. Bureau of Indian Standards, New Delhi, India

IS: 3495–1992a (Reaffirmed 2002). Methods of tests of burnt clay building bricks—part I: determination of compressive strength. Bureau of Indian Standards, New Delhi, India

IS: 3495–1992b (Reaffirmed 2002). Methods of tests of burnt clay building bricks—part II: determination of water absorption. Bureau of Indian Standards, New Delhi, India

Kampf L (1963) Factors affecting bond of mortar to brick. In: Symposium on masonry testing, American Society of Testing Materials, ASTM STP 320, pp 127–141

Lange DA, DeFord HD, Werner AM (1999) Microstructural investigation of mortar/unit interaction. Masonry Soc J 17(1):T31–T42

Lawrence SJ, Cao HT (1987) An experimental study of the interface between brick and mortar. In: Proceedings of the fourth North American masonry conference, Los Angeles, pp 1–14

Olivier M, Mesbah A (1987) Influence of different parameters on the resistance of earth, used as a building materials. In: International conference on mud architecture, Trivandrum, India

Reddy BVV, Gupta A (2005) Characteristics of soil-cement blocks using highly sandy soils. Mat Struct 38(280):651–658

Reddy BVV, Gupta A (2006) Strength and elastic properties of stabilised mud block masonry using cement–soil mortars'. J Mater Civ Eng 18(3):472–476

Reddy BVV, Latha MS (2014) Influence of soil grading on the characteristics of cement stabilised soil compacts. Mater Struct 47(10):1633–1645

Reddy BVV, Latha MS (2018) Mortar shrinkage and flexure bond strength of stabilized soil brick masonry. J Mater Civil Eng 30(5):05018002

Reddy BVV, Lal Richardson, Nanjunda Rao KS (2007) Optimum soil grading for the soil-cement blocks. J Mater Civil Eng 19(2):139–148

Sinha BP (1967) Model studies related to load bearing brickwork. PhD thesis, University of Edinburgh, Edinburgh, UK

Sarangapani G, Reddy BVV, Groot CJWP (2002) Water loss from fresh mortars and bond strength development in low strength masonry. Masonry Int 15(2):42–47

Venu Madhava Rao K, Venkatarama Reddy BV, Jagadish KS (1996) Flexural bond strength of masonry using various blocks and mortars. Mater Struct 29:119–124

Walker PJ (1995) Strength, durability and shrinkage characteristics of cement stabilised soil blocks. Cement Concr Compos 17:301–310

Walker P (2004) Strength and erosion characteristics of earth blocks and earth block masonry. J Mater Civ Eng 16(5):497–506

Walker P, Stace T (1997) Properties of some cement stabilized compressed earth blocks and mortars. Mat Struct 30:545–551

Chapter 4
Bond Strength of Rebars in Cement-Stabilised Rammed Earth

R. Lepakshi and B. V. Venkatarama Reddy

4.1 Introduction

Rammed earth wall is constructed by compacting partially saturated loose soil using a rigid formwork. These walls are generally 200–400 mm thick. Non-load-bearing and load-bearing walls can be constructed using CSRE. The behaviour of CSRE under gravity loads is widely examined by several researchers in the past (Easton 1982; King 1996; Keable 1996; Hall et al. 2004; Walker et al. 2005; Jayasinghe 2007; Jayasinghe and Kamaladasa 2007; Bui et al. 2007; Reddy and Kumar 2009, 2010). The CSRE walls are subjected to in-plane and out-of-plane bending during the action of lateral loads arising due to the wind and seismic loads. The flexural resistance of CSRE walls under out-of-plane lateral bending is poor, especially when the bending is in the vertical direction. Reinforcing CSRE with rebars can improve the flexural strength. The bond strength between rebars and CSRE matrix is essential for better flexural behaviour of CSRE walls. Therefore, there is a need to understand the bond development and the bond strength between rebars and CSRE matrix.

Bond strength of rebars in concrete has been elaborately investigated (Soroushian and Choi 1989; Abrishami and Mitchell 1996; Cairns and Abdullah 1995; Mo and Chan 1996; Yeih et al. 1997). Standard test procedures are available to assess the bond strength of rebars in reinforced concrete (RILEM 1994; IS 2770–2002). Generally, pullout testing procedure is adapted to quantify the rebar bond strength in concrete (Naaman and Najm 1991; Soroushian et al. 1991; Abrishami and Mitchell 1996; Yeih et al. 1997; Pecce et al. 2001; Achillides and Pilakoutas 2004).

R. Lepakshi (✉) · B. V. Venkatarama Reddy
Department of Civil Engineering, Indian Institute of Science, Bangalore, India
e-mail: lepakshi_raju@yahoo.com

B. V. Venkatarama Reddy
e-mail: venkat@iisc.ac.in

B. V. V. Reddy et al. (eds.), *Earthen Dwellings and Structures*, Springer Transactions in Civil and Environmental Engineering,
https://doi.org/10.1007/978-981-13-5883-8_4

Table 4.1 Experimental programme for pullout tests

Density (kg/m^3)	Testing condition	Type of reinforcement		
		4-mm plain bar	5-mm deformed high-strength bar	8-mm HYSD bar
1800	Dry	–	✓	✓
	Wet	–	✓	✓
1900	Dry	✓	✓	✓
	Wet	–	✓	✓

There are limited studies on the bond strength and flexural strength of CSRE. A brief review of the previous investigations contributing to the bond strength of CSRE is as follows. Investigations on the bond strength of steel reinforcement in CSRE were examined by Walker and Dobson (2001) through rebar pullout tests. The pullout test procedure specified in RILEM (1994) was used in this study. The variables considered include soil type, cement content, type of rebar and the embedment length. The pullout specimens were cylindrical in shape with 150 mm diameter and varying height. Dry density was maintained around 2000 kg/m^3. All the specimens were tested in air-dried condition. The study showed that the compressive strength of CSRE and the pullout bond force are linearly related. Two types of failures were observed for plain and deformed rebars: they were pullout failure and splitting failure. The authors report 80% drop in the bond strength in the plain rebars when compared to the bond strength of deformed rebars.

The behaviour of CSRE beams reinforced with steel wires and aramid bars was examined by Walker et al. (2002) and the results showed that the ductility and flexural strength of CSRE were largely improved.

The bond strength of rebars in CSRE is inadequately explored and a comprehensive study on the rebar pullout behaviour in CSRE matrix is essential. The limited literature provides information on the several parameters influencing the bond between the rebar and CSRE matrix. There are no codes or standards prescribing the pullout bond test procedure for reinforced CSRE; therefore, in the present study, the procedure outlined for pullout testing for concrete specified in IS 2770–2002 was adopted.

The scope of the study includes assessing the bond strength of rebars in CSRE matrix through standard pullout test, establishing the bond force–slip relationships and to examine the influence of the type of reinforcement on the bond strength. The variables considered include dry density and type of reinforcement. Pullout testing was conducted in both dry and wet conditions. Table 4.1 provides the experimental details.

4.2 Characteristics of Materials Used

A reconstituted soil with 16% clay was used. The liquid limit and the plasticity index of the soil were 37.2 and 15%, respectively. The soil had a Proctar optimum moisture content (OMC) of 11.36% and maximum dry density of 1917 kg/m^3. The ordinary Portland cement (OPC) used conformed to IS 12269 (2004) specifications. The 7 days and 28 days compressive strengths of OPC were 38.5 and 57.5 MPa, respectively. Three types of reinforcement used conformed to the guidelines listed in IS 1786–2008. The tensile strengths were 380, 580 and 670 MPa for 4-mm GI wire, 5-mm high-strength ribbed bars and 8-mm high-yield-strength ductile bars, respectively.

4.3 Casting and Testing Methodology

Rammed earth cube specimens (150 mm sides) were prepared by adapting the following procedure. The natural soil was sieved through a 4.75 mm mesh and oven dried at 60 °C until constant weight was attained. The soil was later diluted with dry sand in the ratio 1:1 (soil: sand by mass). A known quantity of cement was mixed with the soil. Water content based on optimum moisture content (OMC) was added to the dry mix to obtain a partially saturated soil mix. Metal moulds were fixed with bottom wooden plates with central grooves. These grooves were made to accommodate the rebars and protrude 50 mm out from the base of the rammed earth cube specimens. Figure 4.1 shows the metal moulds with bottom base plate used for casting rammed earth cube specimens. One end of the rebar of 1 m length was welded to square mild steel plate of dimension 30 × 30 mm and thickness of 2 mm. These plates provided support to accommodate the LVDT to measure the slip at the free end. The surface of the rods was cleaned using acetone to remove dirt and any loose particles. The sides of the mould were greased prior to casting to ensure non-adhesion of soil on to the sides of the mould. Contrary to the procedure listed out in IS 2770—Part 1, helical reinforcement was not provided for the rammed earth specimens unlike the concrete specimens because of the interference in compaction process. The helical reinforcement in concrete specimens ensures the failure of specimens in pullout mode. The specimens were compacted in three layers with the rebars held in vertical position. The specimens were de-moulded after 24 h of casting and were subjected to wet burlap curing for 28 days. A rammed earth cube specimen with rebar is shown in Fig. 4.2. The specimens were later shifted to a platform inside the laboratory and were dried for a week. Immense care was taken to keep the rebar-matrix bond intact during the shifting process. It was difficult to cast specimens with 1800 kg/m^3 dry density embedded with 4-mm plain rebar because of its smaller diameter and smooth surface. Also, specimens with 1900 kg/m^3 dry density with 4-mm plain rebar (galvanised iron wire) in wet condition were difficult to handle as they lost bond while handling.

Fig. 4.1 Moulds with wooden base plates

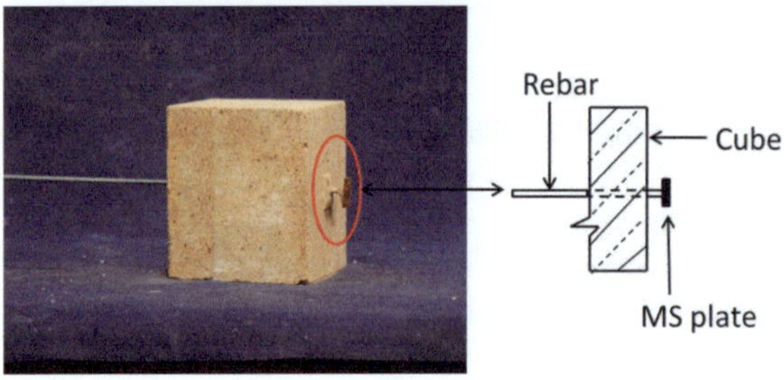

Fig. 4.2 A typical pullout test specimen with rebar

The code emphasises on the measurement of the slip (at the free end) when the rebar is pulled axially, and recording the maximum load taken either until the specimen has failed or the rebar yield point is reached or about 2.5 mm slippage has occurred at the loaded end, whichever is earlier. However, the code lacks in providing an explicit definition for the term bond strength. The code emphasises on monitoring the maximum load taken up by different types of rebars. Eurocode 2 (2005) mentions that the bond strength, just sufficient to prevent the bond failure, is considered as the ultimate bond strength of the material.

Dry testing of rammed earth specimens was carried out using oven-dried specimens and the wet testing after soaking of specimens in water for 48 h. Care was taken while soaking the specimens in water and keeping the rebars upright. All the specimens had an embedment length of 150 mm. Figure 4.3 shows the schematic diagram of the experimental set-up for pullout testing of rammed earth specimens. The testing procedure is as follows.

1. The set-up consisted sandwiching the specimens between the top and bottom plates as shown in Fig. 4.3. The bolts were fastened to hold the specimens firmly in position.

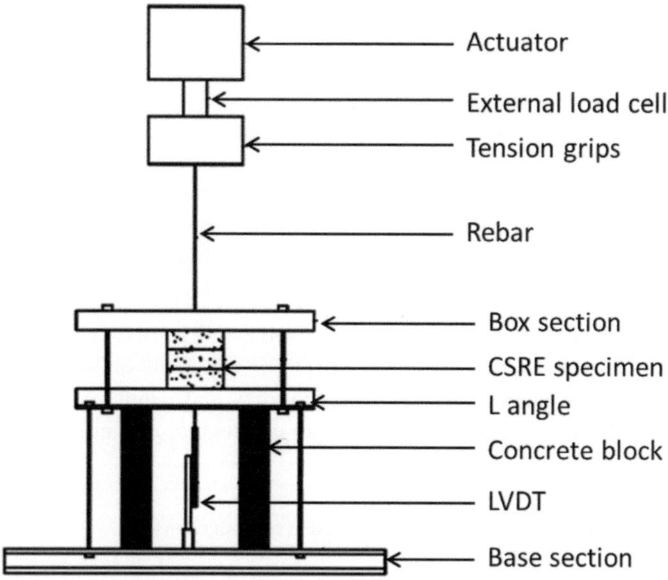

Fig. 4.3 Experimental set-up for pullout testing

2. One end of the rebar was fixed to the tension grips. The slip of the rebar at the free end was recorded by placing a linear variable differential transducer (LVDT). The LVDT was in turn connected to a data acquisition system and the data were recorded.
3. The load was constantly applied to the rebar held in the tension grips. The rate of loading was maintained at 1 mm/min. The specimens were loaded until failure; failure took place either through pullout of rebar or lateral splitting of the CSRE specimen. The ultimate load taken up by each specimen with different types of rebar was noted. Five repeat test specimens were tested in each series.
4. At the end of each test, the moisture content of the samples during the test was determined by collecting the samples of CSRE immediately after the test.

4.4 Results and Discussions

Rebar bond strength of CSRE in dry and wet states, bond force–slip relationships and the different modes of failure of CSRE under rebar pullout were the most significant experimental results inferred from the rebar pullout tests on CSRE.

4.4.1 Bond Strength of CSRE

The code (IS 2770—Part 1) specifies that the bond strength for the case of plain bars is the stress corresponding to the first slip (generally maximum load) and the bond strength for the case of deformed bars is the maximum load that is achieved for a relatively larger slip. Table 4.2 gives details of specimen density, rebar diameter, bond stress at 0.25-mm slip, ultimate bond strength, free-end slip, moisture content of the specimen at the time of testing and the modes of failure. Below are the points inferred from the results provided in Table 4.2.

1. The bond strength in dry condition for 5- and 8-mm rebars is higher than in wet condition irrespective of the dry density. The dry bond strength is 50–60 and 20–45% higher than the wet bond strength for 5- and 8-mm rebars, respectively. The 8-mm bars have higher rib volume per unit length. Higher rib volume corre-

Table 4.2 Pullout test results for CSRE with different types of rebars

Specimen density (kg/m^3)	Rebar diameter (mm)	Testing condition	Bond stress at 0.25 mm slip	Ultimate bond strength (MPa)	Free-end slip at ultimate bond strength (mm)	Moisture content (%)	Mode of failure[a]
1800	5	Dry	0.56	3.14 (0.300)	2.17	3.00	Type A: 5 nos
		Wet	0.58	2.05 (0.353)	2.05	11.30	Type A: 5 nos
	8	Dry	0.37	3.86 (0.485)	2.04	3.20	Type A: 1 nos, Type B: 4 nos
		Wet	0.26	2.40 (0.037)	2.61	11.82	Type A: 2 nos, Type B: 3 nos
1900	4	Dry	0.68	2.78 (0.250)	1.02	3.00	Type A: 5 nos
	5	Dry	0.55	4.59 (0.129)	1.78	–	Type A: 5 nos
		Wet	0.64	3.75 (0.149)	2.19	11.04	Type A: 5 nos
	8	Dry	0.62	5.58 (0.067)	2.10	2.50	Type B: 5 nos
		Wet	0.88	3.82 (0.210)	1.37	11.03	Type A: 1 nos, Type B: 4 nos

[a]Mode of failure: Type A—pullout failure and Type B—splitting failure; values in parentheses indicates standard deviation

sponds to the higher surface area. Hence, the higher bond strength for the 8-mm ribbed bars can be attributed to its higher surface area.
2. Higher density (1900 kg/m^3) CSRE specimens showed higher bond strength when compared to CSRE specimens with a dry density of 1800 kg/m^3 irrespective of the bar diameter and moisture content in the specimen during testing.
3. The range of wet to dry bond strength ratio was between 0.70 and 0.90 for specimens with 1900 kg/m^3 dry density and 0.65 for specimens with 1800 kg/m^3. The difference in bond strength of wet and dry states is attributed to the presence of residual clay minerals in CSRE. The clay minerals loose strength upon saturation.
4. The bond strength of 8-mm deformed bars with CSRE having 1800 kg/m^3 dry density was 3.86 and 2.40 MPa for dry and wet conditions, respectively. Similarly, the bond strength of 8-mm deformed bars with CSRE having 1900 kg/m^3 dry density was 5.58 and 3.82 MPa for dry and wet conditions, respectively. The bond strengths increased as the dry density of the specimens increased and this phenomenon is attributed to the better packing density and lower porosity.

4.4.2 Bond Force–Slip Relationships for CSRE

Bond force–slip relationships for CSRE with plain and deformed rebars are shown in Figs. 4.4, 4.5 and 4.6. The following conclusions can be drawn.

1. Figure 4.4 shows the bond force–slip relationship for CSRE with 1900 kg/m^3 dry density embedded with 4-mm GI wire in dry condition. The force–slip relationship for plain bars is characterised by an increase in bond force with slip and followed by sudden failure (pullout failure), and therefore there is the absence of post-peak profile. In the present study, the maximum load was achieved for a slip corresponding to 1 mm.
2. Figures 4.5 and 4.6 show the relationships for the 5- and 8-mm rebars, respectively with each plot having CSRE specimens with 1800 and 1900 kg/m^3 dry densities. The bond force–slip relationships for deformed bars are characterised by a steady

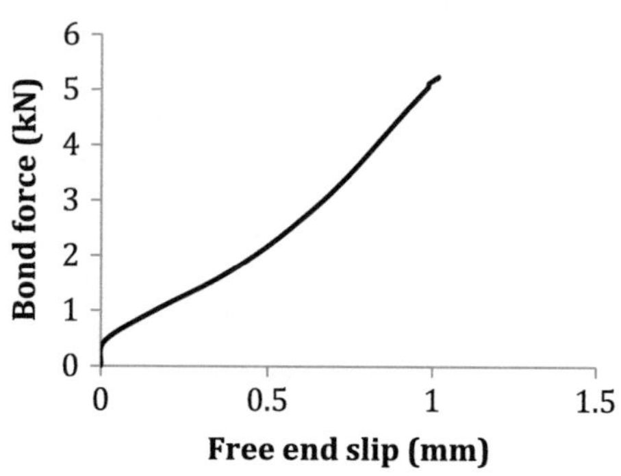

Fig. 4.4 Bond force–slip relationships for 4-mm plain rebar (1900 kg/m^3 dry density) in dry condition

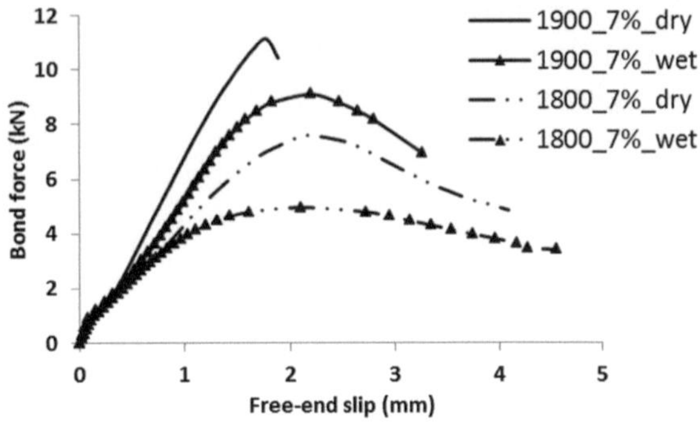

Fig. 4.5 Bond force–slip relationships for 5-mm deformed rebar (1800 and 1900 kg/m^3 dry density). *Note* The legend reads as follows: Dry density_Cement %_Testing condition

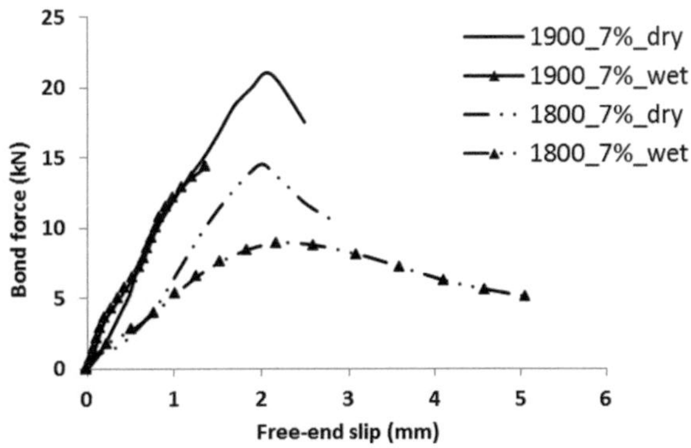

Fig. 4.6 Bond force–slip relationships for 8-mm deformed rebar (1800 and 1900 kg/m^3 dry density). *Note* The legend reads as follows: Dry density_Cement %_Testing condition

increase in the bond force with slip until the ultimate load is reached followed by a drop in the bond force with the slip. The ultimate load was reached for a relatively larger slip when compared to that with plain bars. This behaviour is in accordance with the discussion made in IS 2770–2002. For the case of deformed rebars, the slip corresponding to ultimate bond strength ranges between 1.4 to 3.3 mm for both dry and wet conditions.

4.4.3 Modes of Failure of CSRE

CSRE pullout test specimens exhibited two types of failures. They either failed due to (1) Pullout of rebars or (2) Lateral splitting of rammed earth specimens.

It was observed that all the specimens with plain rebars failed due to rebar pullout as seen in Fig. 4.7, while the specimens with deformed rebars failed either through rebar pullout or lateral splitting of the rammed earth specimens as shown in Fig. 4.8. Table 4.2 lists out the mode of failure of each specimen. Failure due to tensile yielding of rebars was never witnessed during the entire testing regime. This is because the bar experienced very low tensile stress corresponding to the ultimate bond strength.

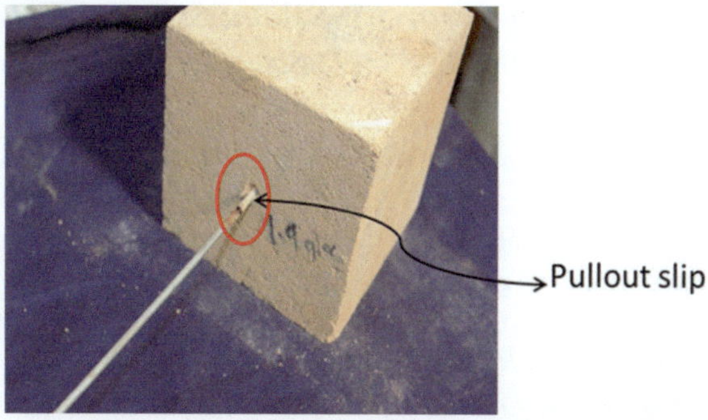

Fig. 4.7 Typical pullout bond–slip failure for plain rebar

Fig. 4.8 Typical splitting failure of deformed rebar during pullout test

The failure modes of CSRE specimens observed in the present study is in accordance with the failure modes of CSRE reported by Walker and Dobson (2001).

The pullout bond strength is dictated by the shear bond strength of the surrounding rammed earth and the surface properties of the rebar. The bond developed between plain rebars and the surrounding matrix is attributed to the adhesion and friction between the rebar and CSRE matrix. The pullout failure occurs when the shear strength of the bond between the rebar and CSRE is exceeded. Lateral splitting mode of failure is common in reinforced concrete specimens embedded with deformed rebar. Embedment length and rebar diameter do not influence this mode of failure for concrete specimens (Cairns and Abdullah 1995). Yeih et al. (1997) mention that the splitting failure occurs when the splitting pressure becomes large and the bond stress is very low comparatively.

4.4.4 Comparison with Bond Strength in Reinforced Concrete

As per the Indian standard code IS 456 (2000), the allowable bond stress for plain bars in concrete is between 1.2 and 1.9 MPa depending upon the compressive strength of concrete. For the case of concrete, the rebar bond strength depends upon a number of factors such as grade of concrete, type of bar and fresh concrete properties. The code also specifies to increase the allowable bond stress by 60% for the case of deformed bars. Investigations of Okelo and Yuan (2005) with 10-mm deformed rebar and concrete of compressive strength 42.8 MPa showed bond strength of 24.4 MPa. The maximum bond strength achieved for CSRE in the present study (1900 kg/m^3 dry density, 7% cement and 8-mm deformed rebars in dry state) was 5.7 MPa. More bond strength tests are required to generalise and to give guidelines for allowable bond strength in rebars for CSRE.

4.5 Conclusions

Based on the investigations on bond strength of rebars in CSRE matrix, the following conclusions can be drawn.

1. The bond between steel rebars and CSRE matrix is sensitive to the density and moisture content of the matrix. Higher the density, higher the bond strength of rebars.
2. The bond strength of deformed rebars is higher than the plain rebars irrespective of the specimen density or cement content. The bond strength using 8-mm high-strength steel bars is 20% more than that of the bond strength using 5-mm bars. This is mainly attributed to high rib surface area in the 8-mm bars.
3. Specimens with plain rebars failed due to rebar pullout. Specimens with deformed rebars failed either due to lateral splitting of CSRE specimens or through rebar pullout mode.
4. The wet and dry conditions significantly affect the rebar bond strength. The residual clay present in CSRE matrix loses strength upon saturation, and thus there is reduction in bond strength.
5. The bond strength in reinforced concrete is much higher when compared to the bond strength in CSRE with rebars.

References

Abrishami HH, Mitchell D (1996) Analysis of bond stress distributions in pullout specimens. J Struct Eng 122(3):255–261

Achillides Z, Pilakoutas K (2004) Bond behaviour of fiber reinforced polymer bars under direct pullout conditions. J Compos Const 8(2):173–181

Bui QB, Hans S, Morel JC (2007) The compressive strength and pseudo elastic modulus of rammed earth. In: Proceedings of the international symposium on earthen structures, Interline Publishers, Bangalore, India, pp 217–223

Cairns J, Abdullah R (1995) An evaluation of bond pullout tests and their relevance to structural performance. Struct Eng 73(11)

Easton D (1982) The rammed earth experience. Blue Mountain Press, Wilseyville, California, USA

Eurocode 2 (2005) Design of concrete structures—part 1-1: general rules and rules for buildings. NF EN 1992-1-1, pp 18–711-1

Hall M, Damms P, Djerbib Y (2004) Stabilised rammed earth and the building regulations 2000, Part A—structural stability. Build Eng 79(6):18–21

IS 456 (2000) Plain and reinforced concrete code of practice. Bureau of Indian Standard, New Delhi, India

IS 12269 (Reaffirmed 2004) (1987) Specification for 53 grade ordinary Portland cement. Bureau of Indian Standard, New Delhi, India

IS 2770 (Part 1) (Reaffirmed 2007) (1967) Method of testing bond in reinforced concrete—(Part 1): pull-out test. Bureau of Indian Standard, New Delhi, India

IS 1786 (2008) High strength deformed steel bars and wires for concrete reinforcement—specifications. Bureau of Indian Standard, New Delhi, India

Jayasinghe C (2007) Shrinkage characteristics of cement stabilised rammed earth. In: Proceeding of international symposium on earthen structures, Bangalore, India, pp 212–216. ISBN 81-7296-051-4

Jayasinghe C, Kamaladasa N (2007) Compressive strength characteristics of cement stabilised rammed earth walls. Constr Build Mater 21(11):1971–1976

Keable J (1996) Rammed earth structures: a code of practice. Intermediate Technology

King B (1996) Buildings of earth and straw: structural design for rammed earth and straw-bale architecture. Ecological Design Press

Mo YL, Chan J (1996) Bond and slip of plain rebars in concrete. J Mater Civ Eng 8(4):208–211

Naaman AE, Najm H (1991) Bond-slip mechanisms of steel fibres in concrete. Mat J 88(2):135–145

Okelo R, Yuan RL (2005) Bond strength of fiber reinforced polymer rebars in normal strength concrete. J Compos Constr 9(3):203–213

Pecce M, Manfredi G, Realfonzo R, Cosenza E (2001) Experimental and analytical evaluation of bond properties of GFRP bars. J Mater Civ Eng 13(4):282–290

Reddy BVV, Kumar PP (2009) Compressive strength and elastic properties of stabilised rammed earth and masonry. Masonry Int 22(2):39

Reddy BVV, Kumar PP (2010) Embodied energy in cement stabilised rammed earth walls. Energy Build 42(3):380–385

RILEM (1994) Technical recommendations for the testing and the use of construction materials. In: International union of testing and research laboratories for materials and structures, E and FN spon, London

Soroushian P, Choi K (1989) Local Bond of deformed bars with different diameters in confined concrete. ACI Struct J 86(2):217–222

Soroushian P, Choi KB, Park GH, Aslani F (1991) Bond of deformed bars to concrete: effects of confinement and strength of concrete. Mater J 88(3):227–232

Walker PJ, Dobson S (2001) Pullout tests on deformed and plain rebars in cement-stabilised rammed earth. J Mater Civ Eng 13(4):291–297

Walker P, Ayala R, Dobson S (2002) Reinforced composite rammed earth in flexure. In: Proceedings 3rd International conference on Non-Conventional Materials & Technologies, University of Bath

Walker P, Keable R, Martin J, Maniatidis V (2005) Rammed earth: design and construction guidelines. BRE Bookshop, Watford

Yeih W, Huang R, Chang JJ, Yang CC (1997) A pullout test for determining interface properties between rebar and concrete. Adv Cem Based Mater 5(2):57–65

Chapter 5
Shear Strength Parameters and Mohr–Coulomb Failure Envelopes for Cement-Stabilised Rammed Earth

R. Lepakshi and B. V. Venkatarama Reddy

5.1 Introduction

Rammed earth technology is vastly used in many parts of the world for wall construction. Since the last several decades, load-bearing walls have been widely constructed using cement-stabilised rammed earth (Verma and Mehra 1950; Easton 1982; Houben and Guillaud 1994; Hall 2002; Walker et al. 2005). The ultimate failure of CSRE structures under compression is observed to be in shear (Walker et al. 2005; Jayasinghe 2007; Jayasinghe and Kamaladasa 2007; Bui et al. 2007; Reddy and Kumar 2009, 2010). Shear failure observed in CSRE is attributed to its monolithic nature of construction. It is essential to characterise the shear strength parameters of CSRE, as it finds its significance in the analysis of CSRE behaviour under compression, in predicting the strength of CSRE, in establishing the failure envelopes for CSRE and in modelling the CSRE failures under different types of loading.

The shear strength of CSRE has been scantily explored. Jaquin et al. (2006) conducted direct shear tests and triaxial tests to determine the shear strength parameters of unstabilised rammed earth. They concluded that Mohr–Coulomb failure criterion can be used to model the rammed earth behaviour. Cheah et al. (2012) investigated stabilised (cement content of 7.7%) rammed earth through triaxial shear tests and triplet shear tests. The shear strength of CSRE obtained through these two methods was considerably different. The study recommends the use of triaxial shear tests with CSRE specimens having aspect ratio of 2:1 to establish the Mohr–Coulomb failure envelopes. The cohesion and angle of internal friction reported in the study were found be 700 kPa and 48°, respectively. Miccoli et al. (2014) conducted diagonal

R. Lepakshi (✉) · B. V. Venkatarama Reddy
Department of Civil Engineering, Indian Institute of Science, Bangalore, India
e-mail: lepakshi_raju@yahoo.com

B. V. Venkatarama Reddy
e-mail: venkat@iisc.ac.in

B. V. V. Reddy et al. (eds.), *Earthen Dwellings and Structures*,
Springer Transactions in Civil and Environmental Engineering,
https://doi.org/10.1007/978-981-13-5883-8_5

compression tests on unstabilised rammed earth wallettes (500 × 500 × 110 mm). The specimens were tilted by 45° and shear force was induced, shear stress–strain plots were obtained. The failure strain reported in the study was about 1.1% and the diagonal shear strength was between 0.65 and 0.84 MPa. Similar diagonal shear tests were conducted by Silva et al. (2014) on unstabilised rammed earth wallettes, the diagonal shear strength reported in this study was about 0.15 MPa. Bui et al. (2016) adopted the discrete element modelling to predict the values of diagonal shear strength of unstabilised rammed earth and recommend discrete element modelling over finite element modelling to predict the behaviour of rammed earth under shear.

The failure envelopes are essential in the analysis and design of rammed earth structural elements. Previous studies do not provide sufficient information on the shear strength of cement-stabilised rammed earth in dry and wet states. Therefore, the present investigation is focused on determining the shear strength parameters for CSRE and establishing failure envelopes for CSRE with two variations in cement content and in both dry and wet conditions.

5.2 Scope of the Study and Experimental Programme

The shear strength parameters of CSRE were determined through triaxial shear tests. CSRE cylindrical specimens were used for the triaxial shear test. The dry density of CSRE specimens was maintained at 1800 kg/m^3. Two cement contents (7 and 10%) and four confining pressures (0, 0.1, 0.25 and 0.5 MPa) were considered. The clay content in the soil was controlled at 16%. The study gives the shear strength parameters and failure envelops for CSRE for two testing conditions: dry and wet. Five specimens were tested in each series.

5.3 Casting and Testing Methodology

The cylindrical specimens of size 76 mm height and 38 mm diameter were cast. The specimens were compacted in three layers with compacted thickness of 25–26 mm. After the compaction of each layer, dents were made to ensure proper interlocking between the layers. Figure 5.1 illustrates the metal mould used for casting the specimens. The specimens were then extruded from the mould using a screw jack set-up. The CSRE specimens were cured under wet burlap for 28 days; air dried for about 7 days and then dried in the oven at 60 °C until the specimens attained constant weight. The specimens were tested after the oven drying process for the case of dry condition and the specimens were soaked in water 48 h prior to the test for the case of wet condition.

Each CSRE specimen was inserted into a protective rubber sheath to prevent the oil in the triaxial cell entering into the specimen during the test. The free ends of the specimens were abutted with two metal discs and sealed with the help of rubber

Fig. 5.1 Cylindrical mould used for casting rammed earth cylinders of height 76 mm and diameter 38 mm

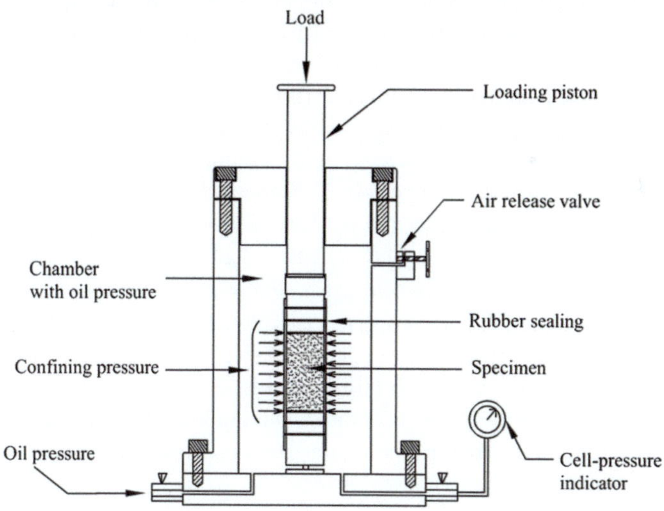

Fig. 5.2 Schematic representation of triaxial test set-up

rings. The CSRE cylindrical specimens with the steel discs were positioned in the triaxial apparatus, and the cell was completely filled with oil through which the lateral pressure was applied. The air release valve was operated for facilitating the escape of any air present inside the triaxial cell. The oil pressure inside the triaxial cell was maintained constant for a particular cell pressure and was controlled by an electrically driven mortar. The axial load was increased until the maximum failure load was reached. Figure 5.2 shows the details of the triaxial test set-up schematically.

5.4 Mohr–Coulomb Envelopes and Shear Strength Parameters

The behaviour of rocks and soils under shear is widely understood through the triaxial tests. The Mohr circles provide a graphical representation of state of stress at a point in terms of normal and shear components. The abscissa contains the possible normal stresses and the ordinate contains the possible shear stresses. A plane in which the shear stresses are zero is called the principal plane. The normal stresses that are acting on this plane are, therefore, called the principal stresses. The confining pressure is the normal stress in the lateral direction (σ_3) that is maintained constant during the triaxial test that was controlled by an electrically driven mortar. The normal stress in the vertical direction was gradually increased until the specimen failure. The maximum normal stress in the vertical direction is nothing but the triaxial compressive strength (σ_1) of CSRE specimens. The principal stresses σ_3 (confining pressure) and σ_1 (triaxial compressive strength) are the intersections of the abscissa that makes the diameter of the Mohr circle. The tangent to the Mohr circles was drawn according to the procedure illustrated by Das (2005) where the Mohr–Coulomb failure criteria is restated in terms of failure stresses given by $\sigma_1 = \sigma_3 \tan^2\left(45 + \frac{\phi}{2}\right) + 2c \tan\left(45 + \frac{\phi}{2}\right)$ where c = cohesion and ϕ = angle of internal friction. Cohesion and angle of friction are the inherent properties of soil, and shear strength is computed from these parameters. The shear strength parameters were obtained for each series and the equation of Mohr–Coulomb failure envelope (tangent to the Mohr circles) in the form of $\tau = c + \sigma \tan \phi$ was generated. Having known the values of cohesion and angle of internal friction for a particular state of stress, the shear strength was computed using the formula $\tau = c + \sigma \tan \phi$.

5.5 Results and Discussions

For different confining pressures, the shear strength and the triaxial compressive strength values for the two cement contents (7 and 10%) in dry and wet conditions were generated. Figures 5.3 and 5.4 show the Mohr circles and the Mohr–Coulomb failure envelops for CSRE with 7% and 10% cement in dry condition. These are

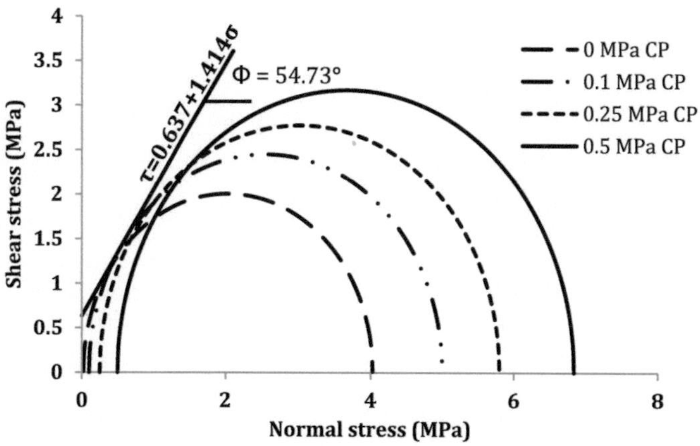

Fig. 5.3 Mohr circles and Mohr–Coulomb failure envelope for 7% cement-stabilised rammed earth in dry condition

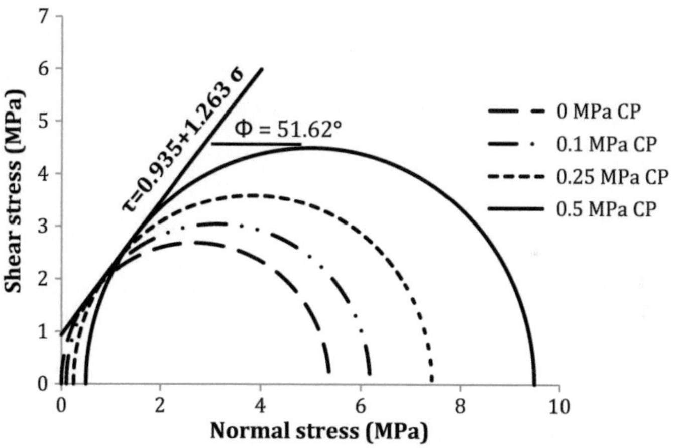

Fig. 5.4 Mohr circles and Mohr–Coulomb failure envelope for 10% cement-stabilised rammed earth in dry condition

typical types. Table 5.1 gives the details of shear strength of CSRE specimens under a given state of stress for both dry and wet conditions. Table 5.2 gives the details of the shear strength parameters and the equations of failure envelopes for the CSRE specimens in dry and wet conditions.

Table 5.1 Triaxial compressive strength and shear strength values for cement-stabilised rammed earth

Confining pressure (MPa)	Testing Condition	Shear strength (MPa)	
		7% cement	10% cement
0	Dry	1.16	1.67
	Wet	0.61	1.07
0.10	Dry	1.42	1.89
	Wet	0.66	1.11
0.25	Dry	1.60	2.23
	Wet	0.91	1.34
0.50	Dry	1.83	2.79
	Wet	1.02	1.45

Table 5.2 Shear strength parameters and failure envelope equations for cement-stabilised rammed earth

Cement percentage (%)	Testing condition	Equation of failure envelope	Shear strength parameters	
			Cohesion, C (MPa)	Angle of internal friction, ϕ (°)
7%	Dry	$\tau = 0.637 + 1.414\ \sigma$	0.64	54.7
	Wet	$\tau = 0.379 + 0.762\ \sigma$	0.38	37.32
10%	Dry	$\tau = 0.935 + 1.263\ \sigma$	0.94	51.60
	Wet	$\tau = 0.69 + 0.672\ \sigma$	0.69	33.89

5.5.1 *Shear Strength Parameters and Failure Envelopes of Cement-Stabilised Rammed Earth*

1. Cement content in CSRE influences the cohesion value irrespective of confining pressure. The cohesion value ranges between 0.6 and 0.9 MPa in dry condition and 0.4 and 0.7 MPa in wet condition. The value of angle of internal friction is not influenced by any parameters and ranges between 50° to 55° and 33° to 37° for dry and wet conditions, respectively.
2. Mohr–Coulomb failure envelops for 7 and 10% cement content stabilised rammed earth in dry and wet conditions are shown in Fig. 5.5. The corresponding equations of these failure envelopes are given in Table 5.2.

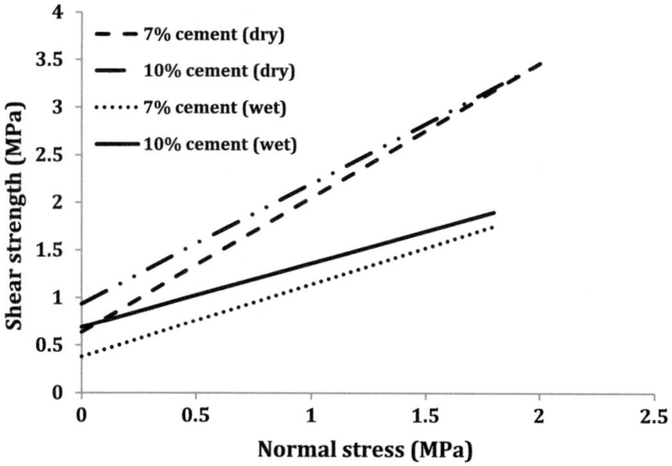

Fig. 5.5 Mohr–Coulomb failure envelopes for 7 and 10% cement content stabilised rammed earth in dry and wet conditions

5.5.2 Shear Strength of Cement-Stabilised Rammed Earth

1. Triaxial compressive strength and the shear strength increased as the cement content was increased in both dry and wet conditions.
2. The triaxial compressive strength and the shear strength of CSRE increased as the confining pressure was increased from 0 to 0.5 MPa. The shear strength of CSRE ranged between 1.2 and 2.8 MPa for dry specimens and between 0.6 and 1.5 MPa for wet specimens. The percentage increase in shear strength of CSRE was about 130 and 150% for dry and wet specimens when the cement content was increased from 7 to 10%, respectively.

New Zealand earth building code (NZS 4297 1998) mentions the design shear strength to be around 0.08 MPa for standard grade construction and 7% of the compressive strength for the case of engineered earth structures. Middleton (1992) gives allowable shear strength in the form of an empirical formula (which accounts for the normal stress) that mentions the shear stress allowance to be 10 kPa per metre depth of rammed earth. The angle of internal friction corresponding to this value is 27°, which was determined computing the stress developed by 1 cubic metre of rammed earth (with dry density = 2000 kg/m^3). The literature does not provide enough information on the allowable shear strength values for CSRE. Cheah et al. (2012) mention the lack of experimental evidence in few codes (such as General Construction Bureau 1991; Standards Association of Zimbabwe 2001) on the allowable design shear strength values for earthquake-prone areas. The present study is one of the preliminary investigations to compute the shear strength parameters and develop failure envelopes.

5.6 Practical Significance and Use of the Results

CSRE is used for the construction of foundations, retaining walls, walls, etc.; therefore, its compressive strength and shear strength are equally important. The shear strength parameters and the Mohr–Coulomb failure envelopes obtained from the present study are used in the following.

1. In the analysis of CSRE behaviour under compression,
2. In determining the in-plane shear strength of CSRE walls,
3. In predicting the CSRE strength and in modelling of CSRE failures under different types of loading.

Mohr–Coulomb failure envelopes give the information of the material strength under the given state of stress. The equations of the form $\tau_{\mathrm{csre}} = c + \sigma \tan \phi$ were generated. Otto Mohr proposed the general strength theory (Lambe and Whitman 1979) and explains the physical meaning of the failure envelope. The general strength theory states that the material is stable if the Mohr circle lies entirely below the failure envelope for a given state of stress. The material has attained full strength on some plane through the soil when the failure envelope is tangential to the Mohr circles. At any given state of stress, Mohr circle cannot intersect the failure envelope, because this represents the failure of the material. Figures 5.3 and 5.4 illustrate the Mohr–Coulomb failure envelops for CSRE specimens with 7 and 10% cement content, respectively, in dry condition, where the practical significance of the failure envelopes can be better understood.

5.7 Conclusions

CSRE cylindrical specimens were subjected to triaxial tests to obtain the shear strength parameters and failure envelopes in dry and wet conditions. Two cement content and four confining pressures were considered as variables. The observations from the tests revealed that the cement content in CSRE influences the shear strength and cohesion irrespective of the confining pressure and the testing condition. The triaxial compressive strength and the shear strength increase as the confining pressure increases. The shear strength in wet condition is almost half of shear strength in dry condition. Failure envelopes of the form $\tau_{\mathrm{csre}} = c + \sigma \tan \phi$ were generated.

References

Bui QB, Hans S, Morel JC (2007) The compressive strength and pseudo elastic modulus of rammed earth. In: Proceedings of the international symposium on earthen structures, Interline Publishers, Bangalore, India, pp 217–223

Bui TT, Bui QB, Limam A, Morel JC (2016) Modelling rammed earth wall using discrete element method. Continuum Mech Thermodyn 28(1–2):523–538

Cheah JS, Walker P, Heath A, Morgan TK (2012) Evaluating shear test methods for stabilised rammed earth. Proc Inst Civ Eng Constr Mat 165(6):325–334

Das BM (2005) Fundamentals of geotechnical engineering. Thomson Brooks/Cole, United States of America

Easton D (1982) The rammed earth experience. Blue Mountain Press, Wilseyville, California, USA

General Construction Bureau (1991) New Mexico Adobe and Rammed earth building code. General Construction Bureau, USA

Hall M (2002) Rammed earth: traditional methods, modern techniques, sustainable future. Build Eng 77(11):22–24

Houben H, Guillaud H (1994) Earth construction: a comprehensive guide. Intermediate Technology Publications/CRATerra-EAG, London

Jaquin PA, Augarde CE, Gerrard CM (2006) Analysis of historic rammed earth construction. In: Proceedings of the 5th international conference on structural analysis of historical constructions, New Delhi, India, pp 1091–1098

Jayasinghe C (2007) Shrinkage characteristics of cement stabilised rammed earth. In: Proceeding of international symposium on earthen structures, Bangalore, India, pp 212–216. ISBN 81-7296-051-4

Jayasinghe C, Kamaladasa N (2007) Compressive strength characteristics of cement stabilised rammed earth walls. Constr Build Mater 21(11):1971–1976

Lambe TW, Whitman RV (1979) Soil mechanics, SI Version. Wiley, p 553

Miccoli L, Müller U, Fontana P (2014) Mechanical behaviour of earthen materials: a comparison between earth block masonry, rammed earth and cob. Constr Build Mater 61:327–339

Middleton GF (1992) Bulletin 5—(revised by Schneider, L. M., 1987). Earth wall construction. Commonwealth Scientific and Industrial Research Organisation (Division of Building Construction and Engineering), 4th edition, North Ryde, Australia. ISBN 0 643 0544 9

NZS 4297 (1998) Engineering design of earth buildings. Standards New Zealand, New Zealand

Reddy BVV, Kumar PP (2009) Compressive strength and elastic properties of stabilised rammed earth and masonry. Masonry Int 22(2):39

Reddy BVV, Kumar PP (2010) Embodied energy in cement stabilised rammed earth walls. Energy Build 42(3):380–385

Silva RA, Oliveira DV, Schueremans L, Miranda TF, Machado J (2014) Shear behaviour of rammed earth walls repaired by means of grouting. In: 9th international Masonry conference

Standards Association of Zimbabwe, SAZS 724 (2001) Rammed earth structures. Zimbabwe standard code of practice, Harare, Zimbabwe

Verma PL, Mehra SR (1950) Use of soil-cement in house construction in the Punjab. Ind Concr J 24(4):91–96

Walker P, Keable R, Martin J, Maniatidis V (2005) Rammed earth: design and construction guidelines. BRE Bookshop, Watford

Chapter 6
Influence of Normal Stress and Bonding Techniques on Shear Bond Strength of Rammed Earth

S. N. Ullas, G. S. Pavan and K. S. Nanjunda Rao

6.1 Introduction

Rammed earth walls are constructed by compacting processed soil–sand mixture with or without stabiliser in layers within a rigid formwork. In general, un-stabilised rammed earth walls are thick and bulky in comparison with brick or stone masonry walls. In order to improve the strength and durability properties of rammed earth, stabilising agents such as cement and lime are added to the soil–sand mixture. Cement-stabilised rammed earth (CSRE) walls have an increased load carrying capacity and are relatively slender. CSRE construction activity has witnessed a steady growth over the last decade, and several structures have been built worldwide. The renewed interest in earth-based materials for construction has necessitated the need for understanding their mechanical behaviour.

CSRE walls, like every other wall system, transmit loads from the overlying beam–slab system to the underlying support system (e.g. foundation). Hence, CSRE walls are primarily subjected to in-plane compressive loading. However, CSRE walls are subjected to in-plane shear loading and out-plane bending during seismic events, wind loading during a hurricane, flooding, uneven settlement of foundation etc. During the in-plane shear action on CSRE walls, shear stresses develop along the layer interface present in CSRE elements. A weak layer interface in rammed earth leads to a sliding shear failure of CSRE wall. Hence, knowledge of joint shear strength, its

S. N. Ullas
Centre for Sustainable Technologies, Indian Institute of Science, Bangalore, India
e-mail: snullas@gmail.com

G. S. Pavan (✉) · K. S. Nanjunda Rao
Department of Civil Engineering, Indian Institute of Science, Bangalore, India
e-mail: pvnstr@gmail.com

K. S. Nanjunda Rao
e-mail: ksn@iisc.ac.in

B. V. V. Reddy et al. (eds.), *Earthen Dwellings and Structures*,
Springer Transactions in Civil and Environmental Engineering,
https://doi.org/10.1007/978-981-13-5883-8_6

behaviour under normal stresses, is essential for the design of load-bearing CSRE wall elements. Techniques to improve the joint shear strength of CSRE may also be required and needs to be explored.

Until now, research studies have largely focused on understanding the mechanical behaviour of CSRE under compressive loading. Studies on the joint shear behaviour of CSRE have been limited. Cheah and da Silva (2007) conducted an experimental study to assess the shear strength of cement stabilised rammed earth reinforced with flax fibres. The test was conducted according to ASTM E519-00. Diagonal CSRE panels of size 1.2 m × 1.2 m were cast and subjected to compressive loading along the diagonal. Shear strength of CSRE was computed from the failure load using the formula suggested by ASTM E5519-00.

Cheah et al. (2012) assessed the performance of existing shear test methods for evaluating the shear strength of cement stabilised rammed earth reinforced with natural fibres. Cheah et al. (2012) considered two test methods: (i) triaxial compression test (ii) masonry triplet test. It was found that the shear strength of CSRE obtained from these two experimental methods was higher than the value of shear strength suggested by existing building codes. Among the two test methods, Cheah et al. (2012) advocated triaxial compression test method for determining the shear strength of CSRE rammed earth material. Nanjunda Rao et al. (2011) assessed the shear strength of CSRE parallel and perpendicular to compacting layers using shear box test apparatus. They have reported that shear strength increases with increase in normal stress in case of shear load parallel to compacting layers. However, there is no influence of normal stress on shear strength in case of shear load perpendicular to compacting layers.

However, several studies exist in literature (Ehsani et al. 1997; Tomaževič 2009; Roca and Araiza 2010; Mojsilović 2012; Alecci et al. 2013; Pavan and Nanjunda Rao 2015) that have adopted triplet test method for assessing the brick–mortar interface shear strength of masonry. Masonry triplet test is best suitable for determining the joint shear strength of both, brick–mortar interface strength of masonry and the interface joint shear strength of rammed earth. Joint shear strength of rammed earth is essential to avoid sliding shear failure mode. Shear strength of rammed earth material may be determined by either, geotechnical trial-axial compression test or the ASTM E519-00 diagonal shear-compression test.

The present study aims at assessing the joint shear strength of CSRE and the influence of normal stress on the joint shear strength of CSRE by adopting masonry triplet test protocol. In addition, three bonding techniques are explored to determine the shear strength of rammed earth joints. The combined influence of the each of these bonding techniques in the presence of normal stress on the joint shear strength of CSRE in wet and dry condition is explored in this study.

6.2 Experimental Program

6.2.1 Characteristic Studies on Materials and Mix Proportion

Locally available red loamy soil and river sand were used in preparation of rammed earth test specimens. The particle size distribution for soil was determined (IS 2386, Part I) to obtain sand, silt and clay size fraction of the soil. This information is essential for reconstituting the soil by adding sand to bring the clay content in the soil within the optimum range. Figure 6.1 shows grain size distribution of soil and sand used in the present study. The clay, silt and sand sized fraction in the soil were about 43, 15 and 42%, respectively. Hence, soil to sand ratio of 1: 1.8 was adopted so that the clay content in the reconstituted soil was about 15%. Ordinary Portland Cement conforming to Indian Standard (IS 8112) was used as the stabiliser for rammed earth and also in cement slurry used between rammed earth layers for few bonding techniques. The casting moisture content was 12.5% (based on optimum moisture content and maximum dry density relationship obtained through standard proctor test), and the dry density targeted was 1850 kg/m^3 for which the specimens were cast with wet density of 2080 kg/m^3.

6.2.2 Preparation of Test Specimens

Triplet specimens similar to that widely used in evaluating the masonry shear strength were used in the present study. The size of the specimens was 230 mm × 230 mm × 75 mm and was cast in three layers of equal thickness. Figure 6.2 shows the dimensions of the triplet specimen. The specimens were cast in a battery of stiff wooden moulds. Three types of bonding techniques, namely (a) making conical

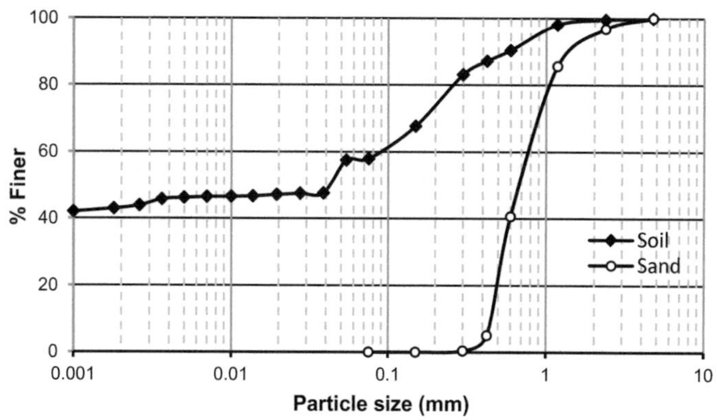

Fig. 6.1 Particle size distribution of soil and sand

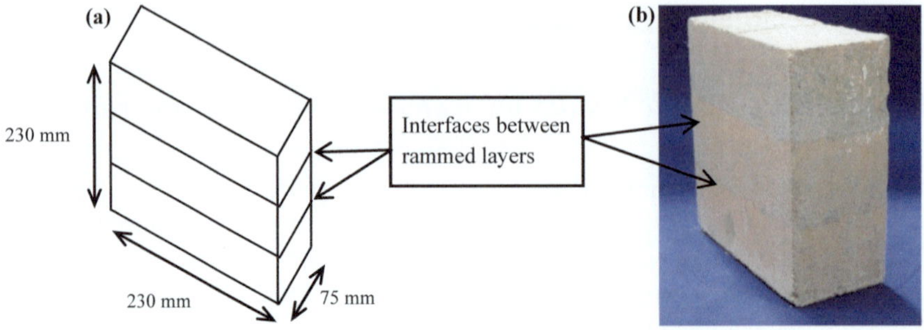

Fig. 6.2 **a** Schematic representation of specimen, **b** actual specimen in dry condition

Fig. 6.3 Casting of rammed earth shear test specimen

dents on freshly made layer before casting the next layer (Tri-A), (b) applying a coat of fresh cement slurry (Tri-B) and (c) combination of both 'A' and 'B' (Tri-C) were examined. Figure 6.3 shows casting of the specimen using stiff wooden mould. The specimens were demoulded after 24 h of casting and cured under wet burlap for 28 days.

6.2.3 Testing of Specimens

The rammed earth triplet specimens were tested in both dry condition (28 days burlap cured specimens stored inside the laboratory for four weeks prior to testing)

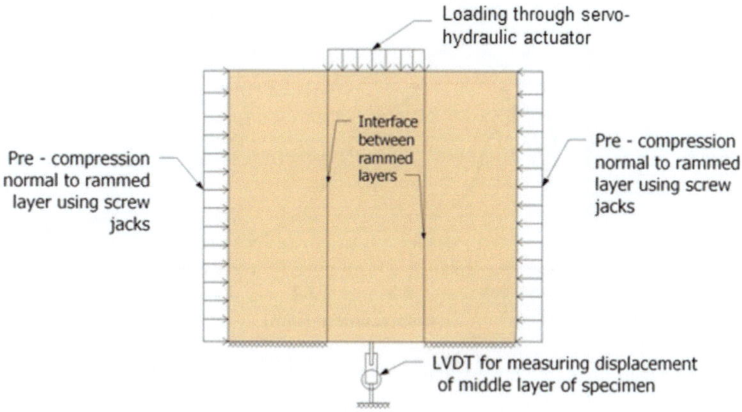

Fig. 6.4 Schematic representation of shear triplet specimen under testing

and wet condition (28 days burlap cured specimens stored inside the laboratory for four weeks and soaked in water for two days prior to testing). The triplet specimens were tested using strain controlled material testing system where the required levels of pre-compressions (0.05, 0.3 and 0.9 MPa) perpendicular to the rammed earth layers were applied through screw jacks. Figure 6.4 shows the schematic of the experimental setup. The extreme layers were firmly supported at the base, and the longitudinal force (parallel to the joints) was applied on middle layer which was free to deform. The longitudinal displacement in the middle layer was recorded using LVDT. The moisture content in the specimens at the time of testing was determined and is reported in Sect. 6.3.

6.3 Results and Discussion

The results of the experimental investigations are presented in terms of shear strength and deformation for various types of bonding between rammed layers in dry and wet conditions. The curves presented are for average of four test specimens. As mentioned in earlier section, the moisture content in the specimens at the time of testing was determined. For CSRE triplet specimens in dry condition, the average moisture content was found to be about 4%, and in the wet condition, the moisture content was found to be about 12%. The influence of pre-compression normal to the rammed layers on the shear strength is also discussed. Figures 6.5, 6.6 and 6.7 presents the load versus displacement response of CSRE triplet specimens in dry condition for various values of pre-compression for three different bonding techniques discussed earlier. Figures 6.8, 6.9 and 6.10 presents the load versus displacement response of CSRE triplet specimens in wet condition for various values of pre-compression for three different bonding techniques. From the above figures, it can be observed that

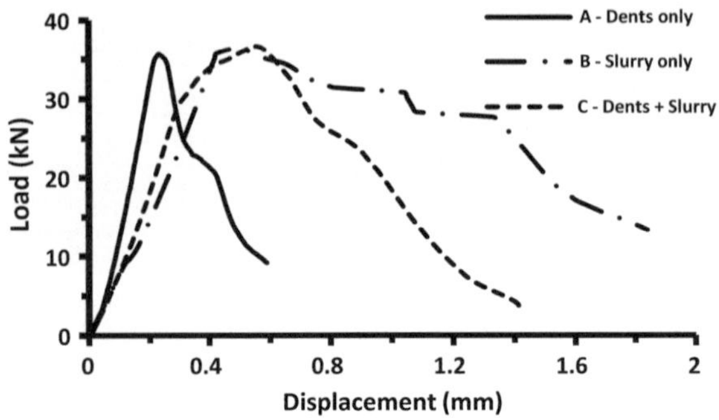

Fig. 6.5 Load versus displacement response in dry condition with 0.05 MPa normal stress

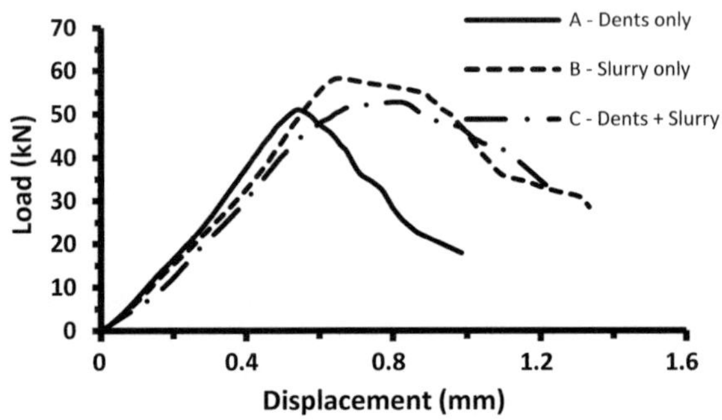

Fig. 6.6 Load versus displacement response in dry condition with 0.3 MPa normal stress

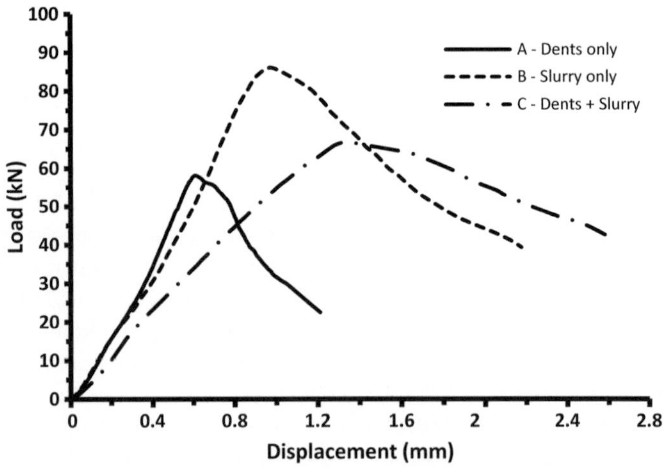

Fig. 6.7 Load versus displacement response in dry condition with 0.9 MPa normal stress

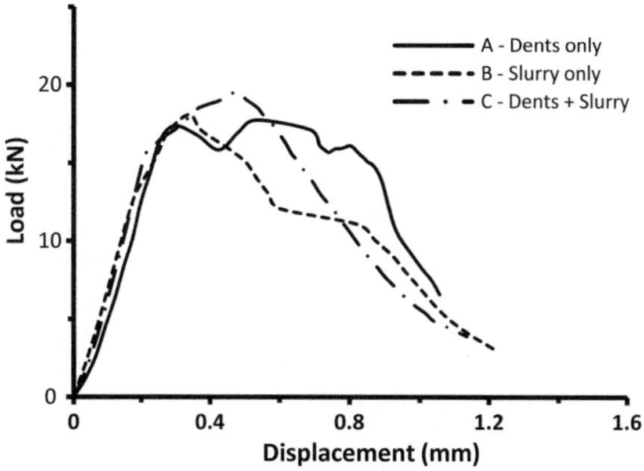

Fig. 6.8 Load versus displacement response in wet condition with 0.05 MPa normal stress

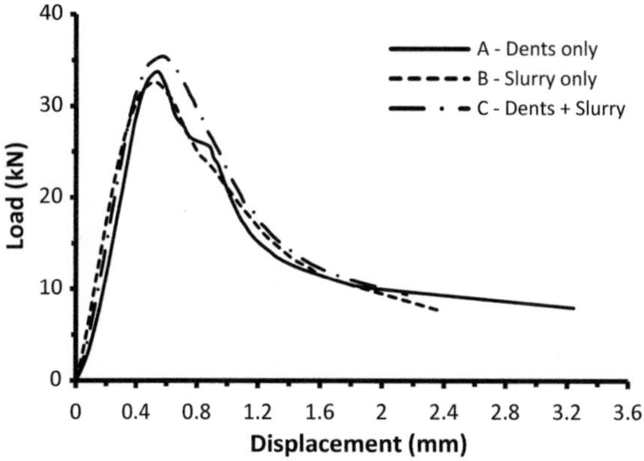

Fig. 6.9 Load versus displacement response in wet condition with 0.3 MPa normal stress

at low values of pre-compression (0.05 MPa) there is no significant influence of different bonding techniques on the shear strength of the triplet specimen in both dry and wet conditions. However, it can be seen that shear strength in the dry condition is more than in the wet condition by about 60–90% for the case of triplet specimens with only fresh cement slurry between rammed layers for all the three values of pre-compression considered in the present study. From Fig. 6.11, it can be observed that in the dry condition for pre-compression values greater than 0.05 MPa, the use of fresh cement slurry has enhanced the shear strength of the triplet specimens. The average value of cohesion (c) in dry condition is 0.781 MPa, and the friction angle (ϕ) ranges between 26.2° and 49.2°. In Fig. 6.12, it can be observed that in the wet condition for all the pre-compression values there is no significant influence of bonding techniques on the shear strength of triplet specimens. The value of cohesion (c) in wet condition is about 0.39 MPa, and the friction angle (ϕ) is about 39.2°. In general, the stiffness

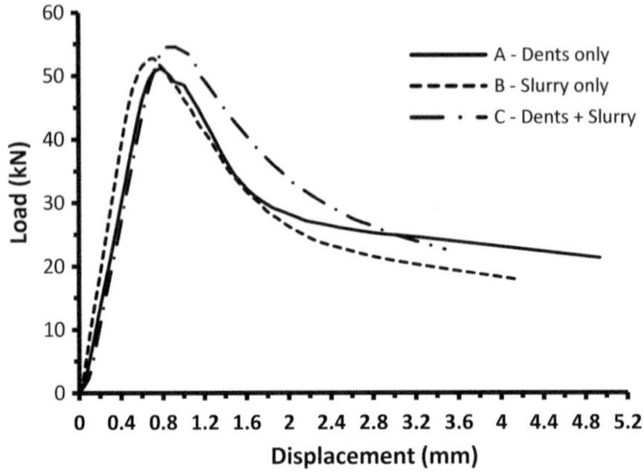

Fig. 6.10 Load versus displacement response in wet condition with 0.9 MPa normal stress

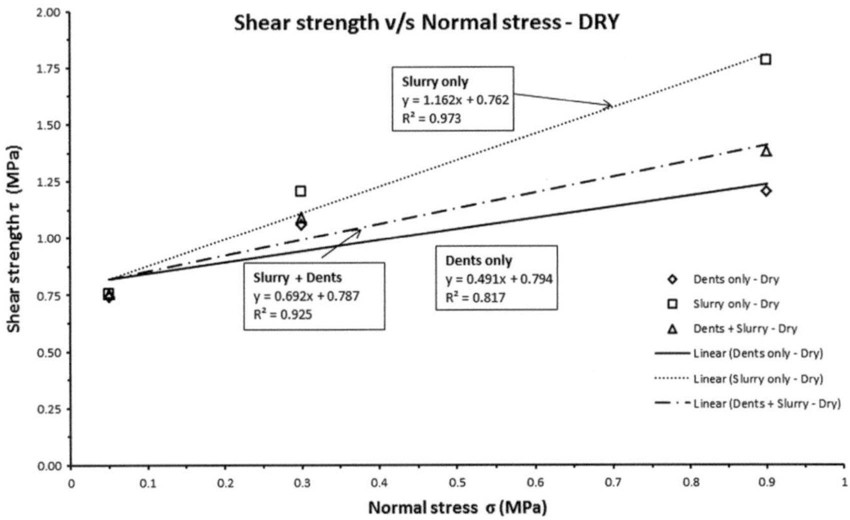

Fig. 6.11 Variation of shear strength as a function of pre-compression in dry state

of the triplet specimen increases with increase in pre-compression in both dry and wet state.

The failure patterns of CSRE triplet specimens are shown in Fig. 6.13. In the case of Tri-A specimens which had only dents for bonding between rammed layers, the failure was essentially along the interface. However, in the case of specimens (Tri-B and Tri-C) with cement slurry only and both dents and cement slurry as bonding between rammed layers, there was material failure in the middle portion of the specimen indicating that shear strength of interface was more than that of the material.

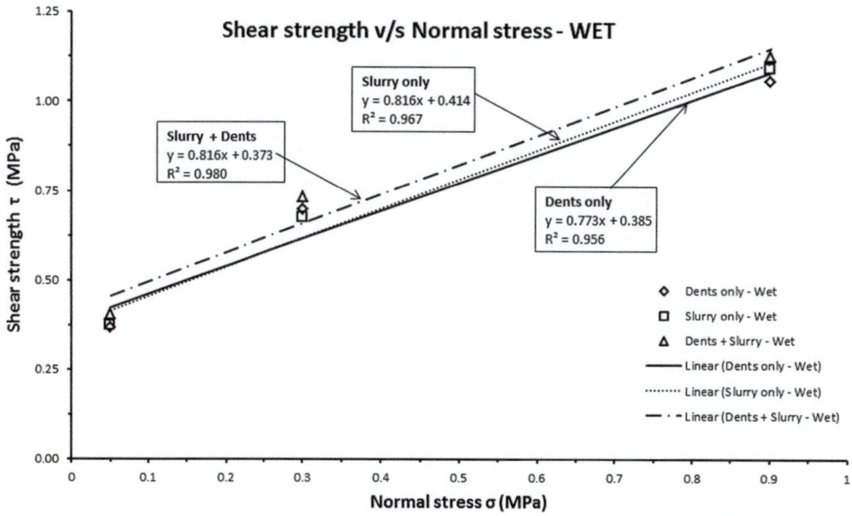

Fig. 6.12 Variation of shear strength as a function of pre-compression in wet state

Fig. 6.13 **a** Interface failure of Tri-A specimen under high pre-compression (0.9 MPa), **b** material failure of Tri-B specimen under high pre-compression (0.9 MPa), **c** material failure of Tri-C specimen under high pre-compression (0.9 MPa)

6.4 Conclusions

Shear triplet specimen test is simple and commonly used procedure to evaluate shear bond strength of brick–mortar interface in masonry and this test procedure can be effectively used to understand the interface shear strength between rammed earth layers. The effect of various bonding techniques on shear bond strength depends on moisture content of the specimen at the time of testing. Amount of pre-compression applied normal to the interfacial layer influences the shear bond strength and stiffness of the CSRE triplet specimens.

Acknowledgements The authors would like to gratefully acknowledge the financial support provided by DST, Govt. of India, New Delhi through the project Ref. No. SR/S3/MERC-0077/2010 for carrying out the research investigations reported in this paper.

References

Alecci V, Fagone M, Rotunno T, De Stefano M (2013) Shear strength of brick masonry walls assembled with different types of mortar. Constr Build Mater 40:1038–1045
ASTM E519-00 (2010) Standard test method for diagonal tension (shear) in masonry assemblages. American Society for Testing and Materials, West Conshohocken, Pa, USA
Cheah JS, da Silva R (2007) Material testing of flax-fibre reinforced rammed earth. The University of Auckland, Auckland
Cheah JS, Walker P, Heath A, Morgan TK (2012) Evaluating shear test methods for stabilised rammed earth. Proc Inst Civ Eng Constr Mater 165(6):325–334
Ehsani MR, Saadatmanesh H, Al-Saidy A (1997) Shear behavior of URM retrofitted with FRP overlays. J Compos Constr 1(1):17–25
IS: 2386 (Part I)–1963 (Reaffirmed 2007) Methods of test for aggregates for concrete: Part I Particle size and shape. Bureau of Indian Standards, New Delhi, India
IS: 8112–1989 (Reaffirmed 2005) 43 Grade ordinary Portland cement—Specification. Bureau of Indian Standards, New Delhi, India
Mojsilović N (2012) Masonry elements with damp-proof course membrane: assessment of shear strength parameters. Constr Build Mater 35:1002–1012
Nanjunda Rao KS, Venkatarama Reddy BV, Maheshreddy G (2011) Strength and elastic properties of cement stabilized rammed earth. In: National conference on recent developments in civil engineering, VVIET, Mysore, Karnataka, India
Pavan GS, Nanjunda Rao KS (2015) Behavior of Brick-Mortar interfaces in FRP-strengthened masonry assemblages under normal loading and shear loading. J Mater Civ Eng 28(2):04015120
Roca P, Araiza G (2010) Shear response of brick masonry small assemblages strengthened with bonded FRP laminates for in-plane reinforcement. Constr Build Mater 24(8):1372–1384
Tomaževič M (2009) Shear resistance of masonry walls and Eurocode 6: shear versus tensile strength of masonry. Mater Struct 42(7):889–907

Chapter 7
Characteristics of Flowable Stabilised Earth Concrete

K. Gourav and S. N. Ullas

7.1 Introduction

Soil has been extensively used for various construction purposes since ages. Soil is one of the most abundantly available materials used for the production of construction products such as adobe, cob, stabilized soil block and rammed earth. Generally, rammed earth technique is used to produce load-bearing monolithic walls (Verma and Mehra 1950; Walker et al. 2005; Easton 1982, 2008; Hall 2002; Kotak 2007; Reddy and Kumar 2009, 2011; Reddy et al. 2017). But, rammed earth construction demands rigid formwork.

Concrete made with soil is being used for pavements and non-structural components. Studies of Arooz and Halwatura (2018) have used mud–concrete for the production of blocks and have checked the durability aspects of the block. Damme and Houben (2017) have attempted to improve the workability and the strength of raw earth by controlling the dispersion of its fine fractions. There are hardly any studies on the workability and strength of lean and rich flowable earth concrete mix proportions.

The scope of the present study focusses on examining the influence of soil content on workability and compressive strength of lean and rich concrete mixes.

K. Gourav (✉)
Department of Civil Engineering, The National Institute of Engineering, Mysore, India
e-mail: gouravmys@gmail.com

S. N. Ullas
Centre for Sustainable Technologies, Indian Institute of Science, Bangalore, India
e-mail: snullas@gmail.com

B. V. V. Reddy et al. (eds.), *Earthen Dwellings and Structures*,
Springer Transactions in Civil and Environmental Engineering,
https://doi.org/10.1007/978-981-13-5883-8_7

7.2 Materials and Methods

7.2.1 Materials Used in the Experimental Investigations

Following materials were used:

i. Ordinary Portland Cement (OPC) confirming to BIS IS 12269-2013 (2013)
ii. Aggregates:

   a. Fine aggregates—Manufactured sand (M-Sand)
   b. Coarse aggregates—Crushed granite

iii. Soil: Red loamy soil having clay, silt and sand size fractions of about 30, 13 and 57%, respectively.

Figure 7.1 shows the particle size distribution curves for soil and M-Sand used in the investigation.

7.2.2 Experimental Programme and Testing Procedures

Two nominal mix proportions of 1:3:5 and 1:5:8 (cement: fine aggregate: coarse aggregate) by mass was chosen as control mix considering the nominal mix proportions given in BIS IS 456-2000 (Reaffirmed 2005) (2005). Two mix proportions were selected, namely Mix 1 (1:3:5) and Mix 2 (1:5:8). The coarse aggregates used were 20 and 12 mm in size, in equal proportions. The fine aggregates in the mix were

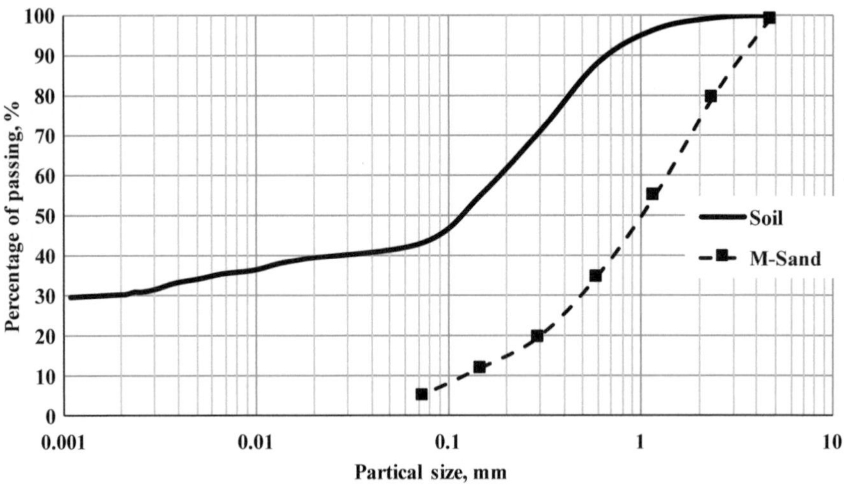

Fig. 7.1 Particle size distribution of soil and M-Sand

Table 7.1 Mix proportions used in the investigation

S. No.	Designation	Soil (%)	Mix proportions: by mass				
			Cement	Fine aggregates		Coarse aggregates (50%—20 mm: 50%—12 mm)	
				Soil	M-sand		
1	Mix 1	A	0	1	0	3	5
2		B	35	1	1.05	1.95	5
3		C	50	1	1.50	1.50	5
4		D	65	1	1.95	1.05	5
5	Mix 2	A	0	1	0	5	8
6		B	35	1	1.75	3.25	8
7		C	50	1	2.50	2.50	8
8		D	65	1	3.25	1.75	8

replaced by 0, 35, 50 and 65% with soil. The soil contents of 0, 35, 50 and 65% are designated as A, B, C and D, respectively, in both the control mixes. The details of the mix proportions used in the investigation are given in Table 7.1.

Workability and compressive strength of Mix 1 and Mix 2 having various percentages of soil were investigated. BIS IS 456-2000 (Reaffirmed 2005) (2005) recommends a range of 50–100 mm slump for medium workability. Hence, water/cement ratio of the concrete mix was controlled to have a slump value in the range of 50–100 mm. The workability and compressive strength of the concrete were determined following the BIS IS 516-1959 (Reaffirmed 2004) (1959) guidelines. The concrete cube specimens of 150 × 150 × 150 mm were used for determining the compressive strength upon 28 days curing period.

7.3 Results and Discussion

7.3.1 Workability and Compressive Strength of FEC

Workability of the mix proportions was examined by conducting standard slump test, by varying the water content. Water/cement ratio corresponding to 80–90 mm slump was noted, and 150-mm cube specimens were cast to determine the compressive strength. The water/cement ratio and the compressive strength of the Mix 1 and Mix 2 for various soil replacements are given in Table 7.2. The slump and the compressive strength values given in Table 7.2 are average of three tests.

Mix 1 A, B, C and D required a water/cement ratio of 0.8, 0.95, 1.07 and 1.13 for a slump value between 80 and 90 mm. The corresponding compressive strengths of the Mix 1 are 24.82, 22.52, 15.56 and 14.81 MPa, respectively. Similarly, Mix

Table 7.2 Workability and compressive strength of FEC

S. No.	Mix proportions		Slump (mm)	Water/cement ratio	Compressive strength (MPa)
1	Mix 1	A	81	0.80	24.82 (0.43)
2		B	88	0.95	22.52 (1.72)
3		C	83	1.07	15.56 (1.27)
4		D	89	1.13	14.81 (1.07)
5	Mix 2	A	82	1.15	14.82 (1.79)
6		B	83.5	1.55	10.27 (0.63)
7		C	84	1.65	8.68 (0.68)
8		D	83	1.8	8.31 (0.33)

*Values given in parenthesis are standard deviation

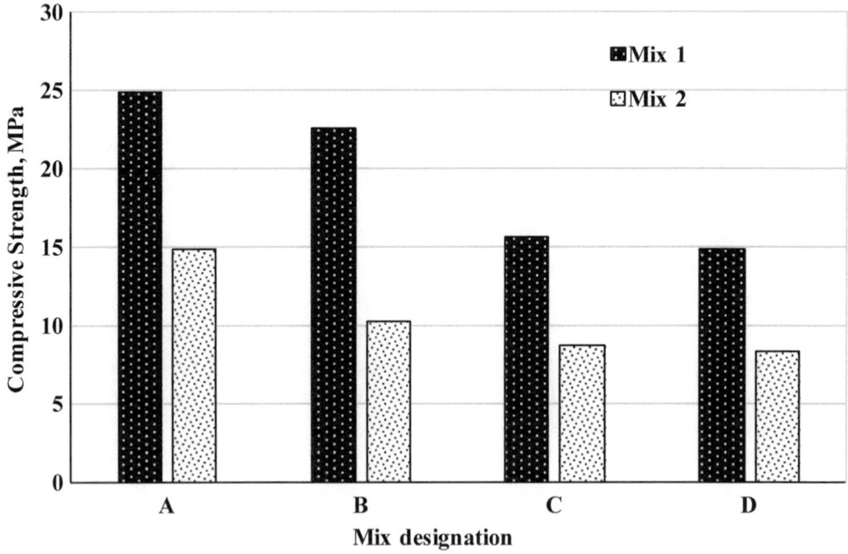

Fig. 7.2 Compressive strength of mix proportions

2 A, B, C and D required a water/cement ratio of 1.15, 1.55, 1.65 and 1.8 for a slump value between 80 and 90 mm. The corresponding compressive strengths of the Mix 2 are 14.82, 10.27, 8.68 and 8.31 MPa, respectively. Figure 7.2 shows the plot of compressive strength of Mix 1 and Mix 2 for various percentages of soil replacement. Mix 2 requires higher water/cement ratio than the Mix 1. In both the mixes, an increase in soil content to replace the fine aggregates demands higher water/cement ratio for achieving required workability. This may be attributed to increase in fine particles such as clay and silt at higher soil content. Hence, strength falls with an increase in soil content.

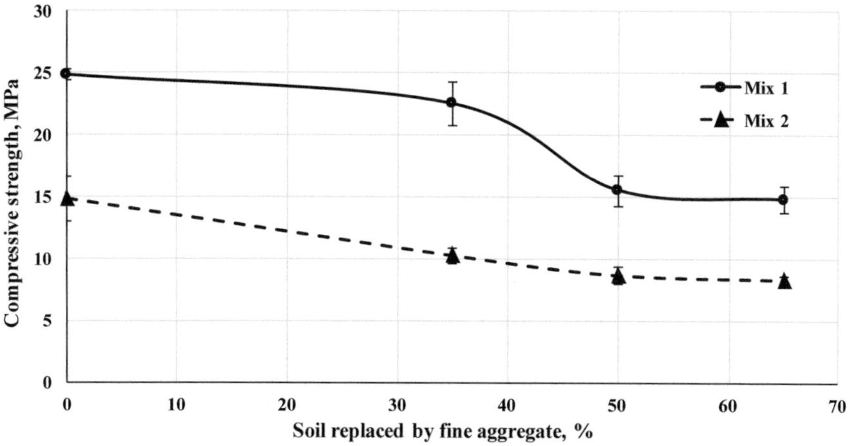

Fig. 7.3 Compressive strength versus fine aggregate replacement by soil

7.3.2 *Influence of Soil on Strength of FEC*

Figure 7.3 shows the plot of compressive strength versus the soil replacement by fine aggregate. Compressive strength of both the mixes (Mix 1 and Mix 2) decreases with increase in soil content in the mix. There is about 40% decrease in compressive strength from 0 to 65% soil replacement in Mix 1 and about 43% decrease in compressive strength in case of Mix 2.

7.4 Comparison with Strength of CSEB and CSRE

The objective of the present investigation was to explore FEC as an emerging material for masonry construction. As discussed in the earlier section, CSEB and CSRE have gained acceptance as alternative masonry materials and the engineering properties of CSEB and CSRE have been widely explored. Since FEC concrete also is a soil-based product, it may be worth comparing some of the properties of FEC, CSEB and CSRE. The compressive strength of FEC, CSEB and CSRE with respect to clay content and cement percentage is reported in Table 7.3. From the results presented in Table 7.3, it may be observed that in CSEB and CSRE, the clay content examined is in the range of 5–20%, cement content for stabilisation is in the range of 5–12%, and the clay and cement content in FEC examined in the present study is also within such range. However, it may be noted that the CSEB and CSRE can be produced with various densities and the strength is sensitive to density achieved, whereas FEC is compacted by vibration to achieve maximum compaction similar to conventional concrete making practice.

Table 7.3 Comparison of properties of CSEB, CSRE and FEC

Type of material		Clay content[a] (%)	Cement content (%)	Compressive strength (MPa)	Source
CSEB		11	5	6.4	Walker (2004)
		11	10	10.3	
		20	5	5.4	
		21–24	4–5	3.3–6.6[b]	
		9	6	3.13	Reddy and Gupta (2006)
		9	8	5.63	
		9	12	7.19	
		5.4	8	4.54	Reddy et al. (2007)
		10.9	8	4.99	
		16.3	8	4.73	
		21.7	8	4.42	
CSRE		9	8	2.32	Reddy and Kumar 2011
		12.6	8	2.45	
		15.8	8	3.2	
		21.1	8	2.82	
		14.3	7	3.44–4.6[b]	Reddy et al. (2017)
		14.3	10	5.01–7.44[b]	
FEC Mix 1	B	10.5	11.12	22.52	Present study
	C	15	11.12	15.56	
	D	19.5	11.12	14.81	
FEC Mix 2	B	10.5	7.14	10.27	
	C	15	7.14	8.68	
	D	19.5	7.14	8.31	

[a]Clay content in mixture of soil and sand; [b]For a range of densities achieved

7.5 Practical Significance of FEC

The FEC was expected to yield compressive strength similar to that of CSEB and CSRE and the results obtained during the present study are promising. Hence, FEC can be a potential material for wall construction in general, and in particular, it can be a convenient material for mass construction needs such as sanitary units and soak pits, pavements and water storage sumps because raw materials are locally available, formwork needed is simple, and construction is quicker. There have been some attempts in using FEC for such applications and are shown in Figs. 7.4 and 7.5.

Fig. 7.4 FEC sanitary unit in IISc Campus, Challakere, Karnataka state, India

Fig. 7.5 FEC sanitary soak pit in Kudapura village, Challakere, Karnataka state, India

7.6 Concluding Remarks

The following conclusions can be drawn from the present study.

(a) Mix 2 (relatively leaner mix) demands higher water/cement ratio when compared with Mix 1 to achieve required workability.
(b) Increase in soil content to replace fine aggregates results in a reduction in compressive strength in both the mixes. This may be attributed to need for higher water/cement ratio at higher soil contents to achieve required workability.
(c) The workability and compressive strength of FEC can be tweaked by varying soil content in the mix for monolithic wall construction.
(d) It is worth exploring other properties of FEC such as flexural strength, creep, shrinkage and durability parameters.

References

Arooz FR, Halwatura RU (2018) Mud-concrete block (MCB): mix design and durability characteristics. Case Stud Constr Mater 8:39–50 (Elsevier). https://doi.org/10.1016/j.cscm.2017.12.004

BIS IS 516-1959 (Reaffirmed 2004) (1959) Methods of tests for strength of concrete. Bureau of Indian Standards, New Delhi, India

BIS IS 456-2000 (Reaffirmed 2005) (2005) Plain and reinforced concrete—code of practice. Bureau of Indian Standards, New Delhi, India

BIS IS 12269-2013. (2013) Ordinary Portland cement, 53 grade—specification. Bureau of Indian Standards, New Delhi, India

Easton, D (2008) The industrialisation of monolithic earth walling for first world applications. In: Proceedings of the 5th international conference on building with earth (LEHM 2008), Koblanz, Germany. Dachverband Lehm EV, Weimar, Germany, pp 90–97

Easton, D (1982) The rammed earth experience. Blue Mountain

Hall M (2002) Rammed earth: traditional methods, modern techniques, sustainable future. Build Eng 77(11):22–24

Kotak T (2007) Constructing cement stabilised rammed earth houses in Gujarat after 2001 Bhuj earthquake. In: Proceedings of international symposium on earthen structures. Interline Publishers, Bangalore, pp 62–71

Reddy BV, Gupta A (2006) Tensile bond strength of soil-cement block masonry couplets using cement-soil mortars. J Mater Civ Eng 18(1):36–45. https://doi.org/10.1061/(ASCE)0899-1561(2006)18:1(36)

Reddy BV, Lal R, Rao KS (2007) Optimum soil grading for the soil-cement blocks. J Mater Civ Eng 19(2):139–148. https://doi.org/10.1061/(ASCE)0899-1561(2007)19:2(139)

Reddy BV, Kumar PP (2009) Compressive strength and elastic properties of stabilised rammed earth and masonry. Masonry Int 22(2):39

Reddy BV, Kumar PP (2011) Structural behavior of story-high cement-stabilized rammed-earth walls under compression. J Mater Civ Eng 23(3):240–247. https://doi.org/10.1061/(ASCE)MT.1943-5533.0000155

Reddy et al (2017) Characteristic compressive strength of cement-stabilized rammed earth. J Mater Civ Eng 29(2):04016203. https://doi.org/10.1061/(asce)mt.1943-5533.0001692

Van Damme H, Houben H (2017) Earth concrete. Stabilization revisited. Cem Concr Res (Pergamon). https://doi.org/10.1016/j.cemconres.2017.02.035

Verma PL, Mehra SR (1950) Use of soil-cement in house construction in the Punjab. Indian Concr J 24(4):91–96
Walker PJ (2004) Strength and erosion characteristics of earth blocks and earth block masonry. J Mater Civ Eng 16(5):497–506. https://doi.org/10.1061/(ASCE)0899-1561(2004)16:5(497)
Walker P et al (2005) Rammed earth: design and construction guidelines. BRE Bookshop Watford, UK

Chapter 8
The Stabilized Rammed Earth Building Technique and its Use in Australia

Rodrigo Amaral Rocha and Pedro Henryque Melo de Oliveira

8.1 Introduction

Building with earth today addresses key questions for contemporary architecture, given its involvement in changing the environment, pollution and the consumption of natural resources in general caused by the construction industry. According to John et al. (2001, p. 92), "the building industry and its products consume approximately 40% of energy and natural resources, and generate 40% of pollution produced by human activity as a whole, reaching up to 75% in the USA."

This article describes and discloses information about the stabilized rammed earth (SRE) technique practised in Australia, more precisely, experiences experienced and developed in Melbourne, Victoria. After discussing the difference between the contemporary Australian practice and the traditional rammed earth practised for centuries in Brazil, for example, the article comments on some particularities and characteristics of the design and execution of this technique.

The vast production of earth buildings currently in Australia has its origins in the second half of the nineteenth century, but it was 100 years later when rammed earth became an extremely popular technique in many regions of Australia, this recent architectural production has been thriving perhaps because of the lack of use history of earth buildings, which differs from other countries where it is related to historical heritage and precarious or low-income buildings.

Although there are other techniques with earth in this country, such as adobe, compressed earth blocks and mixed techniques, it is a fact that rammed earth predominates in contemporary Australian architecture, resulting in one of the countries

R. Amaral Rocha (✉)
Olnee Constructions and Earth House Australia, Melbourne, Australia
e-mail: arqrodrigoamaral@gmail.com

P. H. Melo de Oliveira
UFU, Uberlândia, Minas Gerais, Brazil
e-mail: arq.pedrohenryque@gmail.com

B. V. V. Reddy et al. (eds.), *Earthen Dwellings and Structures*,
Springer Transactions in Civil and Environmental Engineering,
https://doi.org/10.1007/978-981-13-5883-8_8

that has invested most in regulation, university research and in the current implementation of numerous projects (Fernandes 2013, p. 20).

A little of this Australian experience is shared in this paper, which intends to describe and comment a little on the use of rammed earth with stabilized earth.

8.2 About the Production of Rammed Earth Made in Australia

In order to start the discussion about the production of rammed earth, it is necessary to look for a production chain that is structured in three factors emphasized here: in the obtaining of the material, in the manufacture of the formwork and in the expertise of the technicians and constructors. These last two points are related in an inextricable way, since it was from the practice of builders and technicians that the construction system was developed, resulting in the change of the manufacture and design of the forms, a central component for the production of SRE walls. Many Australian architects and engineers now know how to design and calculate for the use of rammed earth, which encourages and expands production, and consequently enables more builders to know how to work, and more products and equipment related to SRE production can be improved to optimize the production of this technique, which can be highlighted as an important contribution of the Australian practice.

On obtaining the construction material, earth, it is important to point out that is not restricted only to that present at the building site. In Australia, suppliers of soil and building aggregates, usually companies that provide paving material for roads and highways, generally located on the urban outskirts, are also responsible for providing material for the production of SRE. These companies provide different types of aggregates, typically types of gravel, rock sediments, sandy loam and clay with various granulometries, grind and reutilize concrete and burnt clay bricks. Diversity and availability of soil and aggregates are essential for the production and propagation of SRE buildings.

After observing the obtaining of the material and briefly pondering its execution, it concentrates on the specificities of the forms used for the execution of the walls, since these are different from those destined to the construction in reinforced concrete. Generally, the forms in question are produced by the constructors themselves entirely in wood or with a metallic structure, as will be seen in detail in the following item, in addition to Figs. 8.1 and 8.2 that illustrate the main instruments and tools necessary for the realization of the work.

Fig. 8.1 Equipment used in the building process of the stabilized rammed earth (*Photo* Oliver Petrovic 2014)

Fig. 8.2 Manufacturing of the steel framework and its use in the construction process (*Photos* Rodrigo Rocha 2015)

8.2.1 Differences Between the Traditional Latin American Method and the Contemporary Australian

The Australian method of SRE has significant differences if compared to the traditional rammed earth technique. Two main points stand out: the stabilization of the earth and the manufacture of the forms, also the mechanized compaction that confers effectiveness and adequacy to the contemporary production.

The traditional procedure for executing SRE buildings is through the use of timber frames locked by fixed or mobile external columns and filled with a mixture of soil with natural stabilizers, the clay already present in the used soil or lime.

Fig. 8.3 Example of the building process of a 12-m-tall reinforced SRE tower (*Photos* Oliver Petrovic 2014)

On the other hand, the Australian practice, besides using cement as the main stabilizer, it is a steel formwork and locked horizontally, avoiding the need for external columns. This allows the builders to build from outside the wall, on platforms over brackets cantilevered on the frames themselves. This change brings important changes to the construction system of rammed earth, as it gives agility to the construction process, enabling a continuous and efficient work. Another change, in aesthetic and economic terms, is the fact that with these forms the walls can be up to 20 cm thick, slenderer than those made by the traditional technique, usually with a minimum of 40 cm.

There is also a change in the design and structural performance of the wall. In the traditional technique, the rammed earth is maintained by its own weight, because the resistance of the wall is in the great inertia of its mass; thus, it resists only the compressive stresses, and thus is self-supporting. In contemporary practice, however, the SRE can function in a different way, similar to a reinforced concrete structure, which is resistant to tensile stress due to the stabilization with cement and with reinforcement in steel bars, as shown in Fig. 8.3 of the execution of a tower of 12 m of height.

(a) Manufacturing and use of the formwork

The formwork predominantly used in Australia was created in Perth, in Western Australia. After many years utilizing a formwork with a timber structure, plywood plates and specific tools and equipment (such as locking columns, screws, spacers, scaffolding and others), builders realized they could increase efficiency by using a formwork with a steel structure and plywood plates.

These plates have standard dimensions between 30 and 60 cm in height and 240, 180, 120 or 90 cm in length and weighing between 10 and 30 kilograms. The forms are structured by pre-fabricated, welded steel parts—a standard created due to the

Fig. 8.4 Different SRE types on the forefront of the Olnee Constructions office (*Photo* Rodrigo Rocha 2015)

need to customize the formwork, as shown in Fig. 8.4. A plywood plate is screwed onto the steel frame, and holes of 1.5 cm in diameter are drilled at every 15 cm on the upper and lower sides of its longest extension. These holes make it easier to assemble the formwork but will be visible on the finished surface of the wall when the formwork is removed. A layer of glass fibre resin is applied to the surface of the plywood to protect it from contact with the soil, ensuring its waterproofing and longevity, and also making it easier to the disassemble the formwork. It also safeguards the quality of the surface on the finished wall.

(b) Ideal mixture and mixing, soil types, and usage of cement and water

The stabilization of the rammed earth with cement allows it to be used in external environments and eliminates the need for plastering or coating.

According to Ciancio and Beckett (2015), the amount of cement in the mix is of 5–15% is required to ensure endurance to structural stress, longevity, stabilization and ability to withstand loads in compression. According to Dobson (2015), the moisture content is related to the type of mixing and compaction, aiming at the maximum dry density to guarantee a compressive strength of 2.5–5 MPa (25–50 kgf/cm^2), enough for using the SRE as sealing masonry, according to Middleton (1992).

The types of soil used in SRE in Australia vary according to the needs of the design and the use of the walls. According to Middleton (1992), usually one or two types of soil are used with not more than 2% of organic matter, free of leaves or branches with 5–20% of clay (including mud and silt), a minimum of 30% of sand, stones no larger than 40 mm and humidity between 4 and 12%. In some cases, crushed concrete or brick are also used as aggregates, materials found in conventional building waste.

Figure 8.5 shows, on the left, SRE wall that supports all loads of the roof structure with thermal insulation embedded in the wall to increase the thermal resistance of the wall and, on the right, the wall assembled over a window span of 1.2 m.

Fig. 8.5 Example of a curved reinforced wall with insulation, 2.5 m^2 K/W of thermal resistance, and a reinforced SRE beam with foam as an expansion joint (*Photos* Oliver Petrovic and Rodrigo Rocha 2014)

Additives are included in the mix to increase the longevity of the wall. It is suggested to use 1% of *plasticure*, a product developed by Techdry[1] for soil–cement monolithic walls. When added to the rammed earth mix during the manufacturing process, this water-repellent admixture will ensure water, salt and mould repellency, and strengthen the structure. It is also used in low concrete *slumps* (low consistency concrete), and its role is to ensure long-term resistance against weathering and deterioration. After the formwork is removed from the finished walls, their external faces receive a layer of a colourless solvent-based silicone binder, a penetrating sealer, which reaches deep into the capillaries of earth surfaces, providing the substrate with dust and waterproof sealer effect. As the strong water repellency reduces the absorption of water substantially, the occurrence of efflorescence and other waterborne staining effects are as good as eliminated, also prevents surface abrasion, that is responsible for most of the degradation in rammed earth structures. The manufacturer guarantees that the surface appearance and breathability are not significantly affected.

A mini loader is used to best prepare the mixture, as it ensures the materials are thoroughly mixed. Mixers are not recommended since the low-humidity mixture tends to stick to its walls and do not mix properly. The mini loader is also capable of placing the mixture into the formwork up to 3 m high, the limit its bucket can reach.

(c) Design details

Knowing how to design for SRE and its possibilities are essential for a plan that respects the technique's needs and potential. It resembles reinforced concrete, since they are both self-bearing and have a structural feature strong enough to support compression loads. According to Ciancio and Beckett (2015), SRE ranges from 1 MPa (10 kgf/cm^2) to 47 MPa (470 kgf/cm^2) of compressive strength, considering

[1]www.techdry.com.au.

Fig. 8.6 Detail of built-in electrical installation, example of built-in steel column and contrast with corten steel (*Photo* Rodrigo Rocha 2015)

the size of the steel structure incorporated inside the walls, granulometry of the aggregates, ramming pressure and quantity of cement and water in its composition.

A steel frame with the desired shape is required when executing fireplaces and ovens in SRE. The frame is incorporated into the building's wall and solid walls are built on its side, underneath and on top of it. The air vent must be centralized and end higher than the final surface of the wall.

The finishing of the top of exterior walls that are exposed to weather must receive a mix of water and *plasticure*, or like is it in other countries, adding a higher concentration of cement in the mix of its top layer.

When the architectural project is finalized, it is possible to plan the electric and hydraulic installations inside the walls. Future changes to them are not viable, as can be seen in Fig. 8.6. Connections to beams and structural columns allow the SRE to interact with other materials, like steel and timber, and other building techniques such as reinforced concrete, as shown in Fig. 8.7.

(d) Construction details

The execution is relatively simple and the building system is comprised of four people in the building site: two professionals with experience in carpentry, masonry and reinforced concrete, and two other workers that can be inexperienced. It is necessary to emphasize the importance of the design stage, which defines the position of the forms and distance between expansion joints, besides resolving architectural details, structural, electrical and hydraulic connections. The SRE execution can be roughly divided into two stages. In the first one, the formwork is assembled and the material mixture is placed in layers between 15 and 20 cm deep and rammed until the desired height for the wall is reached. Each compressed layer leaves a defined line on the wall's face, as visible on Picture 1. The second stage happens on the following day and consists of removing the formwork, triggering the wall's curing process. There is no need for any rendering or waterproofing.

Fig. 8.7 Example of connections to timber floor and steel roof beams (*Photos* Oliver Petrovic 2014)

Fig. 8.8 SRE building process over T-shaped steel beam and built-in steel column (*Photos* Rodrigo Rocha and Oliver Petrovic 2014)

At the start of the construction of a wall, the foundation block receives a strong and thick layer of cement and water, waterproofing it and working as a transition between the surfaces, strengthening the wall's stability. To execute SRE beams, T-shaped steel profiles are used and connected to the steel columns inside the walls. It is recommended to coat the steel column with expandable materials, such as expanded EPS foam, to absorb the movement between the metal structure and the wall. To execute a wall above the steel profile, a formwork is built on top of timber joists temporarily screwed to the wall, and temporary timber columns made *in loco* are used to support the beam until the formwork is disassembled.

The use of T-shaped steel profiles allows for a span of up to 10 m (Figs. 8.8 and 8.9). For spans of up to 4 m, it is recommended the use of only two 12 mm steel

Fig. 8.9 SRE building process over T-shaped steel beams (*Photos* Rodrigo Rocha 2015)

Fig. 8.10 Examples of use of formwork and aluminium platforms resting on cantilevered brackets in the SRE walls building process (*Photo* Rodrigo Rocha 2015)

rebar each 20 cm horizontally, connecting the side walls and supporting the beam's traction movement.

The curing process of a wall will depend on the quantity of cement and clay in the mix as well as weather conditions. Studies reveal that unstabilized rammed earth can take up to two years to complete its curing and retraction process, but this process is reduced to up to 30 days when cement is used. Depending on the amount of water in the mix, and if the external temperatures are high and air humidity is low, it is recommended to water the walls on the days following the framework removal to ensure they will not retract too quickly, causing fissures and cracks due to expansion and chemical reactions between components in the mixture.

To cover cracks and imperfections on the walls, it is recommended to wait until the curing process is finished. When doing it, it is imperative to use the same proportions in the new mixture to avoid changes in colour and texture in the corrected surface. In some cases, colourless water repellent is applied to the external walls after the cure process is over.

In relation to the formwork, it is recommended to standardize the height of the walls to multiples of 30 cm, and to avoid ending the walls halfway through the

Fig. 8.11 Examples of stains on wall caused by bad execution and/or design (*Photo* Rodrigo Rocha 2015)

forms, ensuring the top of the walls are levelled. A choice must be made between a superimposed or interlocked formwork layout depending on the desired appearance, as shown in Fig. 8.10.

8.2.2 Construction Pathologies

Stains are recurrent on SRE walls due to their curing period, which depends on the amount of clay and cement in the mix, as previously mentioned. Because of its water repellence, the use of *plasticure* allows a uniform and stain-free finish. Nevertheless, it is important to make sure no materials resting on top of the walls come in contact with water, as they can release minerals, impurities and chemicals that might interfere in the colour of the wall's surface. Tannin from timber and rust from metals are examples of these substances, and the consequences of their contact with SRE walls can be seen in Fig. 8.11. The tops of external walls must be protected from impurities, as rainwater can spread them along its sides, also causing stains.

8.3 Final Considerations

This article addresses a possibility of building with SRE nowadays. It is important to emphasize that the use of cement as a stabilizer makes it difficult to dispose of the wall in part since in a structured production chain, such as the Australian one, the discarded wall can be crushed and reused as an aggregate for other purposes or even for other SRE. The basic characteristics of SRE construction technique were summarized, gathering information and data that can help in the research about the techniques that use the earth as a construction material.

The earth, despite being proportionally the most used material in the world in construction, in the usual markets of Brazil is still classified as an alternative and primitive material. Most architects, engineers and builders are reluctant to use earth building techniques in projects because they do not know the present applications and possibilities of using this material found in abundance. Presenting how it is possible to build with earth today was, therefore, the main incentive of the article.

This article depicting Australian examples, which have a productive chain developed under construction with SRE, is written under the architect's eye that has experience in the use of SRE in Brazil and Australia, thus understanding the needs of the local market and the potentialities of the construction system in question, providing design and construction information for SRE, as they are being developed and employed in contemporary architectural production in Australia.

In order to promote the use of this technique, theoretical and practical knowledge must be promoted, as well as providing technical information about the use and applicability of this construction technique, which combines traditional knowledge with the current development of technology.

It is essential to build to understand the process of building a SRE wall, as well as designing to optimize the construction details and to improve the development of the technique. It is hoped, with this study, to encourage the research and application of this constructive technique in the academic and professional environment, expanding the use of raw or stabilized earth as a way to mitigate the environmental impact caused by the construction industry.

8.4 Citations

The building industry and its products consume approximately 40% of energy and natural resources, and generate 40% of pollution produced by human activity as a whole, reaching up to 75% in the USA (John et al. 2001, p. 92).

Although there are other techniques with earth in this country, such as adobe, compressed earth blocks and mixed techniques, it is a fact that rammed earth predominates in contemporary Australian architecture, resulting in one of the countries that has invested most in regulation, university research and in the current implementation of numerous projects (Fernandes 2013, p. 20).

References

Ciancio D, Beckett C (2015) Rammed earth construction, cutting-edge research on traditional and modern rammed earth. CRC Press, Perth, Australia

Dobson S (2015) Rammed earth in the modern world. In: Ciancio D, Beckett C (eds) Rammed earth construction, cutting-edge research on traditional and modern rammed earth. CRC Press, Perth, Australia

Fernandes M (2013) The rammed earth in the world. CEAUCP/CAM Center for Archaeological Studies of the Universities of Coimbra and Porto/Archaeological Field of Mertola, Coimbra and Porto, Portugal

John VM, Silva VG, Agopyan V (2001) Agenda 21: a proposal of discussion for the Brazilian construbusiness. ANTAC/UFRGS, II national meeting and I Latin American meeting on sustainable buildings and communities, Canela, Rio Grande do Sul, Brazil, pp. 91–98
Middleton GF (1992) Bulletin 5. Earth wall construction, 4th edn. CSIRO Division of Building, Construction and Engineering, North Ryde, Australia

Chapter 9
Mineralogical, Physical, and Mechanical Properties of Soil for Using in Adobe Blocks

Lucas Miranda Araújo Santos, Aline Figueirêdo Nóbrega de Azerêdo, Givanildo Alves de Azerêdo and Sérgio Ricardo Honório de Assis

9.1 Introduction

There are many types of soil and most are not suitable for use in construction depending on clay type which constitute them. To obtain bricks with adequate strength for building a house, it is necessary to make an improvement of the soil. For this purpose, the soil has to be mixed with a stabilizer material (lime, cement, etc.), and sometimes, it also needs to undergo a granulometric correction. According to Houben and Guillaud (1994), soil stabilization consists of modifying its characteristics in order to obtain greater durability.

Clay is a material found in abundance in nature and it is one of the main raw materials used in the construction industry, as in the manufacture of bricks, tiles, tableware ceramics, etc. In addition, it is a low-cost material and has excellent thermal and acoustic properties (Barbosa and Mattone 2002; Faria et al. 2005; Corrêa et al. 2006). Several studies have been made with respect to the proportions of these stabilizers and granulometric correction for each type of earth (Barbosa et al. 1997; Corrêa et al. 2006; Bouth 2005; Ferreira et al. 2008). In addition to the use of cement and lime, many studies also show that other materials may be added to the mixture

L. M. A. Santos · S. R. H. de Assis
Federal University of Paraiba, PPGECAM, João Pessoa, Brazil
e-mail: lucas.cedro@hotmail.com

S. R. H. de Assis
e-mail: sergioenghcivil@gmail.com

A. F. N. de Azerêdo
Federal Institute of Education, Science Technology of Paraiba, João Pessoa, Brazil
e-mail: alinefnobrega@homail.com

G. A. de Azerêdo (✉)
Department of Civil and Environmental Engineering, Federal University of Paraiba, João Pessoa, Brazil
e-mail: givanildoazeredo@hotmail.com

B. V. V. Reddy et al. (eds.), *Earthen Dwellings and Structures*,
Springer Transactions in Civil and Environmental Engineering,
https://doi.org/10.1007/978-981-13-5883-8_9

for the manufacture of mud bricks (raw ground) in order to improve their mechanical properties, for example, plant fibers, fly ash, and high slag oven (Souza et al. 2000; Mendes et al. 2003; Ferreira et al. 2003, 2008; Mesa-Valenciano and Freire 2004).

This work aims to characterize physically, chemically, and mineralogically two soil samples and to study the dosage of mixtures for adobe using Portland cement as a stabilizer.

9.2 Materials and Experimental Details

9.2.1 Soils

Both soils were characterized for particle size analysis, Atterberg limits, fluorescence X-ray, and diffraction X-rays. The soils were called S1, brown color, and S2, light red. Figure 9.1 shows the soils used.

9.2.1.1 Particle Size Analysis

For particle size analysis, soils were characterized by two methods. The first was based on what is recommended by the standards NBR 7181/84 and DNER-ME 051/94. The second was by laser granulometry. For the laser granulometric analysis, sample was dispersed in 250 ml of distilled water mixed on a mechanical shaker for 20 min. After that, 15 ml of this mixture was separated and dispersed in an ultrasonic bath and finally placed in an apparatus of the type CILAS Model 1064 for the determination of particle size. In Figs. 9.2 and 9.3, granulometric curves obtained by both methods are shown.

Fig. 9.1 Soils S1 (left) and S2 (right)

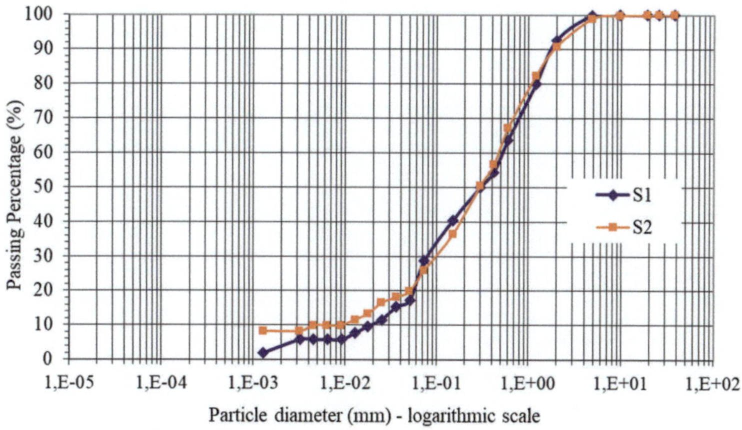

Fig. 9.2 Granulometric curves S1 and S2 obtained by sieving and sedimentation

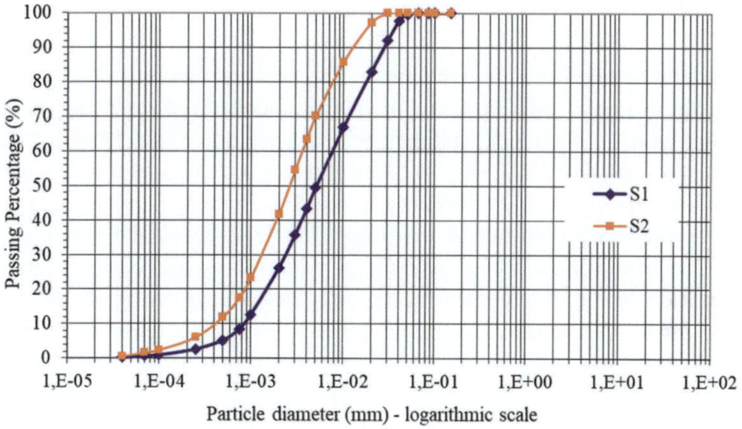

Fig. 9.3 Laser granulometric curves S1 and S2

According to Neves and Faria (2011), Peruvian Standard NTE and 080 (Sencico 2000) proposes the following granulometry composition of earth suitable for use in adobe: clay—10 to 20%, silt—15 to 25%, and sand—55 to 70%. Table 9.1 is a summary of the appropriate granulometry compositions for adobe production according to several authors.

It is noted that S1 and S2 have less than 10% clay. This value is only in accordance with the data proposed by Houben and Guillaud (1994) as shown in Table 9.1. However, according to the granulometric curves shown in Fig. 9.6, S1 and S2 show about 26% clay and 42%, respectively. So considering these figures, it is seen that they are more suitable for use in mud bricks according to Table 9.1. Moreover, at the moment of manufacturing the blocks, it was possible to clearly see the difference between the two soils with regard to fines content. The results so discrepant between

Table 9.1 Granulometry composition of earth suitable for adobe production, according to several authors

Authors	Clay (%)	Silt (%)	Sand (%)
Author 1	35–45		55–65
Author 2	5–29	–	–
Author 3	15–25	–	–
Author 4	10–40	10–30	30–75

Source Neves and Faria (2011)

Table 9.2 Limits of Atterberg S1 and S2

Soil	Limits of Atterberg		
	LL	LP	IP
S1	25.8	19.8	6.0
S2	31.2	25.1	6.1

the clay content came from the two particle size analysis methods raise a question: Which method is best suited to know the clay content of the soil? The method of particle size analysis by sieving and then by sedimentation of the fines described in NBR 7181/84 and DNER-ME 051/94 takes into account the Stokes law, where it is considered that the particles of the clay are rounded. But clay particles are lamellar. In other words, this would not be the most suitable to check the clay content of the soil. One would think then that the laser particle size is the most suitable for the verification of the clay content in the soil. However, there is no standard (or standards) for the execution of laser granulometry test. In the scientific community, there is still no consensus.

9.2.1.2 Limits of Atterberg

The procedures performed are based on what is recommended in the standard NBR 6459/84 for liquid limit and in standards NBR 7180/84 and DNER-ME 082/94 for plastic limit. In Table 9.2, the liquidity limits (LL), plastic limit (PL), and the plasticity index (PI) of the soil are displayed.

According to Neves et al. (2009), the soils are classified into sandy (LL—0 to 30% and 10%—0 IP), silty (LL—20 to 50% and IP—5 to 25%), and clay (LL greater than 40% and IP greater than 20%). According to the study results, S1 and S2 soils can be considered both silty and sand.

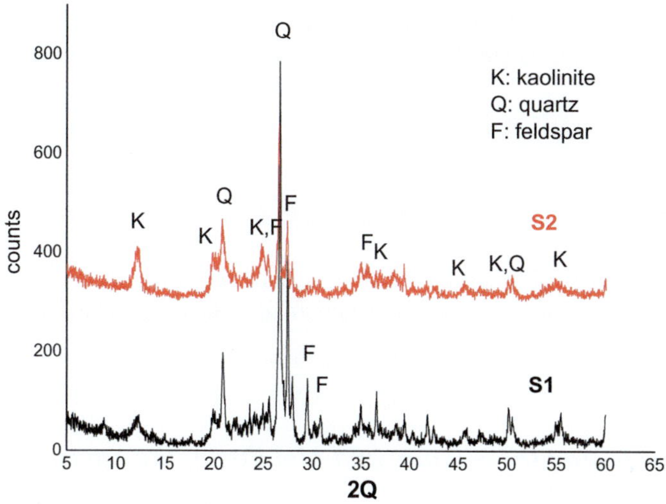

Fig. 9.4 Diffractograms of the two types of soil

9.2.1.3 X-ray Diffraction (XRD)

The mineralogical characteristics of soils were determined by XRD, carried out in a Shimadzu XRD 6000 model equipment under the following test conditions: CuKα radiation of wavelength $\lambda = 1.5418$ Å with X-ray at 40 kV and 30 mA, scanning speed 1°/min in a range of 5°–65° 2θ at a step angle of 0.02° 2θ. Identification of peaks was made by using the JADE 5.0 software MDI. Samples were ground and sieved to 0.075-mm mesh sieve and analyzed as a powder.

According to the result showed in Fig. 9.4, the two soils are the kaolinitic type, with quartz and feldspar peaks. It is noted that S1 has less intense peaks of kaolinite, which may indicate that the soil is less clay content. This result is in agreement with the results obtained from the particle size analysis, both laser and sieving and sedimentation.

9.2.1.4 Fluorescence X-ray (XRF)

The chemical composition of the soil was determined semi-quantitatively by fluorescence X-ray spectrometer (XRF) on a Rigaku RIX device model 3000. In Table 9.3, the data are shown.

Table 9.3 Chemical composition by fluorescent X-ray (wt%) of S1 and S2

Soil	SiO_2	Al_2O_3	Fe_2O_3	K_2O	TiO_2	MgO	CaO	Na_2O	Outros
S1	62.63	16.03	8.93	3.14	1.78	1.22	3.95	0.86	>1%
S2	57.43	22.86	13.16	2.32	1.96	0.54	0.50	0.29	>1%

According to the XRF results, the two soils are mainly SiO_2, Al_2O_3, and Fe_2O_3. The iron content in S1 is less than S2, which explains the difference in color between them.

9.2.2 Mixtures Studied

Portland cement CP II Z was used as stabilizing material in three different proportions (6, 9, and 12%) for each soil type. The amount of cement was added by mass to the dried soil.

9.2.3 Preparing the Mixture and Molding of the Specimens

The material was dry-blended, and then water was added gradually until a good plastic consistency. The mixtures were prepared by hand until the mix becomes homogeneous. The percentage of water added was based on the plasticity limit of each soil being approximately 20 and 25% for S1 and S2, respectively. Then, three specimens were molded into prismatic molds 4 cm $\times$ 4 cm $\times$ 16 cm (Fig. 9.5) for the bending and compression tests. The absorption test was carried out according to standard NBR 8492 (2012), where cylindrical specimens 5 $\times$ 10 cm were cast.

9.2.4 Curing Conditions and Hardened State Tests

After molding, all samples were subjected to the curing process where all samples were placed in plastic bags for a period of 28 days.

After 28 days of curing, the specimens were taken from the plastic bags and left outdoors for a period of 24 h to dry following the recommendation of the document entitled: Guide Compressed Earth Block (2000) and then being subjected to bending, compression, and water absorption tests. The bending and compression tests were carried out on a Shimadzu AG-X AUTOGRAPH 10 kN machine. The bending test was performed with a 0.005 mm/s loading speed on three specimens, and the compression test was run at a speed of 0.01 mm/s in four or six specimens. The results represent an average of three samples for bending and four to six test specimens for compression. Figure 9.6 shows the samples subjected to the bending and compression tests. The water absorption test was performed according to the standard NBR 8492 (ABNT 2012). The result is the arithmetic average of three samples.

Fig. 9.5 Specimens being molded

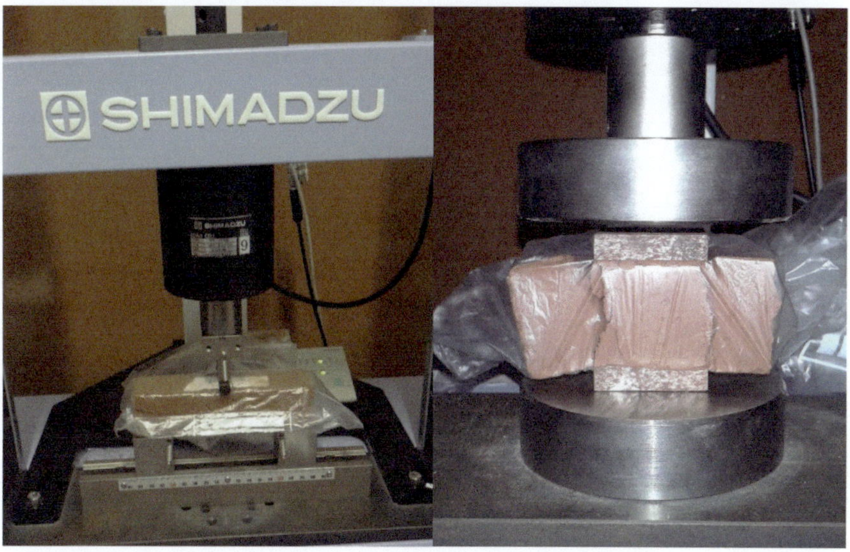

Fig. 9.6 Specimens subjected to bending test (left) and compression (right)

9.3 Results and Discussion

9.3.1 Compressive and Flexural Strength

In Figs. 9.7 and 9.8, the compressive and flexural strength results are given respectively to each mixture and their respective quantity of water used for molding.

According to Figs. 9.7 and 9.8, it is seen that the amount of water was higher for mixtures with soil S2, since this soil contained the higher clay content. It is observed that for mixtures containing 6% Portland cement, greater amount of water to the soil S2 did not influence the strength. The influence of the amount of water in strength occurred only when the cement content was 9 and 12%. It is observed that for both soils, compressive and flexural strength was similar to 6% Portland cement, about 1.30–0.43 MPa, respectively. For higher cement content (9 and 12%), the compressive and bending strength was higher for mixtures with soil S1, that is, the lower the content of clay in the soil is, the higher the strength it promotes. This was expected since the amount of water used to mold the samples with soil S1 is smaller than that for soil S2.

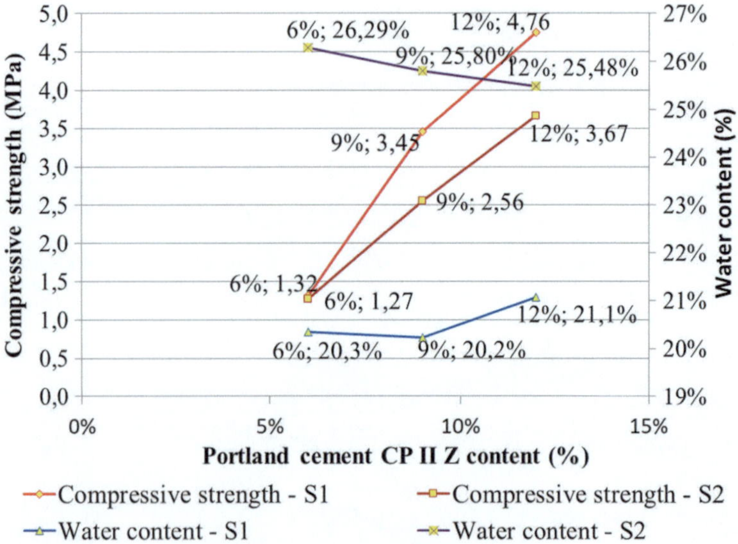

Fig. 9.7 Resistance to compression and quantity of water for molding

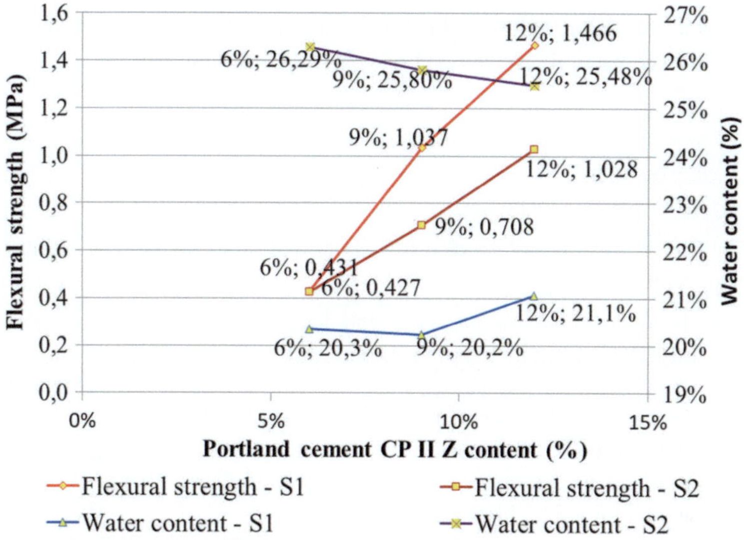

Fig. 9.8 Flexural strength and quantity of water for molding

9.3.2 Water Absorption

Figure 9.9 presents the average values of water absorption of mixtures. It is observed that for mixtures with S1, soil water absorption was about 19% for the three different cement contents, and for soil S2 mixtures, this value was about 24%. This is justified because the soil S2 contains more clay than S1, and hence, it absorbs more water. Also, as for the mixture with the soil S2, greater amount of water is required; this

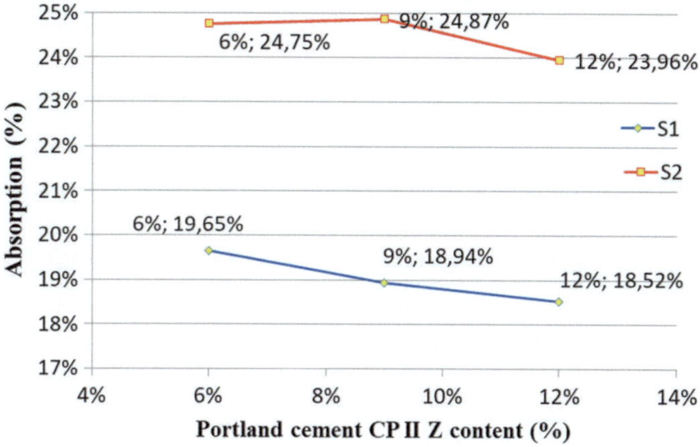

Fig. 9.9 Water absorption

also resulted in a higher porosity. But you can see that there was a slight decrease in water absorption when the cement content increased in both soils.

9.4 Conclusions

According to the results obtained in this study and analysis, we can highlight the following points:

1. Clays showed similar mineralogical characteristics according to the results of XRD and XRF. Both soils contain kaolinite. According to the X-ray fluorescence, the soil S2 presented more FeO_2 content than the soil S2. It is the reason that the soil 1 is brown and the S2 is red.
2. According to the size distribution curves, the soil S2 showed higher content of clay material. This difference was given more significant when size distribution curve was examined by laser particle size test.
3. Due to the higher content of clay material, S2 soil mixtures required more water for molding, and therefore, it had lower resistance values and greater absorption. For the Portland cement content of 6%, there was no difference of mechanical strength at 28 days for soil mixtures S1 and S2.
4. When cement content increased to 9 and 12%, the strength at 28 days also increased. For mixtures with soil S1, the values of mechanical resistances were higher than S2 soil. For 12% of cement, 4.5 and 3.5 MPa (compressive strength) to S1 and S2, respectively, at 28 days of curing were achieved.

Overall, this work has shown that to use these soils for adobe bricks, 9% of Portland cement is sufficient to reach the minimum compression resistance required by standard. Furthermore, this research brings results about the determination of the clay content of the soil, indicating that the traditional method to determine the size distribution curve by sieving and sedimentation may not be the most suitable to check this clay content. However, further studies on how to determine this content by laser particle size need to be performed.

References

Associação Brasileira de Normas Técnicas (ABNT) (2012) NBR 8492: Tijolo de Solo-cimento. Análise dimensional, resistência à compressão simples e absorção de água - método de ensaio

Barbosa NP, Toledo Filho RD, Ghavami K (1997) Construção com terra crua. In: Toledo Filho RD, Nascimento JBW, Ghavami K (eds) Materiais de construção não convencionais. [S. l.]: Sociedade Brasileira de Engenharia Agrícola, Lavras, pp 113–144

Barbosa NP, Mattone R (2002) Construção em Terra. I Seminário Ibero Americano de Construção em Terra, Salvador, BA

Bouth JAC (2005) Estudo da Potencialidade da Produção de Tijolos de Adobe Misturado com Outros Materiais - uma Alternativa de Baixo Custo para Construção Civil. Dissertação (Mestrado em Engenharia Mecância) – Universidade Federal do Rio Grande do Norte, Natal

Corrêa AAR, Teixeira VH, Lopes SP, de Oliveira MS (2006) Avaliação das propriedades físicas e mecânicas do adobe (Tijolo de Terra Crua). Ciências Agrotécnicas, vol 30, pp 503–515

Faria OB, Silva FMG, Ino A (2005) Sistema Construtivo com Paredes Estruturais de Adobe, em Habitação de Interesse Social Rural: Um Estudo de Caso no Assentamento Rural "Fazenda Pirituba" (Itapeva-SP). I Seminário Mato-Grossense de Habitação de Interesse Social. Cuiabá, MT

Ferreira RC, Gobo JCC, Cunha AHN (2008) Incorporação de Casca de Arroz e de Braquiária e seus Efeitos nas Propriedades Físicas e Mecânicas de Tijolos de Solo-Cimento. Engenharia Agrícola, Jaboticabal, vol 28, no 1, pp 1–11

Ferreira R, de C, Freire WJ (2003) Propriedades físico-mecânicas de solos estabilizados com cimento e silicato de sódio avaliadas por meio de testes destrutivos e não-destrutivos. Engenharia Agrícola, Jaboticabal, vol 23, no 2, pp 221–232

Guide Compressed Earth Blocks Testing Procedures (2000) Series technologies Nr 16. CDE, Brussels-Belgium

Houben H, Guillaud H (1994) Earth Construction: a compressive guide. Intermediate Technology Publications, London, p 362

Mendes JUL, Silva LCF, Marinho GS (2003) Fibras de Coco na Composição de Tijolos de Solo-Cimento. Inter American conference on non-conventional materials and technologies in the eco-construction and infrastructure. I IAC-NOCMAT, João Pessoa

Mesa-Valenciano MC, FREIRE WJ (2004) Características físicas e mecânicas de misturas de solo, cimento e cinzas de bagaço de cana-de-açúcar. Engenharia Agrícola, Jaboticabal, vol 24, no 3, pp 484–492

Neves CMM, Faria OB (2011) Técnicas de construção com terra. Bauru, SP: FEB-UNESP/PROTERRA, 79p

Neves CMM, Faria OB, Rotondaro R, Cevallos PS, Hoffman MV (2009) Seleção de solos e métodos de controle na construção com Terra – práticas de campo. Rede Ibero-americana PROTERRA

Souza SMT, Barbosa, NP, Toledo RD (2000) Efeito das fibras de sisal no comportamento de tijolos de terra crua. Anais da Conferência Internacional Sustainable Construction into the Next Millennium. João Pessoa

Chapter 10
A Case Study on Technical and Social Aspects of Earth Houses in Rural India

Y. Kulshreshtha, P. J. Vardon, N. J. A. Mota, M. C. M. van Loosdrecht and H. M. Jonkers

10.1 Introduction

The World Bank has estimated the need for 300 million new housing units for the urban and rural population of the world by 2030 (World Bank 2016). While urban housing projects, especially for upgrading slums, have been given significant attention by international organisations and media, rural housing projects are comparatively neglected and given low importance. However, significantly, 46% of the world population dwells in rural houses. This population is significantly higher in a developing country like India, where the rural population is 67% (World Bank 2016).

The government of India has identified a need for 10 million houses for low-income rural households by 2019 (Ministry of Rural Development India 2016). There is a need for an affordable solution to cater for this shortage of housing. Construction with industrial materials such as concrete or bricks is often considered as a plausible solution, but traditional materials and processes may offer interesting

Y. Kulshreshtha (✉) · P. J. Vardon · H. M. Jonkers
Faculty of Civil Engineering and Geosciences, Delft University of Technology, Delft, The Netherlands
e-mail: Y.Kulshreshtha@tudelft.nl

P. J. Vardon
e-mail: P.J.Vardon@tudelft.nl

H. M. Jonkers
e-mail: H.M.Jonkers@tudelft.nl

N. J. A. Mota
Faculty of Architecture and the Built Environment, Delft University of Technology, Delft, The Netherlands
e-mail: N.J.A.Mota@tudelft.nl

M. C. M. van Loosdrecht
Faculty of Applied Sciences, Delft University of Technology, Delft, The Netherlands
e-mail: M.C.M.vanLoosdrecht@tudelft.nl

B. V. V. Reddy et al. (eds.), *Earthen Dwellings and Structures*,
Springer Transactions in Civil and Environmental Engineering,
https://doi.org/10.1007/978-981-13-5883-8_10

alternatives. Traditional buildings are commonly built by households and/or local communities themselves (Schroeder 2016; Bredenoord 2017), thus saving on labour costs. Therefore, material costs become the major contribution to the expense. Traditional building materials are typically cheap and readily available. The houses made with these materials are built in greater variety as each household builds according to their choice, thereby possibly creating a better social, cultural and psychological environment than that provided by most low-cost mass housing schemes (Agarwal 1982).

Earth (soil) is one of the most abundant resources available on the planet that has been used as construction material for over 9000 years (Minke 2006). Even today, one-third of world population still lives in houses made of earth (UNESCO 2018). In developing countries, this number is much higher. Earth houses are considered environmental friendly and affordable as compared to houses built with concrete or fired clay bricks (Houben and Guillaud 1994). Moreover, it preserves the vernacular social and cultural identity of the community. With rapid industrialisation and the increase in popularity of concrete and brick constructions, a decline in earth houses has been observed especially in the rural areas of India. In 1971, 72.2% of buildings in India were made of earth construction (Houben and Guillaud 1994). According to 2011 census, 21.8% of the houses in India have mud as the predominant material for the walls and 45.5% of houses have mud as the predominant floor material (Census of India 2011).

This rapid decline is caused due to various technical and social factors. In order to understand construction with earth and to investigate whether earth constructions can make a valuable contribution to contemporary dwelling construction, it is important to understand the factors that affect the choice of the earth as a building material. A survey was carried out in five regions of India (Himachal, Orissa–Jharkhand, Gujarat, Tamil Nadu and Sikkim) to understand the technical and social factors favouring or limiting the construction and everyday use of earth houses.

10.2 Research Methodology

A unique non-time-intensive approach was adopted for the survey. A total of 32 unstructured interviews were conducted during a time period of one year. The total duration spent with each individual interviewee ranged from a day to two weeks. These unstructured interviews were based on the development of dialogues between interviewer and interviewee; thus, they were predominantly informal discussions. The motivation was to keep the scope of information provided by the interviewee as broad as possible, in an attempt to encapsulate the underlying philosophy and emotions of each interviewee connected with the everyday use of an earth house. One of the disadvantages of this method is that a lot of details, especially technical details, are missed in the discussion.

The selected locations in 6 different states of India (Himachal, Orissa, Jharkhand, Sikkim, Tamil Nadu and Gujarat) are shown in Fig. 10.1. The locations were selected

Fig. 10.1 Map of India marked with interview locations

based on their geographic location and climatic diversity. Recommendations on locations from experts were also taken into account. Forty per cent of the interviews were conducted in the rural areas of south India.

The interviews were conducted to understand the factors favouring or limiting the construction and everyday use of earth houses. These factors include construction technique, performance of already existing structures, maintenance requirements, affordability, image, personal philosophy, influence of government and policies, and education and training. This article is limited to technical aspects (such as construction techniques), performance and social aspects regarding the image of an earth house. Together with traditional earth houses, modern earth houses constructed in recent times were also considered in this research. The interviewee group consisted of people involved directly in earth construction. This included earth house dwellers with different socio-economic background, earth construction experts, architects, engineers, masons, contractors, consultants, educationalists and volunteers. Interviews were held in English and Hindi, wherever possible, and at other times in regional language. A translator was used in the regions where the interviewer could not speak or understand the regional language.

10.3 Result and Discussion

10.3.1 Construction Technique

The earth construction technique has a strong relationship with the climate of the region. In Himachal, which is a region with a cold climate and regular seismic activities, unstabilised adobe was the most commonly adopted construction technique. Adobe is a technique where rectangular blocks of earth are cast in moulds and joined together with mud mortar. Pine needles and risk husk were added to soil for the production of the adobe. These are known to improve the insulation of earth building materials. The foundations were mostly made up of stone with mud mortar and were raised around 30–50 cm in order to prevent contact of water with the unstabilised adobe blocks. The house walls were usually plastered with cow-dung plaster. A traditional earth house of a low-income household in Bir can be seen in Fig. 10.2a. The house has suffered from significant deterioration over a period of 20+ years and requires significant maintenance. In some modern earth houses, of high-income households, near Dharamshala, significant attention was paid to design and engineering details resulting in durable and aesthetically appealing houses. These houses were constructed with unstabilised adobe and later covered with a stabilised plaster. These modern earth houses consist of lintel and plinth bands (ring beams) that were made with concrete in order to improve their seismic performance. Toilets and bathrooms were constructed with concrete or fired clay bricks which perform well when in contact with water.

The northeast state of Sikkim is cold, cloudy and seismically active. The vernacular construction technique known as “Ikara” is prevalent in this region. Ikara is a type of construction that is a similar technique to wattle and daub. In Ikara houses, timber frames with bamboo weaves act as the structural wall members that are daubed with unstabilised mud. A traditional Ikara house located in Namchi can be seen in Fig. 10.2b. Ikara houses are framed structures that perform structurally well in an earthquake. The lightweight construction also helps to prevent fatal accidents in the case of extreme earthquakes. A modern earth house constructed with compressed earth stabilised blocks (CESB) located in Namchi can be seen in Fig. 10.2c. The construction blocks were stabilised with cement. A stabiliser is added to soil to provide superior mechanical strength and durability to the building. Lintel and plinth bands made of concrete were also used in this building.

The states of Jharkhand and Orissa, located in the central east of India, have a composite to hot and dry climate. Khunti and Sundargarh are both home to tribal communities where living in an earth house has cultural significance. The houses in these locations were constructed with a cob technique. In this technique, a thick

Fig. 10.2 Earth house in **a** Bir (North) **b** Namchi (North east), **c** Namchi (North east), **d** Khunti (East), **e** Sundargarh (East), **f** Tiruvannamalai (South), **g** Sittlingi (South), **h** Pondicherry (South), **i** Khavda (West)

monolithic wall is raised from the foundation. Figure 10.2d, e shows traditional earth houses in Khunti and Sundargarh, respectively. Cow dung has been used for the plastering of cob wall. Local tiles, known as "Khapra", have been used as roofing material. A significant deterioration of the walls was seen in these houses.

Tiruvannamalai and Sittlingi, in Tamil Nadu, in the south of India are located in a hot and humid climate. Cob construction is also commonly seen in this area. Several innovative techniques have been implemented in this area, and a significant rise in modern earth construction has been observed. Figure 10.2f shows an Earthbag building located near Tiruvannamalai. In the Earthbag technique, soil is filled in a jute or plastic bag and these bags are stacked mostly with the help of barb wires. The red soil used in this construction was stabilised with lime. The construction of this building was completed in 4 months. Sittlingi has a mix of modern CSEB (Compressed Stabilised Earthen Blocks) houses and traditional cob houses. Significant abandoned and highly damaged traditional houses were observed in Sittlingi (Fig. 10.2g). The use of local quarry ash was prominent in the modern earth construction houses.

Auroville and Pondicherry, also located in Tamil Nadu, are located in a warm and humid environment. Auroville and the surrounding areas of Pondicherry have many modern earth structures. CSEB blocks stabilised with 5–8% of cement have been commonly used. Figure 10.2h shows a CSEB construction in Pondicherry. One of the architects explained that the soil was dug out from the lowest contour of the site so that the dugout part can be used to collect rainwater. Another architect explained that the most important aspects of modern earth construction are supervisory skills and craftsmanship.

In the western state of Gujarat, a hot and dry climate is prevalent. This zone also has a history of high seismic activities. “Bhungas” are traditional earth houses of this region, and some Bhungas constructed post-earthquake can be seen in the village Gandhi nu Gam located in Khavda (Fig. 10.2i). Many of the houses collapsed during the 2001 earthquake; however, it was stated that many Bhungas survived. The cylindrical shape is considered good for earthquake resistance by locals. The Bhunga shown in Fig. 10.2i was constructed with adobe. The mortar was prepared by mixing cow dung with soil, and there was no foundation. A lintel band was made of locally sourced wood, and the roof was made of bamboo covered with grass.

The climate of a region has a strong influence on the construction technique and the materials used for construction. The type of construction and the recipe of the earth material depended on the availability of raw materials, which are location-dependent. Most of the modern earth houses utilise cement which has standardised properties and quality, and is available almost everywhere. The cost of construction of modern earth houses is comparable to concrete or fired brick houses. Traditional earth houses are constructed with a minimal monetary investment from the dwellers.

It was often mentioned by the interviewee in the regions of high seismic activity that traditional earth houses are inherently earthquake-resistant. These houses and techniques have been developed over a long period of time. Modern earth houses are seismically protected by plinth and lintel bands that are usually constructed of reinforced cement-based concrete. According to this survey, a stabilised earth structure is both labour- and time-intensive. Traditional houses are generally built by the dwellers but modern earth houses require skilled labourers and good supervision. There is not widely available skilled labour for modern earth construction.

10.3.2 Performance

The benefit of earth structures over conventional structure made up of fired bricks and concrete was acknowledged widely. The earth houses were considered environmentally friendly for the following reasons: (1) they utilised locally available resources, thus saving significant embodied energy that is spent on transportation; (2) the material is unfired and therefore reduces energy consumption; and (3) demolition waste can usually be re-used and usually does not end up in landfill disposal.

Almost all interviewees mentioned that the indoor temperature was controlled well in all seasons. This was considered by far the most beneficial aspect of earth houses. An earth house dweller in Tiruvannamalai emphasised “Mud housing is appropriate to the indoor climate, holistic way of leaving. These are live spaces, air is passing through”. A mason in North of India mentioned that several local people were interested in earth houses as they were considered good for health. In another example, an architect, practising in the south of India, mentioned that a doctor was willing to make his house out of mud due to his belief of the healing power of earth.

Earth houses, especially those which are unstabilised, disintegrate in nature and can be re-used numerous times. The dweller of an earth house in Bir mentioned that they constructed their present house from the material of their ancestral house. They believed that earth houses have an infinite life as they can be re-used multiple times. A young architect appreciated the fact that vegetation is possible to be grown on demolished earth material.

Despite the advantages, technical limitations of earth houses led people towards choosing industrialised construction materials. The durability of traditional earth houses was a major concern of all the interviewees. Most of the traditional earth houses faced significant deterioration due to rain and required frequent re-plastering. Sometimes, the rain also resulted in structural weakening of the earth houses. In this specific case, raising the foundation was considered an important precautionary step. The rule of a good hat (roof) and good boots (foundation) was suggested by architects for enhanced durability.

One of the most commonly identified limitations of traditional earth houses was termite infestation. An expert commented that the problem could be solved if the construction is properly detailed and the foundation is treated with pesticides and insecticides. A rise of plinth level was also suggested as a method to prevent termite infestation. The problem of termite infestation was prevalent in the houses, which were not continuously functional for many years. An experienced earth architect living in an unstabilised earth house for over 10 years elucidated the problem "There are so many traditional unstabilised houses that are standing for years. Termite infestation was not a big problem in past. Previously 'Chulas' (wood/coal fired stoves) were used for cooking and the smoke from Chulas functioned as a termite repellent. Now these Chulas are replaced by a gas stove or electric cooking equipment. This has resulted in an increase in case of termite infestation in unstabilised earth building. The building and material techniques are not upgraded to accommodate such changes".

Earth houses are known to regulate indoor temperature and humidity. In a unique case of an Ikara house (North East), the thermal behaviour of an earth house was stated to be poor as compared to a concrete house. This was due to the use of Galvanised Iron (GI) sheet as the roofing material that has a high thermal conductivity and low thermal inertia; thus, the house was hot in the summer and cool in the winter. The GI sheets were also the most widely adopted roofing for fired brick houses in the region. In many cases, traditional earth houses were modified over the time without full consideration and they lost the essential characteristics such as thermal behaviour and aesthetics of an earth building. Other issues such as cleanliness and problems with rodents were also acknowledged. For example, an interviewee mentioned that even though mud flooring kept the house cool but it also result in unhygienic conditions during rainy season, i.e. when a person entered the house with wet feet, the whole house would get dirty.

In the modern earth structures, good design and engineering usually results in a durable structure. One of the contractors mentioned that the CSEB blocks have a longer life than fired brick and they are much stronger. It is thought this stated was considering low-quality "country-fired" bricks where strength could be typically 3 MPa. Termite infestation, deterioration due to rain and frequent maintenance (re-plastering) are not a problem in modern earth construction, due to the presence of cement-based stabilisers. However, in colder regions, the problem of cracking on the exterior surface was observed in few CSEB houses. This was hypothesised to be due to improper curing of cement-stabilised blocks. In a colder climate, the cement takes an extended amount of time to cure and thus results in poor quality of blocks.

In one of the building project which was developed as a community centre for villagers, the architect emphasised on the issue of the weathering of CSEB brick and the rise of water from the foundation due to the absence of an impervious lining under the foundation resulting in the flaking on the wall. This problem was solved by re-plastering the wall, which may have to be undertaken periodically.

Most of the limitations of earth houses has been seen to be able to be overcome by addition of stabilisers and high-quality construction. This, however, results in the increase in construction costs making the structures unaffordable for low- to medium-income households. For the dwellers of traditional earth houses, the issues of termite infestation and frequent re-plastering have resulted in a choice of fired brick over the earth as construction material.

10.3.3 Image

Image of any materials plays an important role in its choice of use. Earth construction in developing countries suffers from a lower societal image.

A Pondicherry-based architect said, "*village people don't want a house which looks like a village house. They want something which urban people aspire for. It may be eco-friendly, or good for the climate or may be good for your health, but status and associations that people have with a concrete house is something which you can't change easily*". An architect from Gangtok (North East India) mentioned, "*Natural building is considered poor man's dwelling. It was once upon a time mass housing for people in Sikkim but then came people from different countries who were influential and powerful. They made houses with foreign material like RCC. Many people were influenced by them to build with this new material. New materials were maintenance free and were much more durable. This changed the perception right away. RCC gave the opportunity to build taller. In a big family, the parents could make a 5-floor house and give an individual floor to their children. Hence, it was definitely an interesting material choice in urban areas. Moreover, the land prices were getting high. RCC became a status symbol in Sikkim. RCC house and a car is the progress in life. People aspire for it. Despite marring the landscape, it is popular because people want to show that they have progressed*" (transcribed from recording).

An expert on earth construction from Bangalore mentioned that the unstabilised mud houses are more of a social problem than the technical one. He gave a different view on the image issue. According to him, in rural areas the walls are thick, and if users are aware of their building, they can live peacefully even though there is erosion on the outside. The user knows that the building will work satisfactorily irrespective of erosion. People looking from outside judge the building and form opinions about it. The problem of image is also a result of aesthetics. A significant amount of earth houses in a deteriorated state can be seen throughout the country.

The users of earth houses also shared their views. A low-income family dwelling in traditional earth house in Bir (North India) mentioned, "*When we see our neighbours, we see houses with bricks and it makes us feel that our house should also be made with bricks. Our kids also say that we want these houses, one with lintels and beams*" (translated from Hindi and transcribed from recording). A community from the tribal village of Sundargarh (central east of India) shared their views, "*Nowadays it has become all about the money in the world. Today we are in an independent India. The mud house days are gone. Before we used to use lungi (traditional pants) and now we use jeans pants. Likewise, slowly people are learning and getting educated and therefore they decided to move to a brick house. When we started earning some more money, we wanted to go for a proper concrete roof. Whoever has a cycle, they think that their life will be better with a motorbike. We see changing from a mud house to concrete building as a positive change. We do it mostly to show to others that we are also modernizing. We do not want to be left behind. When people will see this place changing then they will get a good impression of the people who live here. The mud houses stay strong for 30 years but the brick houses will stay strong for more than 90 years. Hence, we have accepted change and have moved on to brick and concrete houses*" (translated from Oriya and transcribed from recording). Two of the mason interviewed in this survey shared similar views and emphasised that people in the village do not want to build with earth anymore. There are also terms like "Pakka house" and "Kaccha house" that are often used by the government to classify people based on the type of house they own. Houses made with natural materials are considered as Kaccha (weak) houses, and the dwellers of these houses have a lower social status.

The people who can afford and chose to have a modern earth house often derive their motivation from a holistic (ecological) way of living life. Their motivation for choosing an earth house comes from the notion that earth houses are sustainable. They consider industrial building materials as polluting the whole ecosystem. One of the farm owner living in modern cob house near Tiruvannamalai (South India) said, "*People are really searching for a different way of life and they are looking for a different style of architecture. People have been very stifled with the consumeristic and materialistic society and the concrete boxes in which they are living. Cement is not locally sourced and coming through big MNC's (Multi-national companies). It is a material that is really not as durable as people imagine. All the cement construction happening now are also going to fall down someday as they are constructed in an improper way. The idea that an earth construction is not durable need not be true. It depends on how scientifically you are constructing it. The mud houses are living spaces, they breathe. Air can pass through it and the feel of living in these spaces is good and natural*". (transcribed from recording).

The image is the most important factor that favours the choice of building material. For low-income households in the rural area, the image of earth construction is low and something that is outdated. They aspire for a house that urban people possess. For the people interested in a holistic lifestyle, the image of cement and other industrialised material is bad as these materials are known to pollute the environment. Their choice of an earth house is based on living in a sustainable and natural habitat.

10.4 Conclusion

Construction techniques, performance and image govern the choice in favour or against earth construction methods. Earth construction, unlike construction with cement and brick, is dependent on the climate and availability of raw materials. Modern earth houses are labour-intensive, and it is also difficult to find skilled labour in most regions.

Although the advantages of earth houses, such as a better indoor climate, environmental friendliness and recyclability, were widely acknowledged, limitations with regard to termite infestation, poor water resistance and weathering were the major technical drawbacks that motivate the aspiration (for concrete/brick house) of low-income households. The environmental friendliness of earth construction was a major motivation for people interested in a holistic lifestyle. The limitation in traditional earth houses has been overcome by using stabilisers in modern earth houses. However, this modern practice is far away from being affordable for low-income households. In cases where a traditional earth house has been upgraded, a synergy between traditional and modern architecture was missing.

As an outcome of the survey, two specific but opposing motivations were identified: (1) Low-income families living in traditional earth houses aspire for a brick or concrete house; and (2) families that have adopted an alternative and sustainable lifestyle and prefer living in a "natural habitat". The families in the latter group usually have a high income. The social aspects such as a low societal image of traditional earth houses in comparison with other households in the same community were the main reasons behind the choice in favour of modern building materials over earth. The identified negative social, technical and financial aspects place a new requirement and demand re-invention of traditional earth house as a necessary step towards their acceptance.

References

Agarwal A (1982) Research: mud as a traditional building material. In: The changing rural habitat I: 137–146 (Case studies, The Aga Khan Awards)

Bredenoord J (2017) Sustainable building materials for low-cost housing and the challenges facing their technological developments: examples and lessons regarding bamboo, earth-block technologies, building blocks of recycled materials, and improved concrete panels. J Archit Eng Tech 6:187. https://doi.org/10.4172/2168-9717.1000187

Census of India (2011) House. Household amenities and assets. http://www.censusindia.gov.in/DigitalLibrary/TablesSeries2001.aspx. Accessed on 25th Feb 2018

Houben H, Guillaud H (1994) Earth construction: a comprehensive guide. ITDG Publishing, London, UK

Minke G (2006) Building with earth. Birkhäuser-Publishers for Architecture, Basel-Berlin-Boston

Ministry of Rural Development India (2016) PM launches "Housing for All" in rural areas http://pib.nic.in/newsite/PrintRelease.aspx?relid=153931. Accessed on 25th Feb 2018

Schroeder H (2016) Sustainable building with earth. Springer International Publishing, Switzerland

UNESCO (2018) Earthen architecture: the environmentally friendly building blocks of tangible and intangible heritage. http://www.unesco.org/new/en/unesco/resources. Accessed on January 25th 2018

World Bank (2016) Housing for all by 2030. http://www.worldbank.org/en/news/infographic/2016/05/13/housing-for-all-by-2030 . Accessed on 25th Jan 2018

Chapter 11
Identification of Saudi Arabian Soil Appropriate for Stabilised Earth Construction

Mohammad Sharif Zami

11.1 Introduction

Historically, earth has been used as a construction material in every continent of the world. According to Hadjri et al. (2007), stabilised earth is an alternative construction material that is economically and environmentally beneficial compared to the conventional ones. Practice of earth construction in Saudi Arabia nowadays mainly remains in conserving and restoring architectural heritages (Zami 2017). According to Mortada (2016), adobe and sun-dried mud bricks have been in use for construction purposes since ancient time in Arabian Peninsula (King 1998) and were commonly used in building during pre-Islamic eras in the Saudi Arabia. During the time (AD 622) of Muhammad (the prophet of Islam), mud brick (libin in Arabic) was used constructing his residence and the mosque in Medina too (Mortada 2016; King 1998). According to King (1998), mud is used throughout central Arabia Najd, in both the sand desert areas and in the fertile valleys, much of the interior of Yemen and Oman, which also extends northwards into Iraq and the Syrian Desert. According to King (1998), in the recent past, the presence of earth buildings in the central Arabian towns has a long ancestral relation that is proved by the excavations at Al-Rabadha on the western edge of Najd. The buildings in this excavation were found used sun-dried earth blocks from the centuries preceding Islam through to Abbasid times. Amongst hundreds of rich heritage sites, Addiriyah is one of the most prominent earthen architectural heritage sites of Saudi Arabia (Zami 2014). This important settlement was located on an elevated ground which is 700 m above the sea level at Najd central plateau of the Arabian Peninsula on the banks of Wadi Hanifah. This Wadi Hanifah is considered as the most prominent feature of Addir'iyah and an important geographic and topographic natural formation within the Najd region. Throughout the

M. S. Zami (✉)
King Fahd University of Petroleum & Minerals, Dhahran, Saudi Arabia
e-mail: mdszami@kfupm.edu.sa

B. V. V. Reddy et al. (eds.), *Earthen Dwellings and Structures*,
Springer Transactions in Civil and Environmental Engineering,
https://doi.org/10.1007/978-981-13-5883-8_11

past centuries, many tribes were attracted by its suitable and favourable settlement conditions, such as natural features, rich fertile clay soil, and water combined with land formation (Ar-Riyadh City Web Site 2014). Therefore, the Addiriyah settlements during the past centuries utilised soil for construction from the bank of Wadi Hanifah. There are settlements of earth construction that are up to date existing in Saudi Arabia other than Addiriyah but source and type of the soil used in buildings of these settlements are unclear in the literature. Furthermore, there are local soils other than clay that are not commonly used as building material, but existing literature appears to be insufficient to prove that these soils are unsuitable for stabilised earth construction. Therefore, this paper investigates on the typologies of locally available soils appropriate for stabilised earth construction. In order to achieve the aim, this paper critically reviews the literatures, observes on CSEB samples and refines the information to establish whether stabilised earth construction is sustainable considering availability of the soil in Saudi Arabian context.

11.2 Availability of Earth in Saudi Arabia for Construction Purposes

According to Al-Amoudi et al. (2010), there are four types of soil (i.e. sand, marl, clay and sabkha) found in Saudi Arabia's eastern province. Marl soil is unique, most suitable compared to sand, clay and sabkha, therefore extensively used in all types of construction sites including road bases, embankments and foundations (Alshammari and Hamid 2016). Netterberg (1982) defined marl as a soil or similar to rock material that contain about 35–65% calcareous material and varying percentage of clay content (Pettijohn 1975; Qahwash 1989). According to Ahmed (1995), calcareous soils (marl) are commonly used in the highway construction and foundations of buildings. A critical review of the literature reveals that calcareous soils (marl) have drawn very little attention from the geotechnical engineers and researchers despite their wide and extensive use in the construction sites all over the world (Ahmed 1995). Current study shows that calcareous soils (marl) are extremely sensitive to the moulding and testing moisture contents; strength and durability of the calcareous soils can be significantly improved by chemical stabilisation using cement despite the fact that the calcareous soil has poor properties (Ahmed 1995).

According to Aiban (1995), calcareous soil is available in the eastern Province of Saudi Arabia in many places, such as Abqaiq, Dhahran, Dammam, Abu Ali, Hofuf, Berri, Fadhli, Jubail, Abu Hadriyah and Safaniyah areas, as shown in Fig. 11.1. Calcareous soils found in eastern Saudi Arabia can vary depending of their colour, plasticity, physical, mineralogical, chemical composition and also engineering properties. Marl colours could be white, dark or light grey, pink, yellow and brown depending on the locations they are quarried from. Moreover, depending on the composition, especially the clay mineral type and content, the plasticity of marl can vary from

Fig. 11.1 Vicinity map of eastern Saudi Arabia showing locations of major marl quarries in eastern Saudi Arabia. *Source* Al-Amoudi et al. (2010)

none to moderate (Aiban et al. 1998). However, the literature review reveals that a number of studies were carried out on the chemical stabilisation of limited types of marl soil sample blocks; it appears that the compressive strength and durability of marl soil sample blocks significantly improve because of chemical stabilisation. But the ongoing research only considers marls used only for the ground preparation or supporting substructure of the building. Current research in Saudi Arabia does not concentrate at all marls considered for constructing superstructures of the building, such as load-bearing wall, column and floor slabs.

In Fig. 11.2, the types of Saudi Arabian soils are roughly classified. According to Beaumont et al. (1976), the Arabian Peninsula is dominated by desert soils, including lithosols and sand. Lithosol is zonal shallow soil, consisting of imperfectly weathered rock fragments. In general, the characteristics of these soils are: poorly developed soil horizons, very low humus content, predominant direction of the water movement in the soil profile upwards in the sabkah areas resulting in the accumulation of salts and

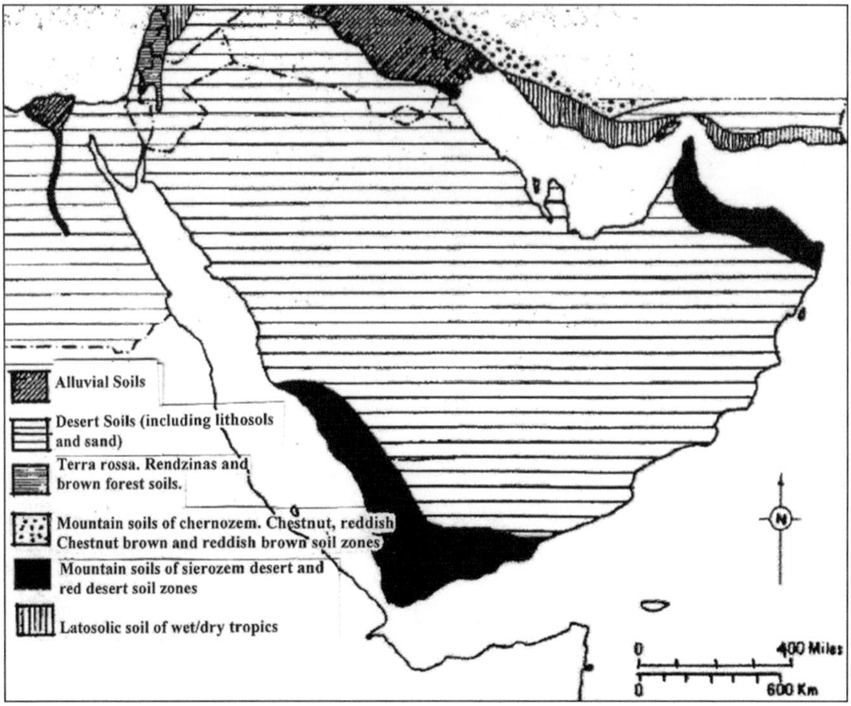

Fig. 11.2 Soil groups of the Middle East. *Source* Beaumont et al. (1976)

other soluble products in the upper layers, rocky stony surface layers of pebbles or moving sand soils (dunes). Other soils found that are not specified in Fig. 11.1 include alluvial soil (soils in river beds and wadis). This soil is sometime fertile or infertile due to high water tables and salt accumulation in the upper soil layers; sometimes, lush vegetation can be found on the higher river banks (Cochrane 1977). Historically, this alluvial soil of Wadi Hanifa was used for the construction purposes in the settlement of Addiriyah. Empirical research on the geotechnical and engineering properties and behaviours of the locally available alluvial soil appears to be very few. Research on modern stabilisation techniques on this type of soil is inadequate; therefore, this needs a long-term research commitment and investments on ways and means to improve the durability and strength of this soil if considered for building construction.

Expansive clay soil is available in different areas of Saudi Arabia and traditionally used for vernacular earth houses. There is a study carried out by Hameed (1991) finding the location of expansive soil in the eastern province of Saudi Arabia and its characteristics. It was found out that expansive soils exist in Al-Qatif and Al-Hofuf areas. Formation and behaviour of these expansive soils are influenced by climate and geology of the area. Clays in both areas are highly plastic, possessing very high swelling potential and rich in smectite, illite, dolomite, palygorskite and kaolinite.

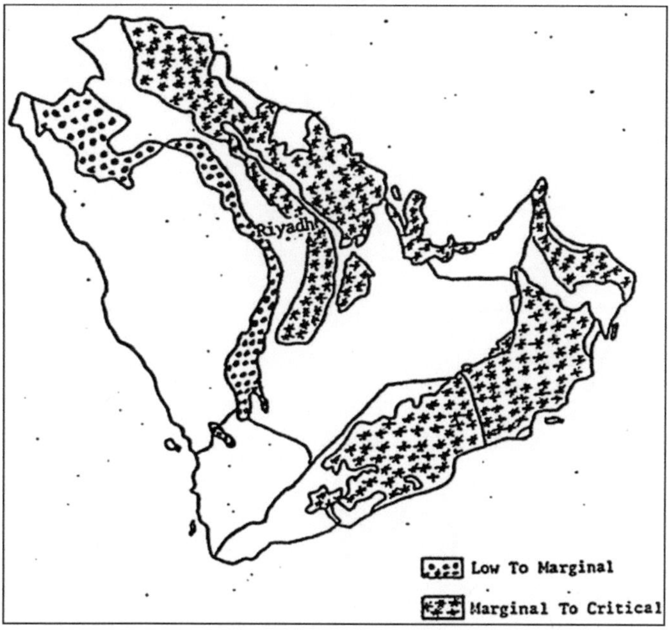

Fig. 11.3 Hazard map for potentially swelling soils of Arabian Peninsula. *Source* Slater (1983)

According to Hameed (1991), the expansive shale zone includes three major development centres, namely Tabuk, Tayma and Al-Ghatt. Expansive soils have also been reported in several other areas, such as Al-Madina and Al-Hofuf (Dhowian et al. 1985) and Al-Qatif area (Abduljauwad and Rafi 1990). Figure 11.3 provides an approximate guide to the suspected distribution and extent of potentially swelling soils (Slater 1983). A study was carried out by Ahmad (1988) whereby engineering properties and behaviour of these expansive clays of Qatif were determined. The results of the investigations showed that Al-Qatif clays are highly expansive and heterogeneous in nature, and 4–8% of commercial lime should be preferred for preconstruction treatment. According to Ahmad (1988), the regions with expansive soil formations in Saudi Arabia are shown in Fig. 11.4.

However, the researches on expansive clays concentrate only for the ground preparation or supporting substructures of the buildings. Current research on expansive clays in Saudi Arabia does not concentrate on constructing superstructures of the building, such as load-bearing wall, column and floor slabs.

Fig. 11.4 Tentative distribution of expansive formations in Saudi Arabia. *Source* Dhowian et al. (1985)

11.3 Suitability of Marl for Stabilised Earth Construction

Two types of marl samples were collected from the field and brought in the laboratory in order to observe and find out the suitability of marl for stabilised earth construction. These two marls are widely, locally used and commonly known as grey and white

Fig. 11.5 Kharj soil (left), grey marl (middle) and white marl (right). *Source* Author, 2017

marl (Fig. 11.5) amongst construction professionals and researchers. It is important to note that marl is not commonly used as building material in Saudi Arabia but found locally in large quantities. As shown in Fig. 11.6, it is a common scenario of Saudi Arabia to see marl transported into any road or building construction sites in large quantities. A third type of soil called Kharj also was collected from the field for comparison purposes as because Kharj soil is commonly used building material in Saudi Arabia. Kharj soil is red in colour (Fig. 11.5), clayey, locally available, queried from Kharj and commonly transported to all over eastern province of Saudi Arabia to construct houses.

Fig. 11.6 A construction site with white marl in Saudi Arabia. *Source* Author, 2017

Fig. 11.7 Belgian handpress machine. *Source* Author, 2017

Fig. 11.8 Making earth block in Belgian handpress machine. *Source* Author, 2017

Fig. 11.9 Crack-free fresh compressed stabilised earth block (CSEB) before curing. *Source* Author, 2017

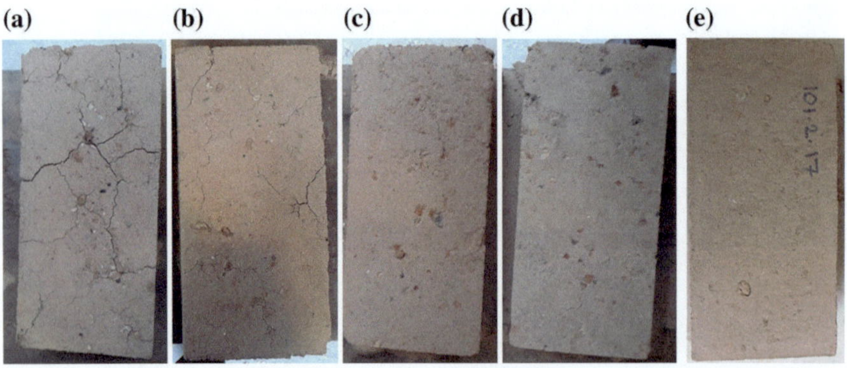

Fig. 11.10 Some of the compressed stabilised earth block samples after curing. *Source* Author, 2017

Table 11.1 Observation remarks on CSEB samples made out of Saudi Arabian soils under fully submerged rainwater

		Types of soil		
		Grey marl	White marl	Kharj soil
Cement as stabiliser	0%	Automatically eroded after curing and completely disintegrated in the submerged rainwater	Automatically eroded after curing and completely disintegrated in the submerged rainwater	Developed a lot of cracks after curing and completely disintegrated in the submerged rainwater
	2.5%	No crack developed but surfaces slightly eroded after curing and partially disintegrated under the rain	No crack developed but surfaces slightly eroded after curing and partially disintegrated under the rain	Developed few cracks, severely eroded after curing and finally disintegrated under the rain
	5%	No crack developed and remained intact after curing and remained integrated under the submerged rainwater	No crack developed and remained intact after curing and remained integrated under the submerged rainwater	No cracks after curing, surfaces eroded slightly and partially disintegrated under the submerged rainwater
	7.5%	No crack developed and remained intact after curing and remained integrated under the submerged rainwater	No crack developed and remained intact after curing and remained integrated under the submerged rainwater	No cracks after curing and remained intact under the submerged rainwater
	10%	No crack developed and remained intact after curing and remained integrated under the rainwater	No crack developed and remained intact after curing and remained integrated under the rainwater	No cracks after curing, intact and remained integrated under the rainwater

(continued)

Table 11.1 (continued)

		Types of soil		
		Grey marl	White marl	Kharj soil
Gypsum as stabiliser	2.5%	No crack developed after curing but completely disintegrated under the submerged rainwater	No crack developed after curing but completely disintegrated under the submerged rainwater	No crack developed after curing but completely disintegrated under the submerged rainwater
	5%	No crack developed after curing but the surfaces eroded significantly; partially disintegrated in the rainwater	No crack developed after curing but the surfaces eroded significantly; partially disintegrated in the submerged rainwater	No crack developed after curing but the surfaces eroded significantly; completely disintegrated in the rainwater
	7.5%	No crack developed after curing but the surfaces eroded significantly; partially disintegrated in the rainwater	No crack developed after curing but the surfaces eroded significantly; partially disintegrated in the rainwater	No crack developed after curing but completely disintegrated in the rainwater
	10%	No crack developed after curing but the surfaces eroded significantly; partially disintegrated in the rainwater	No crack developed after curing but the surfaces eroded significantly; partially disintegrated in the rainwater	No crack developed after curing but partially disintegrated in the rainwater

Source Author, 2017

All three types of soil samples were mixed with two different locally available stabilisers, namely cement and gypsum independently. Mixtures of all types of soil with both stabilisers were independently prepared in five different ratios (0, 2.5, 5, 7.5 and 10%). A Belgian manual earth block making machine (Fig. 11.7) was used to make CSEB out of each soil samples. At total of 27 (9 $\times$ 3) sample blocks were produced manually with the help of the Belgian handpress machine as shown in Fig. 11.8 for preliminary observation. All the freshly made compressed stabilised earth blocks (CSEB) looked crack free (Fig. 11.9). Curing of the sample blocks was then done naturally (stored open-air overhead shaded place under moist condition, i.e. covered one week under soaked sacks) for a period of four weeks.

After curing some of the CSEB samples developed cracks (Fig. 11.10a), surface of the blocks eroded, and some automatically disintegrated in rain water. On the other hand, some samples remained intact and integrated (Fig. 11.10c–e) even under the rainy condition. It is important to mention that the average annual rainfall of Dhahran is 3 mm (Mashat and Abdel Basset 2011) and humidity is 55%. However, for the last 10 years annual rainfall and humidity of Dhahran are not consistent; therefore, all the blocks were fully submerged under the water in a container for a period of 24 h and kept outdoor open air until the water was evaporated and the blocks become dry. It took about 7 days to dry out the water from the container and blocks become dry. Table 11.1 presents observation remarks on the CSEB samples under open-air rainy condition.

11.4 Conclusions

This paper has reviewed, investigated and analysed the existing literature on three types of Saudi Arabian soils appropriate for stabilisation and possibilities to use these as building material. It was found out that research on the appropriateness of all types of Saudi Arabian soils for stabilised earth construction appeared to be none, especially the stabilised marl considered as building material. However, preliminary visual observation of the performance of stabilised Saudi Arabian marls appeared to be very promising as building material. This study also concluded that the application of contemporary stabilised marl as a building material for the superstructure of the buildings is rare in Saudi Arabia. Current practice of earth construction in Saudi Arabia mainly follows traditional methods for the purpose of conserving and restoring architectural heritages. Therefore, a lot of research and practical experimentation are needed on stabilised earth construction. However, research on various indigenous earth samples, such as marl, alluvial, expansive clay soils, is on board but only to use it for the ground preparation of road construction and foundations of buildings. Hence, research and experimentations on load-bearing wall, floor slab and finish materials of stabilised earth are very essential. According to the observations made in this research on local soils confirms that the stabilised earth construction in Saudi Arabia will open up new avenues to solve several prevailing problems, such as housing shortage, environmental pollution, lack of cultural identity and loss of architectural heritage. Soil is locally available in all over Saudi Arabia, they are culturally well known, environmentally sustainable and would be an appropriate alternative to the conventional building materials (fired brick and concrete).

Acknowledgements The author wishes to express their deepest gratitude to King Fahd University of Petroleum and Minerals (KFUPM) for the opportunity to accomplish this work. DSR Project Code: IN161032.

References

Abduljauwad SN, Rafi A (1990) Expansive soil in Al-Qatif area. Arab J Sci Eng 15(2A). http://www.springer.com/engineering/journal/13369/PSE?detailsPage=press

Ahmad R (1988) Engineering properties and mineralogical composition of expansive clays in Al-Qatif area (K.S.A). M.Sc. in Civil Engineering thesis. College of Graduate Studies, King Fahd University of Petroleum & Minerals, Dhahran, Saudi Arabia

Ahmed H-R (1995) Characterization and stabilization of Eastern Saudi Marls. M.Sc. in Civil Engineering thesis. College of Graduate Studies, King Fahd University of Petroleum & Minerals, Dhahran, Saudi Arabia

Aiban SA (1995) Strength and compressibility of Abqaiq marl, Saudi Arabia. Engineering Geology 39 (1995) 203–215, Elsevier Science B.V.

Aiban SA, Al-Abdul, Wahhab HI, Al-Amoudi OSB (1998). Identification, evaluation, and improvement of Eastern Saudi soils for construction purposes, final report. Riyadh, Saudi Arabia: King Abdulaziz City for Science and Technology (KACST AR-14–61)

Al-Amoudi OSB, Khan K, Al-Kahtani NS (2010) Stabilization of a Saudi calcareous marl soil. Constr Build Mater 24:1848–1854. Elsevier Ltd. journal homepage: www.elsevier.com/locate/conbuildmat

Alshammari AM, Hamid AM (2016) Calcareous sediment in Saudi Arabia: a review of marl soil. In: Proceedings of 2016 2nd international conference on architecture, structure and civil engineering (ICASCE'16). London (UK), Mar 26–27, 2016. ISBN 978-93-84422-62-2

Ar-Riyadh City Web Site (2014) High Commission for the Development of Ar-Riyadh. http://www.arriyadh.com/Eng/ADA/Left/DevProj/getdocument.aspx?f=/openshare/Eng/ADA/Left/DevProj/AddiriyaEn093.doc_cvt.htm (Accessed 01.03.2014)

Beaumont P, Blake GH, Wagstaff JM (1976) The Middle East a geographical study. Wiley, London

Cochrane T (1977) Landscape design for the Middle East. Riba Publications Limited, London, p 1977

Dhowian A, Erol A, Ruwiah I (1985) The distribution and evaluation of expansive soils in Saudi Arabia. The second Saudi engineers conference, KFUPM, Dhahran, KSA, Nov 16–19, 1985

Hadjri K, Osmani M, Baiche B, Chifunda C (2007) Attitude towards earth building for Zambian housing provision. In: Proceedings of the ICE institution of civil engineers, engineering sustainability 160, issue ES3. ICE Publisher, UK

Hameed RA (1991) Characterization of expansive soils in the Eastern Province of Saudi Arabia. M.Sc. in Civil Engineering thesis. College of Graduate Studies, King Fahd University of Petroleum & Minerals, Dhahran, Saudi Arabia

King G (1998) The traditional architecture of Saudi Arabia. I. B. Tauris & Co Ltd, Victoria House, Bloomsbury Square, London

Mashat A, Abdel Basset H (2011) Analysis of rainfall over Saudi Arabia. J King Abdulaziz Univ Metrol Environ Arid Land Agric Sci 22(2):59–78. https://doi.org/10.4197/met.22-2.4

Mortada H (2016) Sustainable Desert Traditional Architecture of the Central Region of Saudi Arabia. Sustain Dev 24:383–393. Published online 30 Mar 2016 in Wiley Online Library. (wileyonlinelibrary.com) https://doi.org/10.1002/sd.1634

Netterberg F (1982) Geotechnical properties and behavior of Calcretes in South and Southwest Africa, ASTM STP 777. American Society for Testing and Materials, Philadelphia, pp 296–309

Pettijohn FJ (1975) Sedimentary rock. Harper and Row, New York

Qahwash AA (1989) Geotechnical properties of fine-grained calcareous sediments for engineering purposes. Eng Geol 1989(26):161–169

Slater DE (1983 May) Potential expansive soils in Arabian Peninsula. J Geotech Eng ASCE, USA 109(5)

Zami MS (2014) Earth as a construction material conserving Addiriyah of Saudi Arabia. In: International conference on vernacular heritage, sustainability and earthen architecture VerSus 2014, 2nd MEDITERRA, 2nd ResTAPIA, 11–13 Sept 2014, Valencia, Spain

Zami MS (2017) Contemporary earth construction of Saudi Arabia: a state of art review. In: International conference for sustainable design of the built environment (SDBE 2017). 20–21 Dec, The Crystal, London, United Kingdom. http://newton-sdbe.uk/conferences/sdbe-conference-2017/

Chapter 12
Strength and Cementation in a Termite Mound

Nikita Zachariah, Ramesh K. Kandasami, Aritra Das, Tejas G. Murthy and Renee M. Borges

12.1 Introduction

Animals construct various structures using either collected (birds' nest) or secreted materials (spider web) or a combination of both (termite mounds) (Hansell and Ruxton 2013). These structures perform various functions such as temperature regulation for their nest, protection from predators, attracting mates and therefore play crucial roles in the survival and reproduction of animals (Dawkins 1999). Various aspects of animals construction have been studied in the past such as architecture (Tschinkel 2004), regulation of nest internal environment (King et al. 2015; Kleineidam et al. 2001; Korb 2003) and collective construction behaviour (Werfel et al. 2014) but little is known about the basic building blocks of animal construction and how material properties affect these constructions. Moreover, recent interest in biocementation as an environmentally sustainable and viable technology has drawn further interest from engineers.

N. Zachariah (✉) · R. M. Borges
Centre for Ecological Sciences, Indian Institute of Science, Bangalore, India
e-mail: nikitaz@iisc.ac.in

R. M. Borges
e-mail: renee@iisc.ac.in

R. K. Kandasami
Schofield Centre, West Cambridge Site, University of Cambridge, Cambridge, UK
e-mail: rkk24@cam.ac.uk

A. Das
Centre for Neuroscience, Indian Institute of Science, Bangalore, India
e-mail: aritrad@iisc.ac.in

T. G. Murthy
Department of Civil Engineering, Indian Institute of Science, Bangalore, India
e-mail: tejas@iisc.ac.in

B. V. V. Reddy et al. (eds.), *Earthen Dwellings and Structures*,
Springer Transactions in Civil and Environmental Engineering,
https://doi.org/10.1007/978-981-13-5883-8_12

Termite mounds are conspicuous features of several landscapes, are known to maintain strength and stability for decades to centuries (Erens 2015) and harvest wind energy for ventilation of underground colony (King et al. 2015). They are constructed by insects called termites by mixing soil with their bodily secretions (Papoola and Opayele 2012) and can be three orders of magnitude larger than individual termites. Even though many studies have documented the micro-scale changes in soil due to termite activity (Jouquet et al. 2003), a comprehensive view of the mechanical stability of termite mounds supported by engineering perspective is missing. Moreover, studies on the basic building blocks of these constructions and effect of material properties on these constructions are not available.

In the first part of this study, the strength and stability of termite mounds are explored and in the second part, the process of making the basic building block is explored. Construction is envisioned as involving three stages—material selection, transport and assemblage. The species used for this study was *Odontotermes obesus* (Bose 1984) and the study was conducted at the Indian Institute of Science Campus in Bangalore, India.

12.2 Methods and Results

12.2.1 Stability Analysis of Termite Mound Soil

Cylindrical samples were cored out from different parts of a termite mound (30 mm diameter × 60 mm height) and were tested under unconfined compression at an axial deformation of 0.5%/min. Control soil from the surrounding area was subjected to a similar test. In order to understand the collapse or erodability of termite mound soil, it was subjected to alternate cycles of wetting and drying similar to the crumb test used for dispersive clays (Head, 1984). Samples of different sizes weighing between 150 and 200 g were soaked in water for 2, 4, 8, 16, 32, …, 100 min, were dried and weighed. The immersion/soaking time of soil samples was increased in each cycle till the complete disintegration of the samples. The intergranular contacts failed during this process leading to the loss of weight of soil sample. This was compared to crushed and reconstituted mound soil at its on-site dry density.

In the above experiment, about 20% sample disintegration was seen in termite mound soil when soaked for about 100 min which was followed by rapid collapse (Fig. 12.1). Erosion by water was most rapid at the top of the mound where the soil density was low and porosity was high. Disintegration was much more rapid in case of crushed and reconstituted termite mound soil and 100% collapse was seen in as less as 60 min suggesting that termite-induced cementation in termite mound soil increased resistance to weathering. Since complete immersion of mound soil in water is very unlikely, the actual weathering of soil from a termite mound would be much lesser than observed in this experiment.

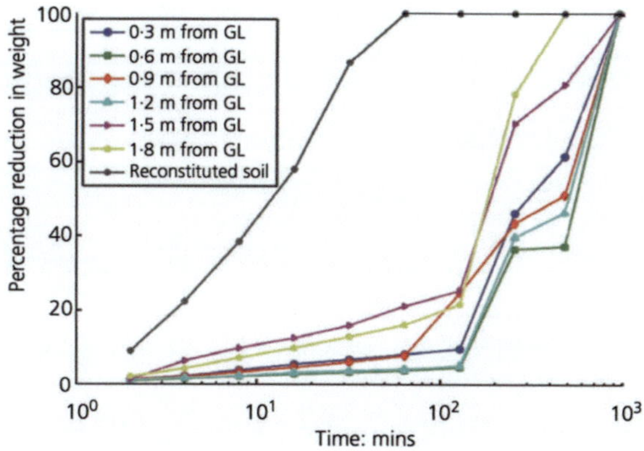

Fig. 12.1 Stability analysis of termite mound soil using alternating wetting and drying and experiments. Samples were collected at different heights from the ground level (GL) of a termite mound and were subjected to alternative cycles of wetting and drying. The results were compared with the weathering of reconstituted termite mound soil

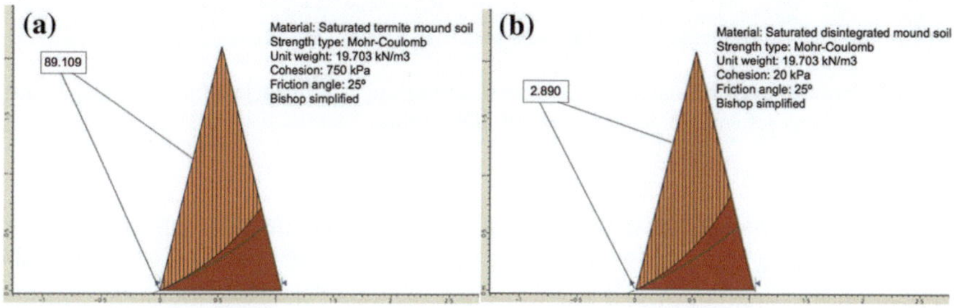

Fig. 12.2 Limit equilibrium slope stability analysis at **a** on-site conditions and **b** with remoulded termite mound soil properties to obtain the factor of safety against failure

The role of biocementation in termite mound was further examined by performing a slope stability analysis under axis symmetric conditions. The termite mound was modelled in two dimensions similar to the field observations (height = 210 cm and base diameter = 105 cm) giving an aspect ratio between 1 and 2 and slopes between 45° and 75°. The stability of such a steep slope may not be very high under saturated conditions considering only frictional parameters. Therefore, the analysis was performed considering both cohesion and friction properties under saturated soil conditions. The method of slices (using Bishop, Janbu and Spencer methods) was carried out where the sliding mass above a failure surface was divided into slices and a force balance was performed to calculate the factor of safety (Salgado 2008). A slope stability analysis programme 'Slide' was used to perform the analysis by considering a circular failure surface and a safety factor was calculated (Fig. 12.2) based on a critical slip surface.

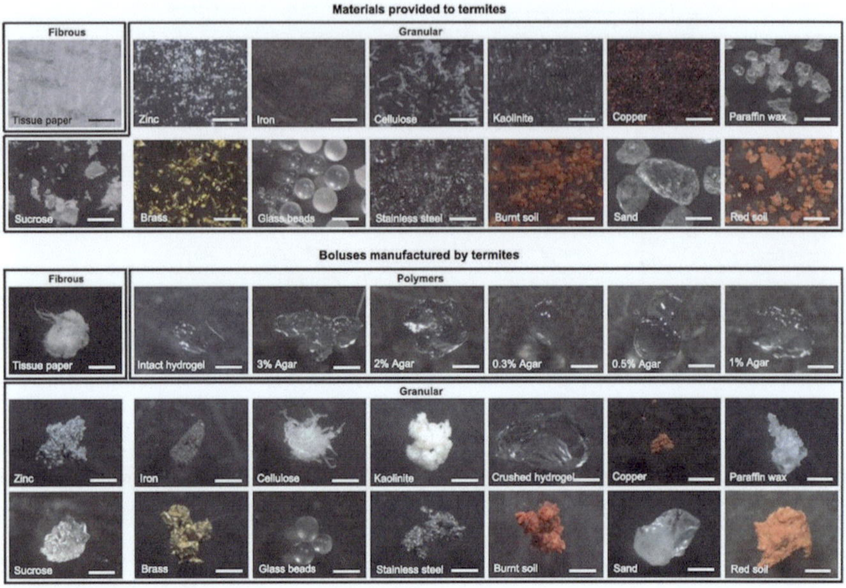

Fig. 12.5 Boluses made by major worker caste of termites with different materials provided (materials indicated on each image). Scale bar represents 0.5 mm

The unconfined compressive strength of termite mound soil was found to be 1200–1800 kPa, while that for the control soil was 125–150 kPa suggesting a tenfold increase in the strength of soil due to termite action. The result for slope stability analysis showed that the safety factor for termite mound on-site properties was 89 while that for the reconstituted soil was as low as 2 (Fig. 12.2) suggesting that termite mound soil is extraordinarily stable owing to biocementation.

12.2.2 Effect of Termite Caste on Bolus Volume

While termite mounds are extraordinarily stable, the basic building blocks of termite mounds are not known. In our preliminary study, we found that termites mix moist soil with their secretions and make small, irregular spheres which we term as boluses (singular bolus; Fig. 12.3a, b). (Termites can make similar boluses with glass beads with the help of their secretions; Figs. 12.3c, d and 12.4a). These boluses act as the basic building blocks of termite mound construction and are analogous to the bricks used in human construction. In order to understand the process of bolus manufacture by termites, a breach was induced at a live termite mound and the process of breach repaired was observed. Three morphologically different castes of termites were found at the site of breach repair—major workers, minor workers and soldiers. Only major and minor workers were found to carry boluses; not the soldiers. The boluses were

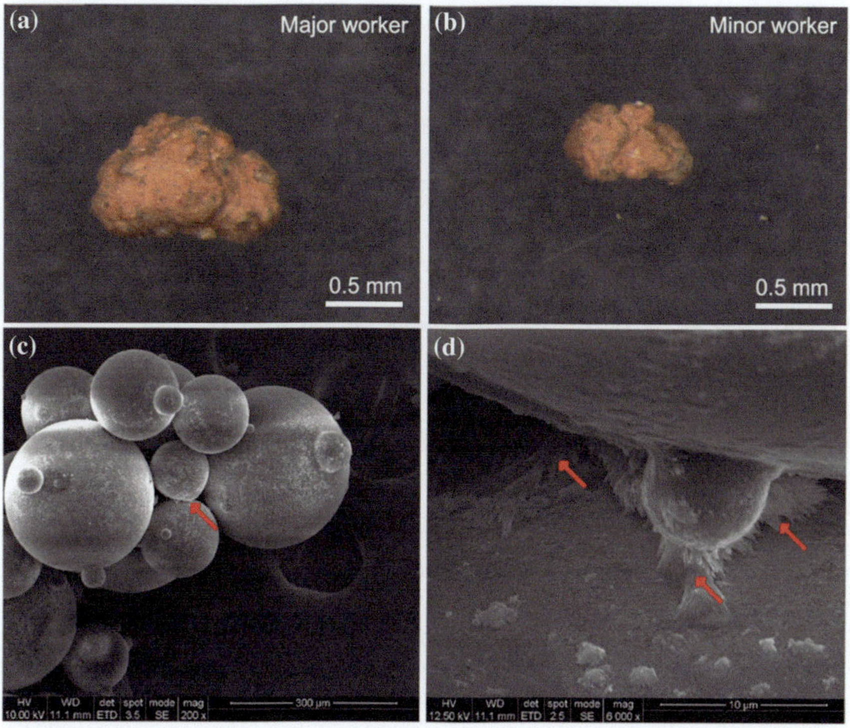

Fig. 12.3 **a** Boluses made with soil by major and minor workers. **b** Electron micrographs of boluses made up of glass beads indicating salivary deposition at the junction of glass beads (arrows)

collected from major and minor workers from five different mounds and their volumes were measured. A generalized linear mixed model (GLMM) analysis (Bolker et al. 2008) was performed with $\log_e$(Bolus Volume) ~ Caste + (1|Mound/Caste).

Results showed a biomodal distribution of bolus sizes between major and minor worker castes such that the boluses made by major workers were 3.7 times larger than those made by minor workers. GLMM results showed very low inter-class correlation for Mound (ICC = 0.039) and Caste:Mound (ICC = 0.056), suggesting that these variables have negligible contribution towards explaining the total variance and the largest predictor of bolus volume is termite caste.

12.2.3 Achieving Optimal Packing of Boluses

From the previous results, it was evident that the boluses of major and minor workers differ significantly in their volumes. Therefore, it was important to understand the packing of these boluses during termite mound construction. Therefore, a breach was experimentally induced in a termite mound and the process of breach repair was video recorded from a fixed location. Boluses deposited by major and minor workers were marked in the frames of the above video. In cases where the termites were not clearly visible, the boluses were marked as caste unknown (Fig. 12.4b).

Results showed that the boluses of major and minor workers, which are different in their sizes, were interspersed during the process of breach repair leading to tight packing and high strength to the mound.

12.2.4 *Effect of Material Properties on Termite Mound Construction*

In order to understand how material properties affect the process of termite mound construction, termites were offered a wide range of materials ranging from fibrous materials to polymers and granular materials. Out of 24 materials offered to termites, 21 were used for making boluses (Fig. 12.5). The materials not used for making boluses were Sodium chloride (NaCl) where 100% termite mortality was observed in less than 90 min and very low viscosity agar (0.1 and 0.2%) on which termites could not locomote. Therefore, termites used all the materials as long as they were alive and were able to locomote on the material surface suggesting that bolus making is a hard-wired behaviour. In order to quantify the ease of handling of these materials, the time taken to start making boluses (T) once a termite is offered a material and the number of boluses made in the first 20 min was recorded. From this, the reciprocal of latency ($1/T$) and the total volume and weight of boluses carried in a fixed time interval was calculated. All the materials were ranked based on these properties and a graph was plotted (Fig. 12.6). Termites were also offered stainless steel balls which they did not use until their surfaces were roughened suggesting the role of surface roughness in material handling.

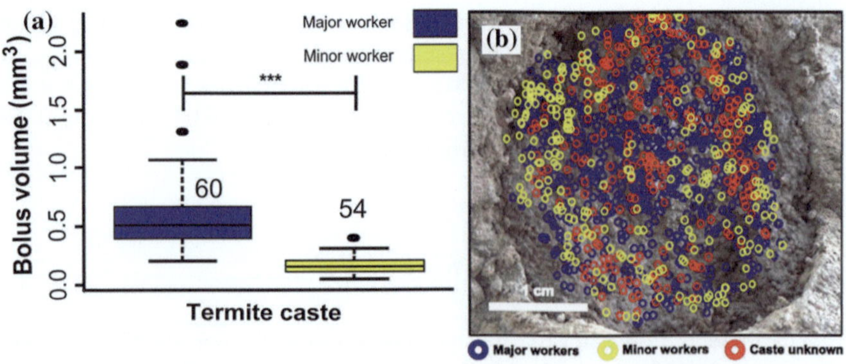

Fig. 12.4 **a** Boluses made by major and minor worker termites at the mound with soil. **b** Packing of boluses made by major and minor workers at the mound

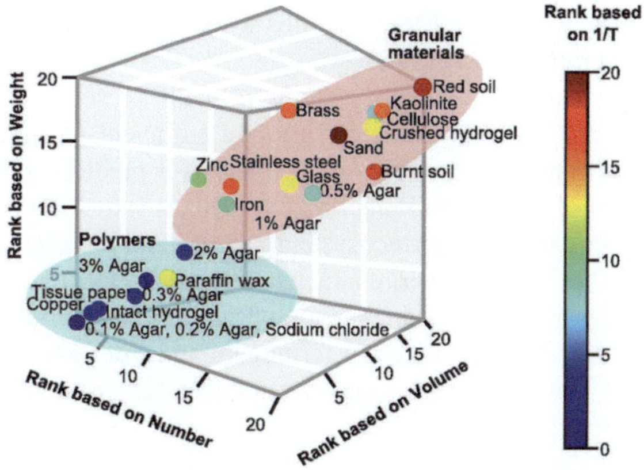

Fig. 12.6 Ease of handling of materials used by termites. Materials were ranked based on four parameters $1/T$, total number, volume and weight of materials carried in 20 min

Results showed that the materials with the highest ease of handling were granular, hydrophilic, osmotically inactive, non-hygroscopic, with surface roughness, rigidity and presence of organic matter.

12.3 Discussion

In this study, we estimated the stability of termite mound soil and characterized the basic building blocks of termite mound construction along with the behaviour associated with material usage for construction. Results showed that termite mound soil had 10 times higher strength than reconstituted mound soil and had a safety factor of 89 as compared to 2.8 for the reconstituted soil. It was observed that immaterial of the slope stability analysis method or failure modes (circular or non-circular), the soil properties modified by termites play a major role in obtaining a higher factor of safety in turn higher stability. Boluses made by termites were bimodal in their size distribution and were interspersed during mound construction such that the smaller boluses filled the voids between the larger boluses imparting tight packing and high strength to the mound. Termites could also use a wide range of experimentally offered materials ranging from fibrous materials, polymers and granular materials but the ease of handling was highest for granular, hydrophilic, osmotically inactive, non-hygroscopic materials, with surface roughness, rigidity and containing organic matter.

Tenfold increase in the strength of termite mound soil and 31-fold increase in the safety factor suggest that termite manipulation imparts very high strength and stability to termite mound soil. The packing of bimodal boluses during termite mound construction ensures optimal packing and high strength to a granular ensemble like a termite mound (Duran 2012). Moreover, this works reasonably well with varying ratios of major and minor worker boluses.

Finally, certain materials properties along with the presence of moisture and favourable climatic conditions are necessary and sufficient for terrestrial mound construction by termites and therefore can be used for predicting the global geographic distribution of mound-building termites. This is valuable since termites are known to modulate the soil-water ecosystem process and thereby climatic effects (Bonachela et al. 2015). Moreover, characterization of biocement used in termite mound construction (which can adhere soil and glass beads among materials) and understanding harvest of wind energy by termite mounds for ventilation (King et al. 2015) will provide earthen materials and technologies for sustainable development in order to meet growing human needs.

Data Availability The data that support the findings of this study are available in Scientific Reports | 7: *4692* | https://doi.org/10.1038/s41598-017-04295-3; Environmental Geotechnics | 3: *EG2*: 99–113| http://dx.doi.org/10.1680/jenge.15.00036 and also from the corresponding author upon request.

References

Bolker BM et al (2008) Generalized linear mixed models: a practical guide for ecology and evolution. Trends Ecol Evol 24:127–135

Bonachela JA et al (2015) Termite mounds can increase the robustness of dryland ecosystems to climatic change. Science 347:651–655

Bose G (1984) Termite Fauna of Southern India (ed. Director, Zoological Survey of India). In Zoological Survey of India, Calcutta

Dawkins R (1999) The extended phenotype: the long reach of the gene. OUP Oxford

Duran J (2012) Sands, powders, and grains: an introduction to the physics of granular materials. Springer Science & Business Media

Erens H et al (2015) The age of large termite mounds—radiocarbon dating of *Macrotermes falciger* mounds of the Miombo woodland of Katanga, DR Congo. Palaeogeogr Palaeoclimatol Palaeoecol 435:265–271

Hansell MH, Ruxton GD (2013) Exploring the dichotomy between animals building using self-secreted materials and using materials collected from the environment. Biol J Linn Soc 108:688–701

Jouquet P, Mery T, Rouland C, Lepage M (2003) Modulated effect of the termite Ancistrotermes cavithorax (Isoptera, Macrotermitinae) on soil properties according to the structures built. Sociobiology 42:403–412

King H, Ocko S, Mahadevan L (2015) Termite mounds harness diurnal temperature oscillations for ventilation. Proc Natl Acad Sci USA 112:11589–11593

Kleineidam C, Ernst R, Roces F (2001) Wind-induced ventilation of the giant nests of the leaf-cutting ant Atta vollenweideri. Naturwissenschaften 88:301–305

Korb J (2003) Thermoregulation and ventilation of termite mounds. Naturwissenschaften 90:212–219

Papoola KOK, Opayele AV (2012) Morphometrics of *Macrotermes bellicosus* (African mound termite) (Blattodea: Termitidae) and the impact of its saliva amylase on the strength of termitarium soil. NY Sci J 5:207–216

Salgado R (2008) The engineering of foundations. McGraw-Hill, New York

Tschinkel WR (2004) The nest architecture of the Florida harvester ant, *Pogonomyrmex badius*. J Insect Sci 4:21

Werfel J, Petersen K, Nagpal R (2014) Designing collective behavior in a termite-inspired robot construction team. Science 343:754–758

Chapter 13
Reviewing the Issue of "Acceptability" of Earthen Structures in Housing

Sujoy Chaudhury

13.1 Introduction

It is not always that one gets to revisit a project that one was involved in, in this case, it was three projects implemented 11, 17 and 23 years back and located in three different states (Sundarbans in West Bengal, Coastal Odisha and Shillong in Meghalaya). In the absence of any institutional monitoring mechanism, it is usually very difficult to know how these projects have performed over time and whether the critical assumptions integrated into the project design are validated. The projects being studied demonstrate the use of various building materials and construction technologies and techniques, the predominant theme being construction with earth. In trying to understand the performance of these earthen structures, the author decided at the very beginning that the study would not just be a technical exercise to assess the performance of the technologies used, but it would look at how the beneficiaries of the structures themselves view the performance of the technology in the context of an entire building system. Four key areas of enquiry were identified, these being

1. State of repair/disrepair of the structure (Structural health).
2. Owner' perception on the efficacy of the building system.
3. Maintenance and issues.
4. Space additions/structural modifications if any to original structure.

Armed with an open mind, a small questionnaire and a camera, the author first travelled to the Indian Sundarbans, West Bengal in mid-January 2018. This was followed by a visit to Jagatsingpur district in Coastal Odisha in mid-February 2018. Unfortunately, due to other commitments, the travel to Shillong could not be made and the author requested a contact to send pictures of the Institution.

S. Chaudhury (✉)
Center for Sustainable Solutions, Kolkata, India
e-mail: sujoy.chaudhury@gmail.com

B. V. V. Reddy et al. (eds.), *Earthen Dwellings and Structures*,
Springer Transactions in Civil and Environmental Engineering,
https://doi.org/10.1007/978-981-13-5883-8_13

Case Study #1

Multi-Hazard resistant houses in the Indian Sundarbans Delta, West Bengal, India.

Constructed while working with GOAL Ireland as a Programme Manager in partnership with a local NGO, about 200 houses were constructed for the "poorest of the poor" families who had lost their homes and hearths to a cyclonic storm followed by tidal surges in 2006.

Traditional wattle and daub construction methodologies were integrated in a flood-resistant core design for the houses reconstructed. The choice of construction was influenced by many factors including appropriateness of building materials and vernacular architecture, it sought to provide a complete, functional and disaster resistant house to the poorest of the poor at a cost which is slightly higher than the cost of a house provided under the IAY Scheme, Government of India (Figs. 13.1 and 13.2).

One of the key considerations in the design was that the completed house had to provide the physical security to its occupants against natural hazards it also had to provide the occupants with social security. Concrete houses are a recent novelty in

Fig. 13.1 Structural framework of the reconstructed house

Fig. 13.2 Completed reconstructed house

the Sundarbans and people continue to live in traditionally built mud houses covered with burnt clay tiles. The targeted beneficiaries of the reconstruction intervention were the "poorest of the poor", and they could not be provided with a concrete house even though it offers more protection against hazards. It would not be socially accepted. The poor could not suddenly own a new concrete house, when the better off neighbors lived in mud houses. A multi-hazard design using traditional construction materials and techniques thus had to be developed.

So were the houses constructed for the poorest of the poor in the Sundarbans actually "poor" houses and how if at all had the new houses benefited the owner, these were the questions that the author was most eager to find out as he visited the randomly selected village of Mohabatnagar in the Nandakumar Gram Panchayat of Mathurapur II block, [S] 24 Paraganas district of West Bengal, India. Beneficiaries of the reconstruction project included several single mothers from this village and who depended on begging and scavenging for their livelihood. Soudama Bibi (name changed) was one such mother, with a brood of children to take care of. She had almost lost her mind when the tidal surge washed away her tiny hut set on an embankment. With most people affected by the tidal surge, families did not have much to spare for Soudama and her family.

On entering the village, the author could see in the distance a woman, bent down and trying to organize a pile of freshly cut bamboo. There was Soudama and her smile of recognition warmed the authors heart. On being asked what did she intend to use the bamboo for, she replied that she intended to undertake repairs to the roof. So where did she get the money from, she replied the money was sent by her children most of whom were working in cities in western India. She no longer begs, and all her children are earning. Other than some problems with the roof, the house had served and her children well. The house itself looked good and it was obvious that

Fig. 13.3 Soudama Bibi inside her house

within her means, Jamila had been looking after the house. The house had made the difference in coping with her poverty and it had also protected her family during cyclone “Aila”. She said that it was within her to take care of the maintenance of the house, herself, but for the repairs to the roof, she would have to call a mason (Figs. 13.3 and 13.4).

The author visited eight other houses and was glad to find all houses occupied and functional. Soudama and another single mother “Feroza Bibi” (name changed) have a similar story. The roof of Feroza’s house located in the crook of an embankment has required attention and she has used money remitted by her son to reconstruct the roof. The wattle and daub house constructed for her, allowed her to cope with her poverty and raise her children (Figs. 13.5 and 13.6).

The author observed that all nine beneficiaries had accepted their houses and to the best of their abilities and resources, had taken care of the repair and maintenance requirements and the general condition or the structural health of the building systems was good, and in the case of the house of Chandi Das (name changed), it was excellent as the owner had upgraded the structure with rendered masonry and has built adjacent spaces that have transformed the entire building system significantly.

Fig. 13.4 Inside of Soudama Bibi’s house

Fig. 13.5 Morgina Bibi’s (name changed) house

Fig. 13.6 Feroza Bibi's house

In terms of efficacy of the building system, beneficiaries perceived the structure to be appropriate to their needs and strong enough to withstand natural hazards such as cyclone "Aila". They also recognized the need for adequate repairs and maintenance (Figs. 13.7 and 13.8).

All households reported problems with the roofing system and most had undertaken partial or complete repairs to the roofing system. The problem with the roofing system can be linked to the quality of timber rafters used for the roofing; some of the hip rafters holding the outer/first level roofing had collapsed or were in the process of collapse. It appears that adequate diligence in selection and procurement of the timber for the rafters was lacking during implementation.

In two cases, while the front and side faces of the structure were properly maintained, the outer windward face was left not taken care of. Although this does not impact on the performance of the building system overall, the health and structural integrity of the building system are compromised. Reasons attributed to this are that the beneficiary could not source building quality clay. Locally available clay is high in salinity and cannot be used for construction.

All houses had settled well in the habitat of the village community, were being lived in and appeared adequate to the needs of the beneficiary families. Most of the functions like sleeping, cooking, storing can be conducted within the house itself. Families have made improvements to the interior of the house based on their requirements and resources. The "houses for the poor" are not "poor" houses, having survived over a decade in the rugged environment of the Sundarbans. The houses have helped families to cope with loss and poverty, their resilience is demonstrated in the way the houses are being looked after and are being strengthened. The case of the house of Tapan Das is testimony to the determination of the families to improve their condition and the acceptance of the technology used to construct their houses (Figs. 13.9 and 13.10).

Fig. 13.7 Houses of beneficiaries in the Sundarbans

Fig. 13.8 Houses of beneficiaries in the Sundarbans [bottom picture–house of Chandi Das name changed]

Fig. 13.9 Entrance to Chandi Das's house

Fig. 13.10 Inside Chandi Das's house

Case Study #2

Multi-Hazard resistant houses in coastal Odisha, India.
Following the Super-cyclone of 1999, CARE in strategic partnership with Development Alternatives, New Delhi implemented a Reconstruction and Rehabilitation Programme aimed at providing reconstruction support to 1450 families along with Cyclone shelters along the affected coastline of the districts of Puri, Jagatsingpur and Kendrapada. Reconstruction was a top priority, but there was also the desire to build better and stronger homes that could withstand future storms and cyclones. There was also the need to work in close partnership with communities and the project "ASHRAYA" was conceived and designed along these lines of partnership and community ownership of processes.

The project "ASHRAYA" was successfully completed in 2002, and was internationally acclaimed as a best practice project and included in the "United Nations Habitat 2002 Best Practices Global 100 list". The project also received appreciation by the Government of India's Ministry of Urban Development and Poverty Alleviation.

The author had a key role in the design and implementation of the reconstruction project. The reconstructed house followed a "core" design philosophy. An anchored structure constructed using interlocking stabilized soil blocks (SSB) resting on a

Fig. 13.11 Completed core house

Fig. 13.12 Building material and services bank

laterite stone plinth and supporting a flat roof, constructed of ferrocement roofing channels. This multi-hazard resistant structure would form the core around which the family could build around to accommodate their functional requirements.

Interlocking SSB's were produced locally in three "Building Materials and Services Banks" (BMSB) using the South African "Hydraform" machine, along with Ferrocement roofing channels and other precast materials. The components of the core house were transported from the 3, BMSB's to villages spread across three districts. The project was logistically challenging and could be completed on schedule largely due to the participation of the community and the intricate nature of partnership between the key stakeholders of the project (Figs. 13.11 and 13.12).

Prior to the visit, the author could not secure any information on the condition of the core houses, constructed between 2000 and 2001. With a little bit of trepidation, the author set out to visit a randomly selected project village. The question predominant in the author's mind was how have the people of coastal Odisha, accepted a radically different building system and how have the SMBs and Roofing Channels, locally produced at scale, performed?

Fig. 13.13 Core houses in Patrapara village, Balikuda, Jagatsingpur, Odisha

The village identified for visit was Patrapara village in the Balikuda block of Jagatsingpur district. A total of seven beneficiaries had received core houses in this village. The author used the criteria used in the Sundarbans project, to assess the performance of the core houses and the technologies used in their construction. The visit was satisfactory, all seven houses were being lived in and appeared to provide the physical and social security to its inhabitants. The owner of one such house was dead and the village had converted her house into a temple, while the owner of another house had demolished his faulty roof and was engaged in integrating the core house into a larger building system, with RCC roof.

The SMBs have performed exceedingly well and the users compare the blocks to “Mankara” which is the local laterite stone used for constructions. It was good to note this as most of the beneficiaries had during construction expressed concerns about soil blocks. Only one family had plastered the internal walls, all walls remained without any rendering. There was minimum "pitting" and the masonry was in very good health.

The roofing system had, however, failed either partially or completely in most cases. Users complained of seepage and in one case one of the roofing channels had crashed after it developed cracks. Cracks in the roofing channels were observed along the reinforcements and the cast-in situ roof beam in three of the seven houses had cracked open in some places exposing the steel reinforcement inside. Other than the roofing system which required periodic care and attention, the homeowners expressed that they did not have to undertake any repairs/maintenance for the walls and other aspects of the core house.

On being asked whether they would use the soil blocks, if it was locally available most answered in the affirmative, however, all complained about the efficacy of the roofing system and cited installation issues and receiving cracked roofing channels during construction.

A couple of houses had integrated the core house into a composite building system, one beneficiary had received a house building subsidy from the Government of Odisha and had already started construction integrating the core house. Others have added a kitchen with traditional construction technology and techniques. The houses suffered no damage during Cyclone "Phailin" in 2013 (Fig. 13.13).

The ferrocement roofing channels were produced locally and during production, and the development of cracks in the channels was observed. It was assumed that the filling of the roofing channels and a waterproof layer of cement concrete at the top would be adequate. This, however, has not been the case. Could the high coastal temperature have anything to do with the cracks and seepages, needs to be investigated.

Case Study #3

Earthen construction technology and alternative construction techniques, demonstration building in Shillong, Meghalaya.

Fig. 13.14 1st SMB building in Shillong, constructed in 1995 and photographed in 2018

Completed in 1995, the 3000 SQFTS building, housing the "Mary Rice Center for Special Education" was a sponsored project of the Meghalaya Council for Science and Technology. It was proposed that the building would incorporate alternative construction techniques and technologies, such that the various construction possibilities could be demonstrated in a single building, a building that was functional and located in an area that is highly visible. The key construction technology that was demonstrated was SMBs, which were produced using the MARDINI Soil Press.

Unfortunately, the author could not visit Shillong to assess the performance of the building; however, photographs of the building sent by a friend indicated that the building was doing well. The Mary Rice Center had expanded and the friend did not have authorization to enter the premises (Fig. 13.14).

13.2 Conclusion

This exercise was very satisfying for the author. Although all three projects could not be revisited, the visits to the homes of the poor and vulnerable in the Indian Sundarbans Delta and Coastal Odisha and to find them doing better was very gratifying. The technology-oriented houses have served their owners well, withstood cyclones and require minimum maintenance. Almost all families have made additions and improvements to the original structures and some have built around the core or integrated into a larger building system. The roofing system in both cases has demonstrated deficiencies. In the Sundarbans, the stepped tile and thatch roof over timber rafters has failed as the quality of the rafters used for construction was inferior. In the case of coastal Odisha, the roofing system failed partially due to production and installation deficiencies and probably the specifications of the roof filling. Five out of seven houses reported problems of seepage and the greatest discomfort in an otherwise comfortable home. All homes

All homeowners in Odisha and in the Sundarbans reported cooler indoors in summers. Earth as a construction material can be used in different ways and in all three case studies, the earthen technology used, be it manually compressed earth blocks or the mechanically compressed earth blocks and the wattle and daub constructions have been used successfully with the technologies performing to their potential. These projects and their performance further the case for the use of earth in constructions including reconstruction projects, as earth constructions greatly outweigh the environmental and social costs of construction when compared with constructions that are made using conventional fired brick and RCC.

Reference

Chaudhury S (2007) Promotion of Earthen structures in housing—the issue of "acceptability". ISES 2007

Chapter 14
Interlocking in Mud Blocks for Improved Flexural Strength

H. G. Vivek Prasad and K. S. Jagadish

14.1 Introduction

Buildings associated with the construction of dwellings up to G + 3 stories can be built as load-bearing masonry structures. An important parameter for load-bearing construction is compressive strength. Another important parameter for such a type of load-bearing wall construction is flexural strength. Flexural strength of a masonry wall is that strength which resists the horizontal components of forces so as to ensure the stability of structure against overturning.

Flexural strength deserves special attention since adequate knowledge on strength parameters can allow structural design engineers to check the adequacy with suitable design methods. Such an approach is gradually becoming a necessity as disaster resistance of dwellings is given considerable attention these days in order to minimize causalities and damages. High lateral loads are generally caused under unusual conditions such as cyclones, floods and earthquakes. Hence, flexural strength is a useful property in resisting lateral loads.

In normal plain blocks, the flexural strength is mainly governed by the bond strength between the mortar and the block. But by providing interlocking in blocks, flexural strength can be improved because here not only the bond strength governs but also the interlocking effect.

Mortar used for masonry structures consists of sand, cement and water. General masonry mortars used are cement mortar, cement–lime mortar, soil–cement mortar, cement–pozzolana mortar, lime–pozzolana mortar, mud mortar-Jagadish (2007).

H. G. Vivek Prasad (✉)
Department of Construction Technology and Management,
Sri Jayachamarajendra College of Engineering, Mysore, India
e-mail: vivekprasad22@gmail.com

K. S. Jagadish
Department of Civil Engineering, Indian Institute of Science, Bangalore, India
e-mail: ksjagadish@gmail.com

B. V. V. Reddy et al. (eds.), *Earthen Dwellings and Structures*,
Springer Transactions in Civil and Environmental Engineering,
https://doi.org/10.1007/978-981-13-5883-8_14

Cement mortars are widely being used for masonry construction. Very often, such mortars are not satisfactory due to lack of plasticity, high suction from bricks and fast-setting character. Combination of mortars like cement–lime mortars and soil–cement mortars eliminates these problems of cement mortar.

14.2 Literature Review

Venu Madhava Rao et al. (1996) examined the effect of mortar composition and strength of masonry flexural bond strength using stabilized mud blocks, stabilized soil–sand blocks and burnt brick, and they concluded that the increase in mortar strength increases flexural bond strength for cement mortar, irrespective of the type of masonry unit. They also found that combination mortars, such as soil–cement mortar and cement–lime mortar, give better bond strength compared to cement mortars.

Anand and Ramamurthy (2000) studied on durability and performance aspects of interlocking block masonry and concluded that the flexural capacity of interlocking block masonry normal to bed joint is higher than parallel to bed joints.

Sarangapani et al. (2005) studied enhancing the bond strength of brick–mortar and masonry compressive strength with and without bond-enhancing parameters, and they have concluded that using regular cement mortars of 1:6 proportion leads to low flexural bond strength of less than 0.10 Mpa; however, it can be increased by coating the surface of the brick with cement slurry/epoxy resin, increasing area of frog.

Venkatarama Reddy et al. (2007) studied on enhancing the bond strength and characteristics of soil–cement block masonry, and they have concluded that by providing a rough textured surface, surface coatings like cement slurry and epoxy resin increase the bond strength.

Konthesingha et al. (2007) attempted to understand bond and compressive strength of masonry for three different types of sand and bricks locally available in Sri Lanka, and also, the effect of soaking time of bricks was studied, and they have concluded that higher shear bond strength can be achieved by increasing the tensile bond strength depending upon type of brick.

Jayasinghe and Mallawarachhi (2008) studied on flexural strength of compressed stabilized earth masonry materials such as compressed stabilized earth bricks, plain solid blocks, interlocking solid blocks, interlocking hollow blocks and rammed earth with low levels of pre-compression load, and they concluded that CSE bricks, blocks and rammed earth can be considered as a safe alternative up to two-storied buildings.

14.3 Experimental Programme

14.3.1 Raw Materials

14.3.1.1 Soil

The locally available soil sample was used which was red in colour with semi-hard lumps, and its grain size distribution was determined. For doing grain size distribution analysis, soil passing through 4.75-mm sieve was used. The sand content was found to be 63% in the soil sample when wet sieve analysis was carried out.

14.3.1.2 Sand/Quarry Dust

For the production of blocks and prisms, sand or quarry dust may be used to reduce the clay content in the soil. Here, quarry dust was used by considering the availability and economy. Ten percentage of total weight of soil taken was added so as to reduce the clay content in the mix. Sand was used in the construction of prisms (mortar joints).

14.3.1.3 Cement

Cement has been used as a stabilizer for preparing the blocks and also as binder for various mortar combinations in mortar joints. Seven percentage of cement was used as stabilizer for the production of blocks. In present work, ordinary Portland cement of 53 grade was used both as stabilizer and binder.

14.3.1.4 Lime

Lime was used only for the mortar joints in preparation of masonry prisms. Locally available fresh lime was procured and slaked. Slaked lime passing through I.S. 1-mm sieve was used.

14.3.1.5 Blocks

Interlocking blocks are shown in Figs. 14.1 and 14.2. The depth of interlocking provided is 20 and 30 mm. These blocks have projections in bottom and recess at the top so that one projection of the block will accommodate at the bottom recess of another block. The area of projection and recess is 100 $\times$ 190 mm and 90 $\times$ 190 mm for 20- and 30-mm-depth interlocking blocks, respectively. The projections and recess

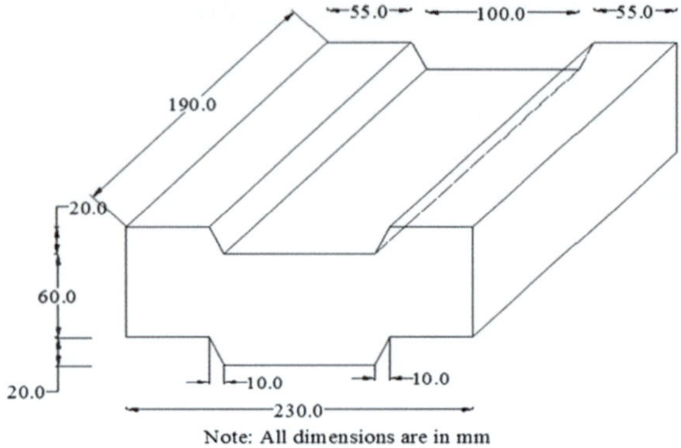

Fig. 14.1 Details of 20-mm-depth interlocking blocks

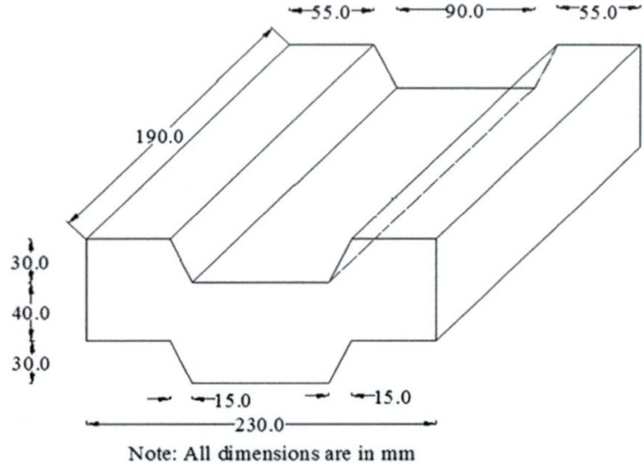

Fig. 14.2 Details of 30-mm-depth interlocking blocks

Fig. 14.3 Wooden pieces used in mould for preparing interlocking blocks

Fig. 14.4 Prepared blocks using Mardini press

are tapered with 10 mm for 20-mm-depth and 15 mm for 30-mm-depth interlocking blocks. These blocks were used to determine both flexural and compressive strength.

Normal plain blocks were made using the machine called "Mardini press". The interlocking blocks were also made using the same machine by making slight modifications to it. These modifications can be done by using steel or wood. In present work, wooden pieces have been used since they were cheaper and the number of blocks production was less. The modifications used in the machine are as shown in Fig. 14.3, and all the three types of blocks produced using the machine and wooden pieces used in the mould are shown in Fig. 14.4.

14.3.2 Casting of Prisms

The prisms for a height of 1 m and 3 block prisms were cast and cured for 28 days in moist condition by covering the prisms with gunny bags. The masonry mortars used for casting of prisms were cement mortar (1:6), cement–lime mortar (1:1:6), cement–soil mortar (1:2:6). For each type of mortar, two sets of prisms were cast. Totally 18 prisms for flexural strength test and 18 prisms for compressive strength test were cast. In order to maintain consistency in the construction of prisms, the prisms were cast by the same mason.

14.4 Results and Discussions

14.4.1 Compression Test

The average compressive strength of individual normal plain blocks was 6.65 Mpa. The average compressive strength of two prisms is given in Table 14.1. Prisms were tested in partially saturated state by pouring water over the prisms 10 min prior to testing.

From the tests, it is observed that the strength of three block masonry prisms is less than that of individual blocks. This difference in strength is being observed because of the presence of mortar between the blocks.

14.4.2 Flexural Strength Test

14.4.2.1 Initial Set-up

For conducting the test, a rigid bottom support was made with reinforced cement concrete of dimensions 24 × 20 × 15 cm. The height of the support was maintained such that only a single joint of prism was restrained inside the support. Arrangements were made to apply load by using wooden bracket, bag and cable wire over a pulley. One end of the cable was tied to wooden bracket which was fixed to the top most block of the prism and the other end to the bag. The arrangement done is as shown in Fig. 14.5.

Table 14.1 Compression test of three block prisms

S. No.	Mortar proportion (by weight) C:L:So:Sa	Type of prism	Compressive strength of prism (Mpa)
1	CM (1:0:0:6)	Plain block	2.63
2	CM	Interlocking block of 8 cm depth	1.95
3	CM	Interlocking block of 7 cm depth	1.95
4	CLM (1:1:0:6)	Plain block	2.63
5	CLM	Interlocking block of 8 cm depth	2.39
6	CLM	Interlocking block of 7 cm depth	2.3
7	CSM (1:0:2:6)	Plain block	2.39
8	CSM	Interlocking block of 8 cm depth	2.39
9	CSM	Interlocking block of 7 cm depth	2.39

Note C cement, *L* lime, *So* soil, *Sa* sand, *M* mortar

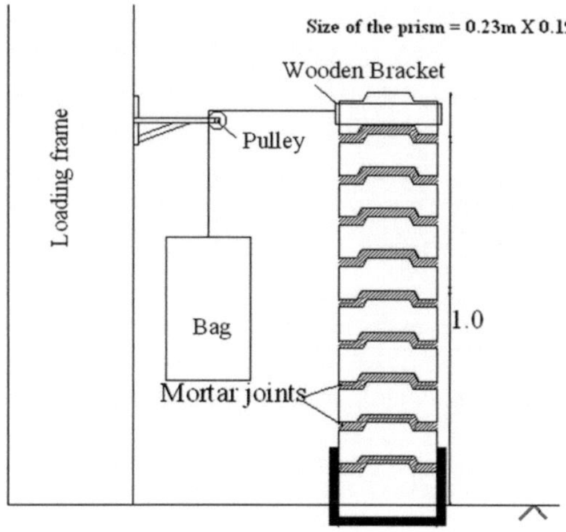

Fig. 14.5 Typical sketch of the arrangement

14.4.2.2 Test Procedure

Prisms were placed as a cantilever over the rigid bottom, and the load was applied (without pre-compression) in the horizontal direction at the free end. Sand was poured into the bag till there was a failure in the prism. The failure load was measured by weighing the quantity of sand in the bag (self-weight of the arrangement was not considered).

Table 14.2 gives the flexural strength of normal plain block prism. The data for cement–lime mortar mix is not available since the specimens failed while handling. The combination of cement–soil mortar mix has given better flexural resistance when compared with cement mortar mix. All the failures have occurred at bond–mortar interface.

The test results of the 1-m-height masonry prisms with 20-mm-depth interlocking are tabulated in Table 14.3. Prisms cast using cement–lime mortar mix proportion gave better flexural resistance compared to cement mortar mix proportion and cement–soil mortar mix proportion which has resulted in better bond strength between the blocks. It can also be observed that there is an improvement in flexural strength

Table 14.2 Flexural strength of normal plain block prisms

S. No.	Mortar proportion	Flexural strength (Mpa)	Average flexural strength in Mpa	Type of failure
1	CM	0.031	0.032	Bond
2	CM	0.033		
3	CSM	0.046	0.048	Bond
4	CSM	0.049		

Fig. 14.6 Failure at bond–mortar interface

Table 14.3 Flexural strength of interlocking blocks with an interlocking depth of 20 mm

S. No.	Mortar proportion	Flexural strength (Mpa)	Average flexural strength in Mpa	Type of failure
1	CM	0.13	0.14	Bond
2	CM	0.14		
3	CLM	0.18	0.25	Bond
4	CLM	0.31		
5	CSM	0.27	0.21	Bond
6	CSM	0.15		

between normal plain block prisms and 20-mm-depth interlocking prisms. All the failures have occurred in bond–mortar interface. These failures can be observed in Fig. 14.6.

The test results of the 1-m-height masonry prisms with 30-mm-depth interlocking are tabulated in Table 14.4. Cement–lime mortar mix gave better flexural strength compared to cement–mortar mix and cement–soil mortar mix which results in better bond strength between the blocks. It can also be observed that there is further improvement in flexural strength of masonry prisms as the depth of interlocking was increased to 30 mm. The failures have taken place at bond and mortar interface, but in the case of cement–lime mortar mix, failure has occurred in the block. The failure is as shown in Fig. 14.7.

Maximum flexural strength of normal block prism in CSM is 0.048 Mpa, whereas in CM is 0.032 Mpa.

Flexural strength with 20-mm-depth interlocking block prism in CLM is 0.25 Mpa, whereas in CSM and CM is 0.21 and 0.14 Mpa, respectively. Flexural strength with 30-mm-depth interlocking block prism in CLM is 0.45 Mpa, whereas in CSM and

Fig. 14.7 Failure in block

Table 14.4 Flexural strength of interlocking blocks with an interlocking depth of 30 mm

S. No.	Mortar proportion	Flexural strength Mpa	Average flexural strength Mpa	Type of failure
1	CM	0.25	0.21	Bond
2	CM	0.17		
3	CLM	0.51	0.45	Block
4	CLM	0.39		
5	CSM	0.22	0.31	Bond
6	CSM	0.27		

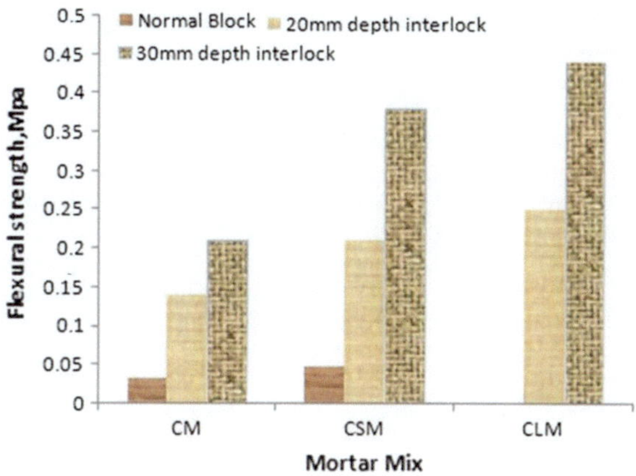

Fig. 14.8 Comparison of flexural strength

CM is 0.31 and 0.21 Mpa, respectively. CLM has given better strength compared to other two mortar proportions. The interlocking blocks have better flexural strength than normal blocks. The details can also be observed in Fig. 14.8.

14.5 Conclusions

Normal plain blocks have a flexural strength of less than 0.1 Mpa which has also been concluded by Sarangpani et al. (2005). There is an interlocking effect on flexural strength which has produced the values of 0.14–0.25 Mpa in case of 20-mm-depth interlocking block and 0.21–0.45 Mpa in case of 30-mm-depth interlocking blocks; thus, interlocking blocks can be used to improve flexural strength of masonry structures.

Bonding between cement–lime mortar mix and stabilized mud blocks functions well when compared to cement–soil mortar mix and cement mortar as the failure has occurred in the block not at bond–mortar interface which clearly indicates the maximum flexural strength. Thus, interlocking blocks with 30 mm depth along cement–lime mortar (1:1:6) can be used in earthquake areas where out of flexure failure occurs.

References

Anand KB, Ramamurthy K (2000) Development and performance evaluation of interlocking block masonry. J Arch Eng 6(2):45–51

Jagadish KS (2007) Building with stabilized mud. I.K. International publishing house Pvt. Ltd., New Delhi

Jayasinghe C, Mallawarachhi RS (2008) Flexural strength of compressed stabilized earth masonry materials. Mater Des 30:3859–3868

Konthesingha KMC, Jayasinghe C, Nanayakkara SMA (2007) Bond and compressive strength of masonry for locally available bricks. Inst Eng Sri Lanka 04:7–13

Sarangpani G, Venkatarama Reddy BV, Jagadish KS (2005) Brick-mortar bond and masonry compressive strength. J Mater Civil Eng 17(2):229–237

Venkatarama Reddy BV, Richardson Lal, Nanjunda Rao KS (2007) Enhancing bond strength and characteristics of soil-cement block masonry. J Mater Civil Eng 19(2):164–172

Venu Madhava Rao K, Venkatarama Reddy BV, Jagadish KS (1996) Flexural bond strength of masonry using various blocks and mortars, Mater Struct 29:119–124

Chapter 15
Earthen Materials as Opportunity for CDW Reduction Results from the EU-Funded Research Project RE[4]

Andrea Klinge, Eike Roswag-Klinge, Christof Ziegert, Caroline Kaiser and Danijela Bojic

15.1 Introduction

In Europe, the building sector contributed in 2008 with approximately 860 million tons of construction and demolition waste (CDW), which accounts for 32% to the total waste generation (European Commission (DG ENV) 2011). Data on generation and recovery and recycling rates, however, are not fully reliable due to some data gaps and the quality of data (European Commission (DG ENV) 2011). According to (http://ec.europa.eu/environment/waste/construction_demolition.htm), recovery rates vary greatly across the Union (between <10 and >90%) and can overall be considered as fairly low, first of all as recovered materials are mainly used for low-grade applications (back filling, unbound road base, etc.). A large percentage of material is still diverted to landfill, as existing buildings were not designed for disassembly or reuse and applied deconstruction strategies focus on efficiency with regards to time and cost rather than on maximisation of material recoveries. Valuable resources and energy are often wasted, while the pressure on a decreasing number of disposal sites is constantly rising. In addition, landfill sites can cause significant impact on air, water and soil and the incineration might generate emissions of air pollutants.

In the EU-27, CDW consists of numerous materials including concrete, bricks, gypsum, wood, glass, metals, plastic, solvents, asbestos and excavated soil. Waste

A. Klinge (✉) · E. Roswag-Klinge
ZRS Architekten, Berlin, Germany
e-mail: klinge@zrs.de

C. Ziegert · C. Kaiser · D. Bojic
ZRS Ingenieure, Berlin, Germany
e-mail: bojic@zrs.berlin

B. V. V. Reddy et al. (eds.), *Earthen Dwellings and Structures*, Springer Transactions in Civil and Environmental Engineering, https://doi.org/10.1007/978-981-13-5883-8_15

figures are not always available and differ due to different qualification frameworks. Statistics from different years provide the following quantities:

- Concrete (2008) 320–380 Mt
- Bricks; tiles and ceramics (2008): no data available
- Asphalt (2008): 47 Mt
- Wood (2004): 70.5 Mt
- Gypsum: minimum of 4 Mt.

Although these figures do not include excavated soil (European Commission (DG ENV) 2011), such waste also occurs in large quantities, as it can be observed from figures in Germany, where 85% of the mineral waste consisted of excavation and gravel (Bundesverband Baustoffe—Steine und Erden e. V. 2017). A clear up-cycling or reuse strategy for high-grade applications, first of all for the fine fraction, is currently missing. In addition, the majority of the CDW is not so easy to recycle and the amount of embodied energy has to be assessed carefully in order not to deteriorate the situation.

The European Waste Framework Directive (2008/98/EC) sets out ambitious goals until 2020, and Member States are asked to minimise the generation of CDW through reuse, recycling and material recovery by 70% (by weight) (https://www.umweltbundesamt.de/daten/ressourcen-abfall/abfallaufkommen#textpart-2). However, economical (low price for raw materials) and technical (sorting qualities, occurrence of harmful substances) barriers are still dominating so that these goals will be difficult to achieve.

The EU-funded RE[4] (REuse and REcycling of CDW materials and structures in energy-efficient pREfabricated elements for building REfurbishment and construction) research project develops manifold concepts for minimising, specifically CDW, and aims to bring back waste into construction. This study focuses on the development of the following earthen materials for interior application:

• Earth plaster	For non- and load-bearing internal partition walls
• Earth mortar	For non- and load-bearing internal partition walls
• Rammed earth for building blocks	For non- load-bearing internal partition walls and façades

Initial steps at material level have been undertaken and developments demonstrate promising results with recycling rates of up to 87%. Furthermore, the project looks into the design development for reversible construction elements for residential and non-residential buildings for new construction and refurbishment. Strategies focus on opportunities for prefabricated internal partition walls but also façade elements, where earthen materials can be applied. The aim is not only to maximise the CDW content for the respective construction but also to promote the reuse and up-cycling of materials and components for future applications and therefore to promote circular construction. In addition, the study intends to illustrate the potential of materials demonstrating an increased ability for circular construction in comparison with materials that require high amounts of embodied energy in order to be brought back into the construction cycle.

15.2 Materials, Screening Tests (Extract), Methodology and Test Methods

15.2.1 Material Provision

Earth as a construction material is well known for its recyclability. The major challenge concerning a material development on the basis of CDW, however, consists in the likely diversity of the incoming materials and associated questions as the occurrence of unwanted contaminations. Besides the fact that earthen materials have only a small market share, it can be assumed that similar to other construction materials especially easy access to raw materials and the variety of the base material constitute the main reasons why CDW until now is hardly ever used for the production of earthen building products. To address this issue, it has been decided to work with material from different regions in Europe (Northern (NE) and Southern Europe (SE)) so that potential differences in the base material, resulting from different construction materials and methods applied in the past, can be identified at an early stage.

CDW batches were provided in different sorting qualities, and initial tests have been conducted with material undergoing only a basic degree of separation. Visual inspections identified a variety of aggregates such as glass, bricks and tiles within the mineral fractions. Further mixtures will be prepared based on improved sorting of the basic material. Table 15.1 provides an overview about the CDW fractions and virgin material that were used for the development of earth plaster, earth mortar and

Table 15.1 Incoming CDW and virgin material and their intended purpose

CDW	Virgin material	Corn size	Source	Purpose
Silt and clay		<0.063 mm	SE/NE	Binder for rammed earth
Sand		0–2 mm	SE/NE/Berlin	Aggregates for earth plaster, earth mortar and rammed earth
Fine gravel		2–8 mm	SE/NE	Aggregates for rammed earth
Medium gravel		8–16 mm	SE/NE	Aggregates for rammed earth
	Clayey soil	<0.5 mm	Germany	Binder for plaster, mortar and rammed earth

rammed earth. It has to be noted that due to the processing of CDW, MF 0–2 mm contained also parts of silt and clay, although only in small amounts. Furthermore, a ready-made binder (clayey soil), <0.5 mm, dry, from Germany has been purchased to complement the silt and clay press cake, where required.

15.2.2 Methodology

Two different strategies were applied for the material development. For the earth plaster and mortar, a ready-made binder, tested by the manufacturer with regards to adhesive forces according to Dachverband Lehm e. V. (2009), was used, which allowed placing the focus on the maximisation of the amount of CDW aggregate within the mixture. To increase the efficiency of the material development, it was aimed to prepare one mixture suitable for both, application as plaster and mortar.

As the incoming CDW material demonstrated an appropriate corn-size distribution similar to what is required for the development of rammed earth, the strategy for this material was amended. To maximise the content of CDW also for the binder, the silt and clay press cake was used and the ready-made clayey soil was only added as required. This strategy could be applied as the adjustment of the optimum mixture for rammed earth is less complex and the impact of the binder is less pronounced.

15.2.3 Screening Test—Adhesive Force

As a first step, the general suitability of the incoming silt and clay press cake for the development of a binder was determined by means of Achterlingsprüfung (test to determine the adhesive force) according to the Lehmbau Regeln (Dachverband Lehm e. V. 2009). The material was rammed into an eight-shaped test mould, and the adhesive force was determined as an average value from three tensile tests performed. The test is considered as an orienting assessment.

15.2.4 Screening Test—Carbonate Content of Clayey Soil

To determine the suitability for use as building material, the lime content of the silt and clay press cake was carried out according to (Dachverband Lehm e. V. 2009; DIN EN 14688-1: 2013-12) as a semi-quantitative determination of the carbonate content (natural lime or added lime as binder). Diluted hydrochloric acid was sprinkled on crushed sample material and the effervescence due to carbon dioxide as a product of the reaction between acid and carbonate was assessed.

15.2.5 Sieving Curves

To characterise the CDW aggregates for the earth plaster and mortar development, the particle-size distribution of the sand fraction (0–2 mm) was obtained and compared with one another. The final sieving curves provide indications with regards to mechanical properties and the influence of the aggregates on the strength properties of the developed earthen building materials.

15.2.6 Test Campaign for Earth Plaster

Developed plasters were tested against DIN 18947 (DIN EN 14688-1: 2013-12, (mandatory and voluntary tests)), which sets out the requirements for earth plasters to be applied in construction. Strength class II (S II) was set as target to enable unrestricted use of the material.

15.2.7 Test Campaign for Earth Mortar

Developed mortars were also tested against DIN 18946 (Deutsches Institut für Normung: DIN 18947:2013-08 2013, (mandatory and voluntary tests)), which sets out the requirements for earth mortars to be applied in construction. Tests with regards to shear strength are still outstanding.

15.2.8 Test Campaign for Rammed Earth

The developed rammed earth mixtures were tested with regards to drying shrinkage and the bulk density was determined. Tests with regards to compressive strength are still outstanding, however, as the material is intended for non-load-bearing application this test is not mandatory.

15.3 Results

15.3.1 Screening Test—Adhesive Force

The eight-shaped test sample demonstrated an average binding force with a value of 157 g/m^2, which means that the press cake is nearly fat and suitable for the production of earthen building materials.

15.3.2 Screening Test—Carbonate Content of Clayey Soil

The test demonstrated a strong, long-lasting effervesce, indicating an increased carbonate content in different particle sizes.

15.3.3 Sieving Curves

CDW from NE and SE contained a small proportion of oversize grain larger than 4 mm. Even though, the particle-size distributions of the different CDWs are similar, as shown by the course of the grain distribution curves, the biggest difference lies in the proportion of grain size bigger than 0.25 mm, which was about 8% higher for deliveries from NE than for deliveries from SE. The oversize fraction >4 mm of the sand from Berlin was significantly higher, whereas the content of the grain fractions 2, 1 and 0.5 mm significantly lower.

The fraction of fine particles <0.063 mm is very small for all CDW aggregates and accounts for approximately 1.4% of the material. Figure 15.1 provides an overview about the most representative sieving curves for materials from the different sources.

15.3.4 Test Campaign for Earth Plaster

For the material development, different mixing ratios of CDW aggregates and binder were tested in a first step using CDW from SE, NE and Berlin. In addition, the workability was assessed during the production of the mortar mixtures and the test

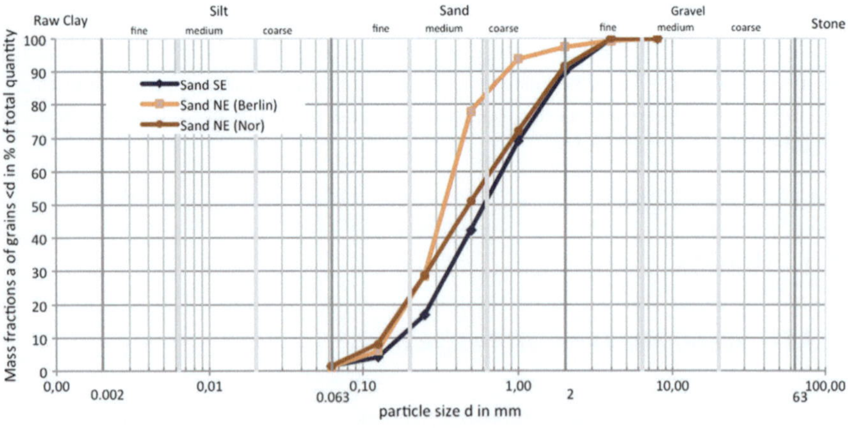

Fig. 15.1 Comparison of particle-size distributions for 0–2 mm fraction

specimens. The most promising mixture, which met all requirements of the screening tests according to Deutsches Institut für Normung: DIN 18947: 2013-08 (2013), was undertaken the full test campaign for earth plasters and mortars. This mixture contained two parts (by volume) of the 0–2 mm CDW fraction and one part (by volume) of the clayey soil and thus in total 67% of CDW content (by volume). In addition, repeating tests were carried out to determine whether results can easily be reproduced or not. Table 15.2 shows test results of mandatory and voluntary test for the optimised mixture according to Deutsches Institut für Normung: DIN 18947: 2013-08 (2013, Deutsches Institut für Normung: DIN 18946: 2013-08 (2013) and demonstrates that overall requirements are met.

The drying shrinkage shows little contingency, and was not met in the repeating tests. However, this property could easily be improved through the addition of fibres or further optimisations of the mixtures. This would also improve the tensile bending strength, which could not be repeated either. Although suitable CDW fibres have not been provided within the project yet, it is aimed to develop them from CDW timber in the near future. The compressive strength of all tests was easily met. For the adhesion strength, unusually, high standard deviation arose. Interestingly, the results for the voluntary water vapour sorption test achieved outstanding results with 94 g/m^2, which exceeds market products by 30–50%.

Table 15.2 Results of mandatory and voluntary tests according to Deutsches Institut für Normung: DIN 18947: 2013-08 (2013), Deutsches Institut für Normung: DIN 18946:2013-08 (2013) with optimised mixture (1 test)

Property	UOM	Mix	Required*/target value		Result	Comments
		SE-2	Plaster	Mortar		
Drying shrinkage	%	1.9	<2*	<2.5*	Ok	Little contingency
Bulk density	kg/dm^3	1.67	>1.41*	>1.41*	Ok	
Bending tensile strength (S II)	N/mm^2	1.0	>0.7*	/	Ok	
Compressive strength (S II)	N/mm^2	2.7	>1.5*	>2.0*	Ok	
Adhesive strength (S II)	N/mm^2	0.12	>0.10	/	Ok	
Abrasion (S II)	g	0.1	<0.7	/	Ok	
Water vapour adsorption test (12 h)	g/m^2	94	60 (WS III)	/	Ok	Excellent result
Adhesive shear strength	N/mm^2	/	/	≥ 0.02	N/A	Test outstanding

15.3.5 Test Campaign for Earth Mortar

As the aim was to develop a mixture suitable for application as plaster and mortar the majority of tests relevant for the earth plaster are also relevant for the earth mortar. Respective test results can be found in Table 15.2. As requirements for the earth mortar are lower than those of plasters, results from all performed tests (initial and repeat tests) exceeded the targeted values. Due to the lack of testing capacities, the shear strength test is still outstanding.

15.3.6 Test Campaign for Rammed Earth

The raw density of the mixtures was higher than 1.41 kg/dm^3. The drying shrinkage achieved a value of 0.25%, which is below the threshold value of 0.5% for monolithic, visible components and well below the value of 2% for monolithic, non-visible components.

15.4 Discussion

15.4.1 Screening Tests—Adhesive Forces—Carbonate Content of Clayey Soil

All screening tests, also those ones not reported in this study (e.g. water content upon delivery, olfactory test) demonstrated promising results. The silt and clay press cake was generally suitable for the development of earthen materials, provided that:

- the content of organic matter is very low to avoid mould growth,
- the material is sufficiently dry to enable evenly mixing with the other components,
- suitable aggregates are added to make the material leaner to avoid cracking.

15.4.2 Sieving Curves

Although the incoming CDW (0–2 mm fraction) demonstrated different particle-size distributions, earthen building materials could be produced from all CDW aggregates matching the required properties. The effect caused by the range of different grain distributions can be considered as neglectable.

15.4.3 Test Campaign for Earth Plaster

The investigations carried out proved that an earth plaster can be produced on the basis of CDW aggregates instead of virgin sand and can be used as earth plaster and as earth mortar. However, a fixed mixing ratio, in which the requirements of the German standards DIN 18946 and DIN 18947 can be reliably met, could not be determined yet. The proven mixing ratio of two volume parts CDW aggregates and one volume part of building clay can be used as a starting point for the adaptation required for each new development. The ratio of CDW in this dry mortar was thus 67%.

15.4.4 Test Campaign for Earth Mortar

Although the shear strength test is still outstanding, it can be stated that the best-performing mixture of the screening test fulfilled all remaining requirements set out in the DIN 18947.

15.4.5 Test Campaign for Rammed Earth

The mixture for rammed earth was actually the easiest one to manufacture. The developed material mixture met all requirements for non-load-bearing application and contained with 87% the highest amount of CDW with. Further investigation should enable to increase the CDW content also for the binder. For load-bearing applications (which is not intended for the European market), compressive strength tests are still outstanding.

15.4.6 Material Application

The intended use for the developed materials consists mainly in the application for non- and load-bearing, internal partition walls, which can either be constructed as dry-lining walls or as solid walls, depicted in Fig. 15.2. However, the application for non-load-bearing timber façade elements is also under development within the RE[4] project.

Earth plasters can be applied as a finishing material, offering a high-quality surface, while improving the IEQ of spaces (Roswag-Klinge et al. 2016). Earth mortars can be used for dry-lining systems, which contain earth blocks as an infill or for masonry walls either made out of earth blocks or recycled bricks. For the rammed earth material, it is intended to manufacture rammed earth blocks that in relation to the

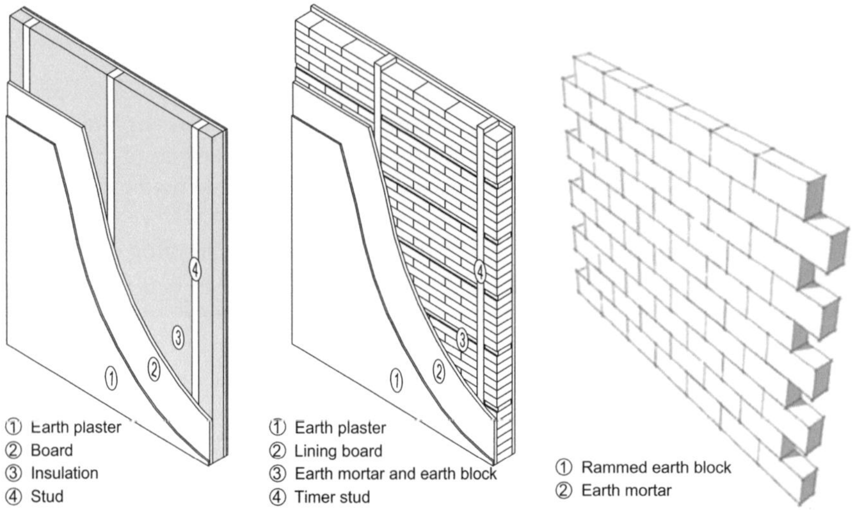

Fig. 15.2 Design concepts for non-load-bearing internal partition walls

final dimensions can either be brick laid or glued. Initial tests have been performed, however, the final dimensions have to be determined.

In Europe, the spread of skeleton structures after World War II triggered the development of drywalls that were able to overcome the weak points of solid internal partition wall systems. As a result, earthen building materials have been largely replaced by industrialised products such as gypsum plasterboards in combination with metal studs and mineral wool insulation. The materials applied in these systems offer almost no potential for a direct reuse or up-cycling and very limited potential for recycling. Furthermore, the lifespan of such wall systems, first of all in non-residential buildings, has significantly been decreased and the cost-efficient erection and changing aspiration with regards to floor layouts contributes to an increased waste generation.

In light of current efforts to minimise the waste generation and to maximise the reuse of systems, the earthen building materials developed in the RE[4] project offer great potential for a significant reduction in this field. In addition, the inherent material properties would enhance the IAQ of spaces and also contribute to an improved performance against overheating in summer, an important topic not only for Southern Europe.

15.5 Conclusion and Outlook

Overall, it can be stated that the material developments of the various different earthen building materials presented in this study required very little resources and requirements could be met without major adjustments of the mixtures. Although the silt and clay press cake has to be watered and separated with hand tools, the sand fraction incorporated into the earthen plaster and mortar could be used as provided; no further sifting was required. Even if there is an intrinsic heterogeneity, the typical quality and average composition are comparable, regardless of the source. Due to time constraints, the earth and silt CDW was only used for rammed earth mixtures, further investigation and testing would be required to establish whether the ready-made binder could be replaced by silt and clay from the sorting of CDW—provided that there are no pollutants—to produce an earthen material entirely made up of CDW.

For each silt and clay batch, the suitability as a binder would have to be tested, as individual batches might not be suitable because they contain too little or too much cohesive ingredients, so that they need to be adjusted, respectively.

A range of suitable mixing ratios cannot be defined accordingly. If earthen materials should consist entirely of CDW, an individual determination of the mixing ratios must be made for the materials available in each case.

Although a load-bearing application for the rammed earth blocks is not anticipated, compressive strengths tests are planned. In addition, it would be desirable to extend the material development and application to other products such as earth blocks and earth dry boards out of CDW.

As all developed materials are derived from more than 67% of CDW and in addition offer a high level of either direct reuse or recycling their potential for CDW reduction is outstanding. LCA studies have only started recently and will provide more insight into this specific evaluation in due course. These advantages, in addition to other benefits should overcome current barriers in Europe, where capital cost are still the main driver for material selection instead of truly sustainable construction.

Another issue that has to be addressed for future application of this approach is the absence of harmful substances, a topic that all material developments on the basis of CDW have to tackle. Although additional research will be required it is felt that control at source, respectively, at site would be the most effective and appropriate strategy. This approach would go in line with a necessary change to building demolition, where selective dismantling strategies should be applied instead of mechanical techniques that only offer very little control about the single waste streams and minimise material purities and in hence the outcome of material recycling.

This study might also be relevant and potentially applicable for countries located in the hot climate, where a rising trend for concrete construction becomes noticeable.

Acknowledgements This research study was made possible with the support of the European Union's Horizon 2020 Programme for research and innovation programme under grant agreement no. 723583 ([RE^4], www.re4.eu).

References

Bundesverband Baustoffe—Steine und Erden e. V. (2017) Mineralische Bauabfälle Monitoring 2014, Bericht zum Aufkommen und zum Verbleib mineralischer Bauabfälle im Jahr 2014

Dachverband Lehm e. V. (2009) Lehmbau Regeln Begriffe—Baustoffe—Bauteile, Vieweg Teubner 3, revised edition. ISBN 978-3-8348-0189-0

Deutsches Institut für Normung (Ed.): DIN 18947: 2013-08 (2013) Earth plasters—terms and definitions requirements, test methods, edition, Aug 2013

Deutsches Institut für Normung (Ed.): DIN 18946: 2013-08 (2013) Earth mortars—terms and definitions, requirements, test methods, edition, Aug 2013

DIN EN ISO 14688-1: 2013–12 Geotechnical investigation and testing—Identification and classification of soil—part 1: identification and description

European Commission (DG ENV) (2011) Service contract on management of construction and demolition waste—SR1

http://ec.europa.eu/environment/waste/construction_demolition.htm. Accessed 26 Jan 2018

https://www.umweltbundesamt.de/daten/ressourcen-abfall/abfallaufkommen#textpart-2. Accessed 26 Jan 2018

http://ec.europa.eu/environment/waste/framework/. Accessed 26 Jan 2018

Roswag-Klinge E et al (2016) Gesund bauen mit Naturbaustoffen. Published in Wohnung und Gesundheit, Nr. 161, 12.16, pp 25–27

Chapter 16
Organic Stabilisers in Traditional Mud Homes of India

Rosie Paul and Sridevi Changali

16.1 Introduction

India is at the wake of a new era as the country rushes towards an idea of modernity and development. This hurried attitude towards development is reflected in the construction sector as well wherein everyday a new product is introduced to the market and it is almost immediately used in construction with little or no thought given about the material in question—embodied energy, origin, industrial processing, health aspect are seldom valuable concerns. Most often cost, aesthetics, speed of construction carry more weightage in material selection. This model of construction is unsustainable and will have dangerous implications on the environment. The Sustainable Development Goals by the UNDP that came into effect in January 2016 lists out 17 venues to ensure a sustainable development path to developing nations. In the domain of construction, the most relevant would be—sustainable cities and communities, responsible consumption and production, affordable and clean energy, industry infrastructure and innovation.

In our hurry to catch up with the developed nations what was unfortunately left behind or taken for granted was our past, rich in knowledge and skill. The traditional building practices of ancient India also fall under this cap of forgotten past—mud homes varying across different landscapes from palaces to quaint mud dwellings. This led to scarcity of such mud dwellings due to negligence and eventually loss of know-how.

Mud construction is seeing a revival in the construction sector today through its earth warriors and natural builders who strive to promote and create awareness on the

R. Paul (✉)
CRA terre-ENSAG, Grenoble, France
e-mail: rosiepaul87@gmail.com

S. Changali
Masons Ink, Bangalore, India
e-mail: sridevichangali87@gmail.com

B. V. V. Reddy et al. (eds.), *Earthen Dwellings and Structures*,
Springer Transactions in Civil and Environmental Engineering,
https://doi.org/10.1007/978-981-13-5883-8_16

various advantages of living in a mud structure. As a result, there are an increasing number of contemporary mud structures being built that act as the catalysts for change.

16.2 Life Cycle Assessment of Mud Buildings

LCA or life cycle assessment is quite an easy and simple way of looking at the environmental impact of any material or a built structure. It gives a view of the "cradle to grave" energy cycle of the built structure (Khasreen et al. 2009) (Image 16.1).

A life cycle assessment of an unstabilised mud building in comparison with concrete structures brings to light the reason why mud is one of the most sustainable materials to use in construction if it is easily available and usable.

The LCA is altered when mud is stabilised, taking the commonly used stabilisers of cement and lime. The mud used in the building becoming soil again at the end of its life is taken away as once these stabilisers are added the effect is irreversible. However, in comparison with the conventional materials used nowadays, it still performs better. A 2001 study in India focused on embodied energy in load-bearing masonry buildings. A brickwork building and a soil–cement block building were compared, and the study showed that the total embodied energy can be reduced by 50% when energy efficient building materials are used (Venkatarama Reddy and Jagadish 2003).

The goal, however, should be to further minimise the impact of mud buildings in present day construction even if the mud needs to be stabilised for site-specific reasons. This can be achieved by looking into organic additives as stabilisation using these additives can further lower the embodied energy in the earthen structure.

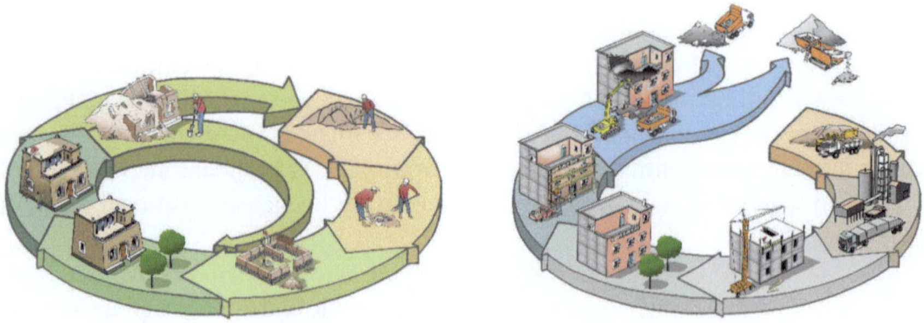

Image 16.1 Comparison of life cycles—figure by Sebastien Moriset, CRATerre

16.3 Stabilisation of Mud

16.3.1 Need for Stabilisation

The use of mud in construction does not necessarily mean that the soil requires reformulation or stabilisation of any kind. There are many examples across the world wherein mud has been used as is found when it is of ideal composition. Stabilisation is not compulsory and can be avoided if soil is found to be satisfactory. Stabilisation is carried out to achieve better mechanical characteristics, improve tensile strength, better cohesion, reducing porosity and improvement of resistance to wind and rain erosion (Houben and Guillaud 2001).

16.3.2 Types of Stabilisers

In case of very sandy or very clayey soils, soil reformulation is required which uses mineral additives such as sand and gravel to attain a good grain size distribution. Depending on availability, often natural fibres are also added to strengthen the structure of the soil. These fall into the category of inert stabilisers that help mainly in densification and reinforcement. The other category of stabilisers is physio-chemical stabilisers that improve cementation, linkage, imperviousness and water proofing. These physio-chemical stabilisers could be natural or synthetic, for example cement, lime, bitumen, asphalt, gypsum, plant or animal product. (Houben and Guillaud 2001) (Fig. 16.1).

16.3.3 Stabilisation in the Indian Context

India being a tropical country with its fair share of torrential rainfall and harsh summers—most of the traditional and contemporary earthen constructions in areas with high rainfall require certain additives to protect it against water, abrasion and general wear and tear or a final exterior coat/skin to protect the unstabilised interior mass.

The earlier earth dwellings when stabilised used mainly organic additives which later moved to using lime. Industrialisation led to cement becoming the additive that is added to protect/strengthen the mud mix against water and other factors. Today, the most common stabilisers used are lime and cement. Very often due to the unavailability and cost of lime, cement is the preferred choice of stabiliser for the common man. This is a matter of concern as cement may augment the mechanical characteristics of the soil but it greatly reduces the breathability of the soil. The use of lime or cement increases the embodied energy of the material and interferes with the cradle to grave lifecycle of mud buildings. In the hope of revival of earthen

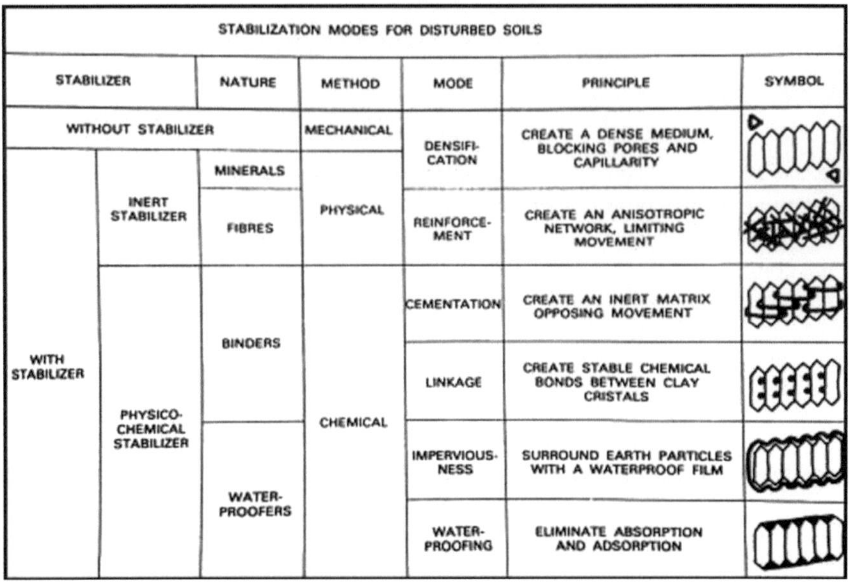

STABILIZATION MODES FOR DISTURBED SOILS						
STABILIZER		NATURE	METHOD	MODE	PRINCIPLE	SYMBOL
WITHOUT STABILIZER			MECHANICAL	DENSIFI-CATION	CREATE A DENSE MEDIUM, BLOCKING PORES AND CAPILLARITY	
WITH STABILIZER	INERT STABILIZER	MINERALS	PHYSICAL			
		FIBRES		REINFORCE-MENT	CREATE AN ANISOTROPIC NETWORK, LIMITING MOVEMENT	
	PHYSICO-CHEMICAL STABILIZER	BINDERS	CHEMICAL	CEMENTATION	CREATE AN INERT MATRIX OPPOSING MOVEMENT	
				LINKAGE	CREATE STABLE CHEMICAL BONDS BETWEEN CLAY CRISTALS	
		WATER-PROOFERS		IMPERVIOUS-NESS	SURROUND EARTH PARTICLES WITH A WATERPROOF FILM	
				WATER-PROOFING	ELIMINATE ABSORPTION AND ADSORPTION	

Fig. 16.1 Stabilisation modes for disturbed soils (Houben and Guillaud 2001)

construction, it is imperative that we look not only towards the future but also to our past to learn and see how the ancient recipes can be used with modern technology to result in mud constructions that are truly sustainable in its holistic sense.

16.4 Organic Stabilisers in Traditional Recipes

16.4.1 Research Methodology

The methodology followed a tripartite approach to find traditional recipes used in mud construction. In order to study the recipes in greater depth, the premier research was restricted to southern India with special focus on the South Canara region. Future research will extend to other regions following the same methodology with a vision of covering the country in the next few years (Fig. 16.2).

16.4.2 Classification of Organic Stabilisers

The stabilisers of animal and plant origin can be broadly classified based on their molecular nature into polysaccharides, lipids, proteins and others (Vissac et al. 2017)

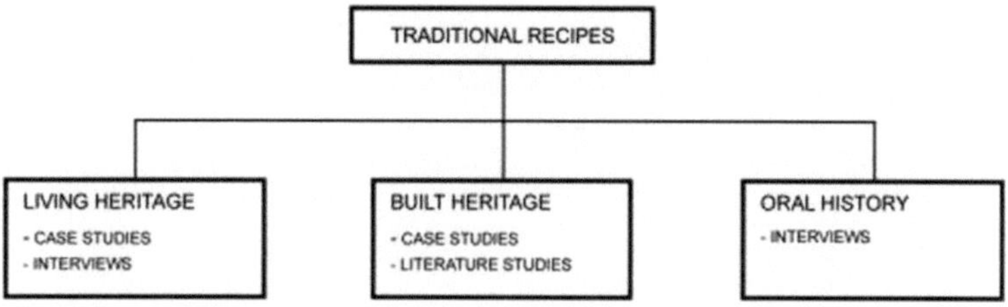

Fig. 16.2 Research methodology chart

Polysaccharides—cellulose, gums, starch, etc.
Lipids—wax, oils
Proteins—casein, albumine, etc.
Others—resins, tannins.

16.4.3 Recipes Found Through Living Heritage

Research carried out in living heritage refers to learning from present examples and recipes that are still used today by the people of that region. This involves field studies conducted in remote villages and tribal communities in the Western Ghats like Wayanad, Nagercoil, etc., where people continue to live in mud homes. It throws light on mixes of mud and natural ingredients that were passed on from generation to generation and provides insight into the type of additives mixed into mud to form durable and resistant vernacular mud houses.

The constraint here however is that information of the exact quantities of the ingredients was not found as the tribals added the ingredients through intuition and experience, than calculated precise quantities.

Many recipes included natural additives commonly heard of like cow dung, soil from termite hills, egg whites, natural fibres—rice straw, rice husk. The lesser known ingredients are mentioned are.

16.4.3.1 Glue from Bark and Leaves of Kulamavu (Scientific Name—*Persea macrantha*)

Slices of the bark of this tree give a gluey mix when water is added to it. It is left to rest overnight and mixed with the soil. This when added into the soil visibly increases the viscosity of the mix and is said to render it stronger and more resistant to mechanical shock (Thirumalini et al. 2011) (Images 16.2 and 16.3).

Images 16.2 and 16.3 Kulamavu Tree and Glue

16.4.3.2 Juice from Kadukkai Seed (Scientific Name—*Terminalia chebula*)

The seeds of Kadukkai or common name Gall Nut are wrapped in a cloth bag and left in water and left to macerate. After 24 h, the water becomes blackish in colour. This is then added into the mud mix (can be added to the entire mass or given as a final coat). It has been proved to have water proofing qualities (Vissac et al. 2017).

Kadukkai water is still used in construction today for water proofing in conventional buildings as well. The old Madras terrace roofing also uses this water for water proofing in its final layers.

16.4.3.3 Paste of the Hibiscus Leaf (Scientific Name—*Hibiscus rosa-sinensis*)

Hibiscus leaves and flowers are drenched in hot water, ground and strained, and this oily paste is taken rubbed over as a final coat in 2–3 layers. This gives a shine to the wall and renders it water resistant. This finish with hibiscus paste is seen in the final finishing of the walls and floors of traditional mud homes in Wayanad. This same mix has been used extensively as a herbal shampoo in the Kerala region even today.

16.4.4 Additives from Recipes Found Through Built Heritage

Under this section, the main source would be from ongoing best practice restoration projects that use mixes that are true to its original fabric and does not incorporate any modern additives. Thirumalini et al. (2011) who talks about a herbal juice mix used to improve performance of lime found on an ancient manuscript in the Padmanabhapuram temple detailed below.

16.4.4.1 Herbal Juice Used in Restoration Work in East Fort—Trivandrum

Aqueous extract of herbs such as oonjalvalli (*Cissus glauca* Roxb.), pananchikaai (*Cochlospermum religiosum*), kulamavu (*Persea macrantha*), gallnut (*Terminalia chebula*) and palm jaggery (from *Borassus flabellifer*) was used in this herbal mix. 0.25 kg (wet weight) of each herb and palm jaggery is taken and crushed well. Crushed herbs and jaggery were soaked together in water for 15 days. At the end, the juice is separated and used. For preparation of one kg of lime putty, equal amount (one litre) of herbal cocktail juice is used. Thirumalini et al. (2011).

16.4.4.2 Herbs Used in Restoration Work at Vadakkunnathan Temple—Thrissur

The UNESCO award winning restoration project of the Shri Vadakkunnathan Ksethram Temple used 9 different herbs along with lime in the external plaster restoration. Interactions with the architect Vinod Kumar co-ordinator of the renovation project and artisan Vijay Kumar, we found that similar herbs were used as in the earlier example—oonjalvalli, gallnut, kulamavu and jaggery.

There are already examples of most of these herbs being used in mud mixes. Hence, a well-grounded hypothesis is made that these combination of herbal mixes should work in earthen plasters, and this will have to be tested as part of the future research in order to be found conclusive.

16.4.5 Additives from Recipes Found Through Oral History

Unlike the western world that passes on information through written documented facts, India has a rich tradition of knowledge transfer from one generation to another through apprenticeship. It is very rare to find written texts on practical skills that are used. Thus, any research in India would be incomplete without a section on Oral history which needs to be validated through test samples and laboratory analysis.

Image 16.4 Varaal Fish

16.4.5.1 Mucous-like Residual Water from the Bral Fish (Scientific Name—*Channa striatus*)

Bral or Varaal is the local name given to the fresh water fish, which releases a sticky mucous as it swims. The mucous content in the water if added to mud is said to increase viscosity of the mix. The approximate proportions would be about 5 fish in 200 L of water—left to swim in it for 2 days. The fish releases a mucous as it swims making water sticky. This water is then used to mix the mud used in plasters. It was used extensively in Kerala especially in Thrissur and Alleppey regions 50 years ago. The varied benefits of this are still unknown till further testing is carried out (Image 16.4).

16.4.5.2 Arabic Gum (Acacia Senegal)

Arabic gum was the other ingredient commonly used in the context of Mural wall paintings—a coat is applied on the wall before the mural is done to improve the durability of the paintings (Mini 2010) Though examples of it being used in India have not been found, it is used extensively in mud plasters in the African continent as a stabiliser and acts as water proofing (Vissac et al. 2017).

16.4.5.3 Karingota (Scientific Name: *Quassia indica*)

The leaves are mixed in water and this water is used in the mud mix which is said to work against termites and hence were also used in the attic of the traditional mud roofs as a layer between the wood and the mud filling to prevent the rotting of wood.

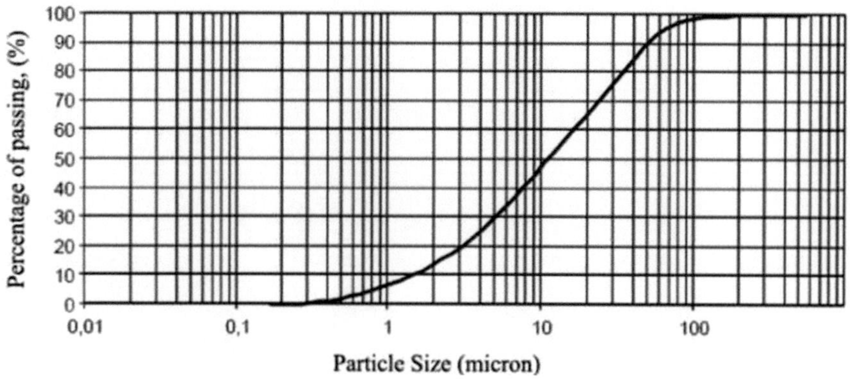

Fig. 16.3 Particle size distribution of soil

16.5 Performance Tests on Organic Additives

Certain laboratory tests were conducted taking specific ingredients from the broad classification of polysaccharides, proteins, lipids and others (Vissac et al. 2017). This was done to understand in detail how ingredients from each of these families perform and the specific property they bring to the mud mix in terms of water resistance, resistance to mechanical shock etc.

The testing was done in a laboratory in France and hence is constrained to the availability of the ingredients. The soil that was used is the residue from washing sand, locally known as FAC—Fine Argilo Calcaire. The soil can be better understood from its grain size distribution and composition chart (Fig. 16.3 and Image 16.5).

Test Samples—The ingredients of each sample have been mentioned (Table 16.1). The ingredients taken to be tested are organic stabilisers used around the world (Vissac et al. 2017). Though the ingredients differed, the size of the samples was kept similar and base ingredient the same.

Base Ingredient—1 Soil: 4 Sand.

Table 16.1 Test samples

Sample No.	Description
0	Base ingredient (pure)
0A	Base ingredient with two coats of linseed oil
1A	Base ingredient with 1% Egg white
1B	Base ingredient with 5% Egg white
1C	Base ingredient with 10% Egg white
2A	Base ingredient with 1% Arabic gum
2B	Base ingredient with 5% Arabic gum
2C	Base ingredient with 10% Arabic gum

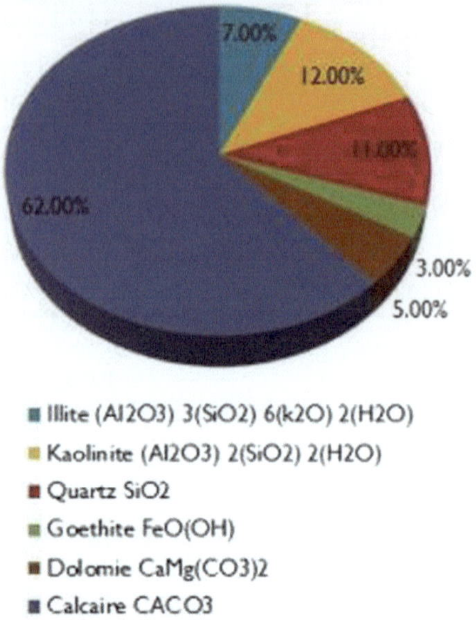

Image 16.5 Mineral composition

16.5.1 Water Erosion Test—Drip Test

The principle of this test is to take the test samples and subject them to the falling of water droplets from a height and in a certain time. This test makes it possible to observe the behaviour of each mixture to continued exposure to water and therefore deduce its behaviour in the wall and water resistance (Images 16.6, 16.7 and Table 16.2).

16.5.2 Abrasion Test

A metal brush weighted to 3 kg is used to scrub the face of the test sample. A single back and forth motion of the brush is regarded as one cycle of abrasion. Brushing is continued for twenty cycles. The measurement consists of weighing the material before and after brushing (Table 16.3 and Images 16.8, 16.9).

Image 16.6 Drip test

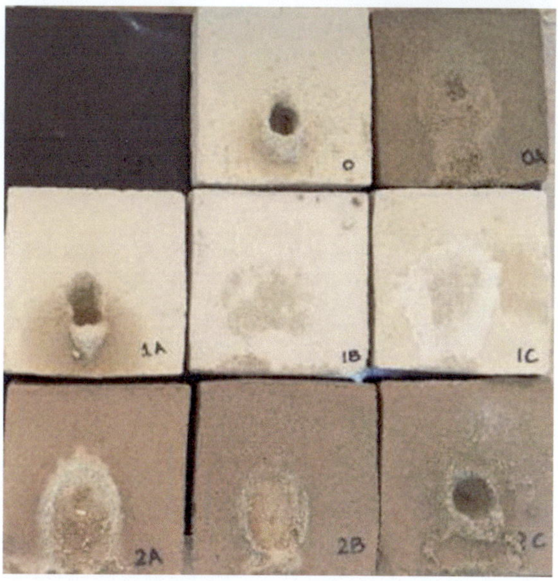

Image 16.7 Test samples—drip test

Table 16.2 Drip test for water erosion

Sample No.	Dry weight (in gm)	Wet weight (in gm)	Size of the hole (in cm)	Observation
0	365.57	367	Diameter: 2	A hole through and through is formed
			Depth: 2	
0A	375.95	After 10 min: 376.75	Diameter: –	
			Depth: –	
		After 30 min: 377.64	Dimension: 1 × 1.5	Earth easily comes off. Wet edges
			Depth: 0.2	
1A	375.68	379.78	Dimension: 1.2 × 2	A depression is formed
			Depth: 1.5	
1B	385.76	After 10 min: 387.64	Diameter: –	No change
			Depth: –	
		After 30 min: 389.62	Diameter: –	No change. Water has spread over the surface making the edges sensitive. But the wet surface was not soft or deformed
			Depth: –	
1C	377.21	After 10 min: 378.33	Diameter: –	No change
			Depth: –	
		After 30 min: 379.05	Diameter: –	No change. Lesser amount of water has spread over the surface. The sample was resistant to deformation
			Depth: –	
2A	374	After 10 min: 378.33	Diameter: –	The damage is concentrated and the wet portion easily comes off
			Depth: –	
		After 30 min: 382.66	Dimension: 1 × 1.2	
			Depth: 0.3	
2B	360	After 10 min: 360.56	Diameter: –	The damage is spread out
			Depth: –	
		After 30 min: 360.38	Dimension: 2 × 3	
			Depth: 0.2	
2C	343.77	After 10 min: 344.03	Diameter: 1.2	There is a very large hole. It is very delicate and sensitive
			Depth: 0.5	
		After 30 min: 344.69	Diameter: 2	
			Depth: 1.2	

Image 16.8 Abrasion test

Image 16.9 Test samples—abrasion test

Table 16.3 Abrasion test

Sample	Weight before	Weight after	Observations
0	365.55	354.6	It is not resistant
0A		361.7	
1A	375.77	366.89	Did not show a great difference between 0 and 1A
1B	380.78	378.03	It was much more resistant compared to 0 and 1A
1C	369.46	368.98	It is very resistant
2A	372.1	372.05	It is very resistant
2B	368.89	368.86	It is very resistant
2C	346.2	346.14	It is very resistant

16.6 Conclusion

To summarise the above technical paper, the main objective of all the studies conducted is to be able to bring to the forefront the use of these natural stabilisers into conventional practice. The main constraint that was observed in the use of these natural stabilisers was the availability of some of the ingredients—either they were in remote locations and were inaccessible or they were not available in the required quantities. This constraint could be addressed by cultivating required ingredient in large quantities and make it available for commercial use. Alternatively, a study of existing indigenous fauna can be done and suitable replacements could be found and further tested for performance.

There is a huge potential to further the research already conducted in this paper. Firstly, the testing of the ingredients has to be continued for all ingredient identified in the traditional recipes and compared it in terms of performance with those stabilised with lime/cement. This would further inform the performance of each ingredient used in the mix. Secondly, to ensure success of bringing the natural additives onto a larger platform for accessibility to more people, an inventory can be generated with area/region specific additives, with information on location of availability and the particular performance enhancing properties it would bring about.

References

Houben H, Guillaud H (2001) Earth construction. ITDG Pub., London

Khasreen M, Banfill P, Menzies G (2009) Life-cycle assessment and the environmental impact of buildings: a review. Sustainability 1(4):674–701

Mini PV (2010) Preparation techniques of pigments for traditional mural paintings of Kerala. Indian J Tradit Knowl 9(4):635–639

Thirumalini P, Ravi R, Sekar SK, Nambirajan M (2011) Study on the performance enhancement of lime mortar used in ancient temples and monuments in India. Indian J Sci Technol 4(11)
Venkatarama Reddy BV, Jagadish KS (2003) Embodied energy of common and alternative building materials and technologies. Energy Build 35:129–137
Vissac A, Bourges A, Gandreau D, Anger R, Fontaine L (2017) Argiles & biopolymères. CRAterre éditions, Villefontaine

Chapter 17
Advances in the Use of Biological Stabilisers and Hyper-compaction for Sustainable Earthen Construction Materials

Sravan Muguda, George Lucas, Paul Hughes, Charles Augarde, Alessia Cuccurullo, Agostino Walter Bruno, Celine Perlot and Domenico Gallipoli

17.1 Introduction

The construction industry is one of the largest contributors to carbon emissions, and therefore, there is considerable interest in research to transform the current industry to be more sustainable, partly by developing new eco-friendly construction materials and techniques. The use of unbaked earth bricks as a building material has clear advantages in the field of sustainability over conventional construction reducing carbon emissions and energy consumption throughout the lifetime of buildings (Morel et al. 2001; Gallipoli et al. 2017). "Raw Earth" consists of a compacted mix of soil and water which is put in place with the least possible transformation (Jaquin et al. 2009) and because of its hydrophilic nature, exhibits a strong tendency to adsorb or release moisture, and therefore to emit or store latent heat, depending on current levels of ambient humidity. Two key barriers to the wider use of "Raw Earth" are poor mechanical properties and questions over durability, and both are traditionally tackled by using stabilisers. However, when the stabiliser is cement, as is most common, the material produced is really a weak concrete and has the carbon footprint approaching that material (Lax 2010). Here we present findings from two avenues of research under the TERRE project, a European Commission funded project training early stage researchers in the development of eco-friendly construction, including earthen materials. Delivering improved mechanical properties and good durability is tackled here in two ways. Firstly, the use of biopolymer stabilisers is presented

S. Muguda · G. Lucas · P. Hughes · C. Augarde (✉) · A. Cuccurullo
Department of Engineering, Durham University, Durham, UK
e-mail: charles.augarde@dur.ac.uk

S. Muguda · A. Cuccurullo · A. W. Bruno · C. Perlot · D. Gallipoli
Laboratoire SIAME, Fédération IPRA, Université de Pau et des Pays de l'Adour, Anglet, France

B. V. V. Reddy et al. (eds.), *Earthen Dwellings and Structures*, Springer Transactions in Civil and Environmental Engineering,
https://doi.org/10.1007/978-981-13-5883-8_17

with an emphasis on durability. Secondly, a new means of manufacture using hyper-compaction is presented where the focus is on the mechanical properties.

17.2 Durability of Earthen Construction Materials

Durability has always been a key issue to the acceptance of earthen materials. The earliest known earthen construction material was used in Mesopotamia and consisted of hand-moulded alluvial deposit mixtures (Deboucha and Hashim 2011). With time, organic compounds such as animal dung and plant extracts were added to these soil mixtures to improve their erosional resistance (Ngowi 1997). One class of modern, widely available organic products are biopolymers which are receiving attention as stabilisers for earthen materials due to their potential green credentials (Chang et al. 2016). Recent work reported in Aguilar et al. (2016) and Nakamatsu et al. (2017) has investigated the use of biopolymers (namely chitosan and carrageenan) as stabilisers and has reported that the addition of these biopolymers improved mechanical and durability performance of earthen materials. Very recently, the mechanical behaviour of earthen construction materials stabilised with the biopolymers guar gum and xanthan gum was studied by Muguda et al. (2017) which showed that the addition of these biopolymers improved compressive and tensile strengths. These biopolymers sequester CO_2 during production (Chang et al. 2016; Krishna Leela and Sharma 2000) in contrast to cement, which leads to the opposite; however, energy required in production of the gums may be much greater than for an equivalent amount of cement (e.g. see Lo et al. 1997), so it would be good to see a full life cycle assessment of these biopolymer-based stabilisers, which is not yet available. At present, durability performance of earthen construction materials is assessed via different tests as described in various international standards, all of which measure the resistance of the earthen material against the erosional action of water. For unstabilised earthen construction, the standard tests are immersion, contact, drip and suction tests, while for stabilised materials, accelerated erosion, spray and wire brush tests are conducted to assess durability. Here we examine the durability properties of the materials studied in Muguda et al. (2017).

17.3 Materials and Methods: Durability Testing

17.3.1 Materials

For this study, an engineered soil mixture comprising 20% Kaolin, 70% sharp sand and 10% gravel by mass was used. This mix complies with the requirements for earthen construction materials given in Oliver and Mesbah (1987) and Houben and Guillaud (1994) and is a combination widely investigated in earthen construction.

Table 17.1 Physical properties of the unamended soil mixture used in this study

Index property		*Atterberg limits*	
Standard compaction tests (BS 1377-2 1990; BS 1337-4 1990)		Liquid limit (%)	36
Maximum dry density (kg/m^3)	1870	Plastic limit (%)	18
Optimum moisture content (%)	9.8	Plasticity index (%)	18
Grain size distribution			
Gravel content (%)	10		
Sand content (%)	70		
Silt content ($\leq$63 μm, %)	04		
Clay content ($\leq$2 μm, %)	16		

Atterberg limits and compaction characteristics for the unamended soil mixture are given in Table 17.1. Commercially available guar gum and xanthan gum were chosen as biopolymer stabilisers in this study. The biopolymer stabiliser content was maintained at 2.0%.

17.3.2 Methodology

Stabilisation using biopolymers is achieved through "hydrogels" which are formed through the interaction of soil, biopolymer and water particles. Unlike cementitious bonds formed due to hydration of cement, these "hydrogels" bind soil particles through a combination of chemical bonds and soil suction (Muguda, et al. 2017). As these hydrogels become susceptible to weakening on contact with water, durability tests such as accelerated erosion tests, spray tests and wire brush tests were considered to be too vigorous and hence an alternative test was chosen, namely the "Geelong" test (NZS 4298, 1998). Samples in the form of 150 × 150 × 20-mm tiles and 38 mm diameter and 76-mm-length cylinders were tested. In both cases, the required bulk mass of the sample was placed in a mould and statically compacted. All blocks were compacted to achieve an initial dry density equivalent to the maximum dry density, i.e. 1870 kg/m^3. Once the sample was compacted, it was carefully removed from the mould and left to air cure at a relative humidity of 50% and temperature of 21 °C. The durability tests were then performed on samples cured for 7 days.

The test procedure involves the dripping of 100 ml of water for up to 60 min from a height of 400 mm on to the surface of the sample. For the tile samples, the surface was kept at an inclination of 2H:1V, while for cylindrical specimens the surface of erosion was held perpendicular (Fig. 17.1). As well as noting the final erosion at 60 min as recommended by the code, the erosion depths were also noted at intermediate 15 min intervals. The results presented herein are the average values of

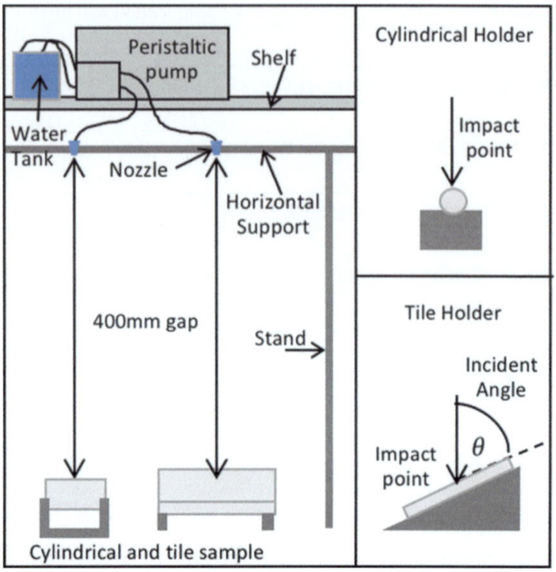

Fig. 17.1 Schematic representation of durability test set-up

five replicates. These results are compared with those of the unamended soil mixture and 8.0% cement-treated specimens.

17.4 Results and Discussion: Durability

Table 17.2 presents the final erosional depth after 60 min for both tile and cylindrical samples for unamended, cement and biopolymer-stabilised material. It can be observed from the results that unamended samples failed against the permissible limit for both tile and cylindrical samples, while cement-stabilised samples had negligible erosion. In the case of the biopolymer-treated samples, both guar- and xanthan gum-stabilised samples had erosional depths within 5 mm and passed the durability tests satisfactorily, with xanthan gum-treated samples performing better. The erosion rates for the biopolymer-treated samples are presented in Fig. 17.2. For both the biopolymers, the observed rates of erosion for tile samples are higher than for the cylindrical samples. This higher rate of erosion may be due to the sample orientation with the drip direction. It is also notable that the rate of erosion for xanthan gum-treated samples was less than guar gum-treated samples. In order to assess the time required to achieve an erosion depth of 5 mm, linear extrapolation was carried out indicating that guar gum-treated samples would require 118 and 200 min for tile and cylindrical samples, respectively, while xanthan gum samples would require 165 and 235 min.

It can be concluded from the above findings that biopolymer-treated earthen materials appear to have much improved durability properties as compared to unstabilised

Table 17.2 Final erosional depths after 60 min for both sample types (tile and cylinder)

Sample	Eroded depth (mm)		Remarks
	Tile	Cylinder	
Unamended	8.00	10.00	Permissible limit as per NZS 4298 is 5 mm
Cement treated	0.10	0.01	
Guar gum—2.0%	2.65	1.51	
Xanthan gum—2.0%	1.86	1.25	

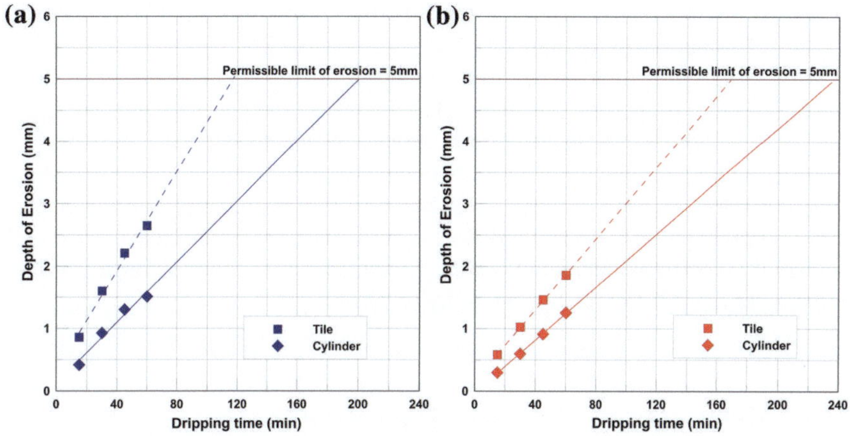

Fig. 17.2 Depth of erosion versus dripping time for both tile and cylindrical samples **a** guar gum, **b** xanthan gum

materials. While the performance of biopolymer-treated earth material cannot match that of cement-treated material, it can still provide an acceptable level of durability as measured by this test. These results support the further investigation of biopolymers for this purpose.

17.5 Hyper-Compaction

It is well known that one of the keys to achieve high strength and stiffness of earthen materials is the compactive effort used in creation of the in situ or unit-based materials. In particular, the application of a pressure significantly higher than that applied during production of conventional earth bricks increases dry density and consequently stiffness and strength, thus resulting in mechanical characteristics similar to those of conventional building materials. An innovative static hyper-compaction method has been developed by the authors using compaction effort corresponding to a 1D stress level of 100 MPa. Table 17.3 summarises previous studies using this

Table 17.3 Comparison in terms of compressive strength

Material	Compressive strength (MPa)
Compressed earth bricks (Bruno et al. 2016)	14.6
Compacted unstabilised and stabilised soils (Guetlala and Guenford 1997)	From 5.2 to 12.9
Standard masonry bricks (ASTM C270 2014)	From 6.9 to 27.6

method (Bruno et al. 2016) indicating that unstabilised hyper-compacted earth bricks are competitive with standard masonry construction according to ASTM C270 (2014) in terms of compressive strength.

In the study presented below, hyper-compaction was applied to soil mixtures containing large proportions of fine materials. Finer soils are able to retain more water than coarser soils thus resulting in stronger hygroscopic behaviour. However, a larger fine fraction may weaken mechanical characteristics and undermine durability. Properties such as stiffness and strength were measured by performing unconfined compression tests on cylindrical raw earth samples compacted at very high pressure (100 MPa) at the optimum water content and after equalisation at the same temperature (25 °C) and relative humidity (62%).

17.6 Materials and Methods: Hyper-Compaction

17.6.1 Soil Type and Index Properties

The earth used in this work has been provided by the Bouisset brickwork factory from the region of Toulouse in France. The grain size distribution is an influential parameter for assessing the suitability of earthen materials for construction and its role affecting the soil behaviour make it central to most existing recommendations (Delgado and Guerrero 2007). The grading curve of the soil used here has been determined by means of wet sieving and sedimentation tests to French standards. The grain size distribution of the Bouisset soil lies close to the upper limit of current recommendations by AFNOR (2001)/CRATerre-EAG (1998) and MOPT (1992) relevant to the manufacture of earth bricks (Fig. 17.3). In order to investigate the role of the grain size distribution, the Bouisset soil was mixed with a sandy soil to obtain three different earth mixes. The percentage of sand added was established looking at the recommended area to obtain three earth mixes with a clay content, respectively, equal to the minimum, the maximum and the average between the maximum and minimum suggested by the guidelines, and the first earth mix is the Bouisset soil itself. Table 17.4 shows the calculated percentages of Bouisset and sand that were mixed together in order to obtain the desired clay content of the resulting earth mix.

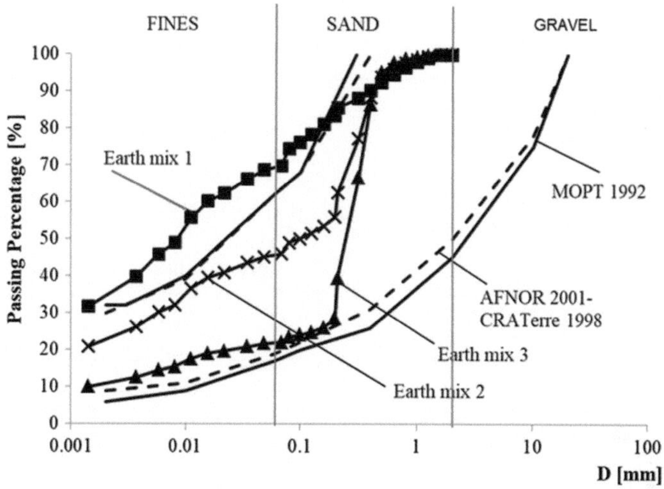

Fig. 17.3 Grain size distribution of earth mixes analysed in relation to recommendations for the manufacture of compressed earth bricks by CRATerre-EAG (1998) and MOPT (1992)

Table 17.4 Physical composition of earth mixes

Sample ID	Bouisset percentage (%)	Sand percentage (%)	Clay content (%)
Earth mix 1	100	0	≈32
Earth mix 2	66	34	≈20
Earth mix 3	32	68	≈10

Figure 17.3 shows the grain size distribution curves of the earth mixes presented above and the discussed guidelines relevant for compressed earth bricks.

The properties of Bouisset soil are summarised in Table 17.5, which indicates that the Bouisset soil can be classified as a well-graded silty clay. The plasticity properties of the fine fraction (i.e. the fraction smaller than 0.400 mm) of the Bouisset soil have been measured in agreement with French standards. The liquid limit, plastic limit and plasticity index, determined as the average of four independent tests, classify the material as a low plasticity clay.

17.6.2 Hyper-Compaction

Prior to compaction, the dry soil was mixed with the desired amount of water and subsequently placed inside three plastic bags to prevent evaporation. After that, the wet soil was left to equalise for at least one day so that moisture could redistribute prior to compaction. The soil was placed inside a stiff cylindrical steel mould with a diameter of 50 mm and vertically compacted by using a load-controlled Zwick

Table 17.5 Bouisset measured index properties

Index property			
Grain size distribution		*Atterberg limits*	
Gravel content (> 2 mm, %)	0	Plastic limit (%)	18.7
Sand content (≤ 2 mm, %)	31	Liquid limit (%)	29.0
Silt content (≤ 63 μm, %)	35	Plasticity index (%)	10.3
Clay content (≤ 2 μm, %)	34	*Mineralogical composition* Goethite, Muscovite, Orthose Kaolinite, Quartz	
Specific gravity	2.65		

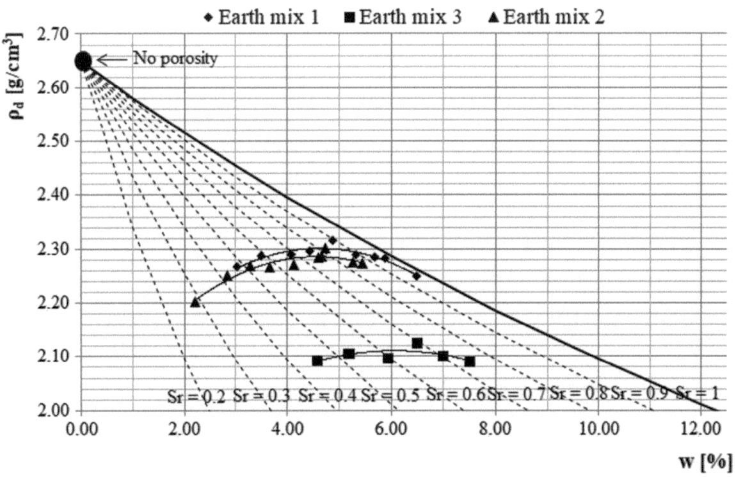

Fig. 17.4 Compaction curves for the static pressure of 100 MPa

press with a capacity of 250 kN. Pressure was applied by two cylindrical aluminium pistons acting at the top and bottom extremities of the specimen. Additional details about the compaction procedure are available in Bruno (2016). Figure 17.4 presents the experimental values of dry density, ρ_d plotted against the corresponding water contents, w together with the respective interpolating curves for each earth mix used in the study.

Figure 17.4 also shows the theoretical "no porosity" point corresponding to an extremely high-compaction effort, which produces a dry density equal to the density of the solid particles.

Table 17.6 Results from strength and stiffness testing

Earth mix	Dry density g/cm^3	Young's modulus (MPa)	Compressive strength (MPa)
1	2.31	2851	7.19
2	2.28	1795	3.44
3	2.12	1320	0.45

17.7 Results: Hyper-Compaction

In order to measure the stiffness and strength of the material, unconfined compression tests were performed at the scale of small cylindrical samples of 50 mm of diameter and 100 mm of height. A set of two samples were manufactured for each earth mix considered. A constant displacement rate of 0.001 mm/s, which is the slowest rate that can be applied by the Zwick/Roell Amsler HB250 press, was chosen in order to obtain a regular stress–strain curve without instabilities (Bruno 2016). Young's modulus was measured based on five unconfined loading–unloading cycles performed at a loading rate of 0.005 MPa/s between one-ninth and one-third of the estimated compressive strength of the material. Axial displacements were measured between two points along the height of the cylindrical samples at a distance of 50 mm by means of two transducers placed on diametrically opposite sides. Based on the assumption that material behaviour is elasto-plastic during loading but essentially elastic during unloading, Young's modulus was determined by considering only the unloading branches of the five cycles. In particular, Young's modulus was determined as the average slope of the five unloading branches in the axial stress–strain plane. Table 17.6 shows Young's modulus and compressive strengths of the three soil mixes.

The earth mix characterizsed by the highest value of dry density exhibits the highest Young's modulus, and compressive strength is consistently higher for the more compacted and denser soil. Interestingly, despite the negligible difference in terms of dry density between earth mix 1 and 2, a significant augmentation of the material stiffness and strength is noticeable. An explanation of this result might be the different physical composition of the two earth mixes. Earth mix 2 is, in fact, a combination of a silty clay and a sandy soil characterised by a lower amount of clay. Inspection of Fig. 17.3 indicates a bimodal grain size distribution (gap-graded soil). It is suggested that not just the density but also the inclusion of a coarser soil to a fine-grained soil or addition of fines to sand strongly affects the mechanical behaviour of the material.

17.8 Conclusions

The use of raw earth as sustainable construction material is being explored as one of the most promising possibilities to replace conventional options, but this is only likely to be successful if key issues such as durability and mechanical properties are improved. Biopolymers and hyper-compaction both show promise in this regard, and in this paper, we have focussed on some aspects of their performances. However, further investigations are necessary to understand how to improve not only mechanical properties but also hygroscopic and durability properties developing a sustainable stabilisation method that could not negatively impact one of these performances. In addition, a full LCA for the proposed stabilisers is needed to truly prove the green credentials discussed above.

Acknowledgements The authors wish to acknowledge the support of the European Commission via the Marie Skłodowska-Curie Innovative Training Networks (ITN-ETN) project TERRE 'Training Engineers and Researchers to Rethink geotechnical Engineering for a low carbon future' (H2020-MSCA-ITN-2015-675762).

References

AFNOR (2001) XP P13-901; Compressed earth blocks for walls and partitions: definitions–Specifications–Test methods–Delivery acceptance conditions

Aguilar R, Nakamatsu J, Ramírez E, Elgegren M, Ayarza J, Kim S, Pando MA, Ortega-San-Martin L (2016) The potential use of chitosan as a biopolymer additive for enhanced mechanical properties and water resistance of earthen construction. Constr Build Mater 114:625–637

ASTM C270 (2014) Standard specification for mortar for unit masonry. American Society for Testing and Materials International

Bruno AW (2016) Hygro-mechanical characterisation of hypercompacted earth for sustainable construction. Ph.D. thesis, Université de Pau et des Pays de l'Adour

Bruno AW, Gallipoli D, Perlot C, Mendes J (2016). Effect of very high compaction pressures on the physical and mechanical properties of earthen materials. In: E3S web of conferences, vol 9, p 14004. EDP Sciences

BS 1377–2 (1990) Methods of test for soils for civil engineering purposes—Part 2: Classification tests. BSI, London

BS 1377–4 (1990) Methods of test for soils for civil engineering purposes—Part 4: Compaction tests. BSI, London

Chang I, Im J, Cho GC (2016) Introduction of microbial biopolymers in soil treatment for future environmentally-friendly and sustainable geotechnical engineering. Sustainability 8(3):251

CRATerre-EAG (1998) CDI, Compressed earth blocks: Standards–Technology Series No.11. CDI, Brussels

Deboucha S, Hashim R (2011) A review on bricks and stabilized compressed earth blocks. Sci Res Essays 6(3):499–506

Delgado MCJ, Guerrero IC (2007) The selection of soils for unstabilised earth building: a normative review. Constr Build Mater 21(2):237–251

Gallipoli D, Bruno AW, Perlot C, Mendes J (2017) A geotechnical perspective of raw earth building. Acta Geotech 12(3):463–478

Guetlala A, Guenfoud M (1997) Béton de terre stabilisé: Propriétés physico-mécaniques et influence des types d'argiles. La technique moderne 89(1–2):21–26

Houben H, Guillaud H (1994) Earth construction: a comprehensive guide. Intermediate Technology Publications

Jaquin PA, Augarde CE, Gallipoli D, Toll DG (2009) The strength of unstabilised rammed earth materials. Géotechnique 59(5):487–490

Krishna Leela J, Sharma G (2000) Studies on xanthan production from Xanthomonas campestris. Bioprocess Eng 23:687–689

Lax C (2010) Life cycle assessment of rammed earth (Doctoral dissertation, M.Sc. thesis. University of Bath, United Kingdom

Lo Y-M, Yang S-T, Min DB (1997) Ultrafiltration of xanthan gum fermentation broth: Process and economic analyses. J Food Eng 31(2):219–236

MOPT (1992) Bases Para el Diseño y Construcción con Tapial. Madrid, Spain: Centro de Publicaciones, Secretaría General Técnica, Ministerio de Obras Públicas y Transportes

Morel JC, Mesbah A, Oggero M, Walker P (2001) Building houses with local materials: means to drastically reduce the environmental impact of construction. Build Environ 36(10):1119–1126

Muguda S, Booth SJ, Hughes PN, Augarde CE, Perlot C, Bruno AW, Gallipoli D (2017) Mechanical properties of biopolymer-stabilised soil-based construction materials. Géotech Lett, 1–6

Nakamatsu J, Kim S, Ayarza J, Ramírez E, Elgegren M, Aguilar R (2017) Eco-friendly modification of earthen construction with carrageenan: water durability and mechanical assessment. Constr Build Mater 139:193–202

Ngowi AB (1997) Improving the traditional earth construction: a case study of Botswana. Constr Build Mater 11(1):1–7

NZS 4298 (1998) Materials and workmanship for earth buildings [Building Code Compliance Document E2 (AS2)]. New Zealand Technical Committee

Olivier M, Mesbah A (1987 Nov) Influence of different parameters on the resistance of earth, used as a building material. In: International conference on mud architecture, Trivandrum, India

Chapter 18
Stress–Strain Characteristics of Unstabilised Rammed Earth

Holur Narayanaswamy Abhilash and Jean-Claude Morel

18.1 Introduction

Adobe, compressed earth blocks (CEBs), rammed earth and Laterite stones are commonly used natural earth-building materials. These building technologies have evolved over a very long period of time, the early usage of raw earth could possibly be dated back to first-ever dwelling built by mankind. Currently, the modern tools and techniques have significantly aided these materials to improve their strength and durability (Bruno et al. 2017). Alternatively, cement or lime additives are added to raw earth mixture to enhance their mechanical properties (Reddy et al. 2007; Walker and Stace 1997). From late twentieth century, the utilisation of CEBs and rammed earth has been significantly increased, thanks to their energy-efficient and sustainable performance.

CEBs are manufactured by static compaction technique, which involves double compaction through top and bottom plates of the press (Rigassi and Craterre 1995). Thus giving a more homogenised product, in which, the density of the material is uniform throughout the section. Fabrication of rammed earth involves dynamic compaction process within a formwork. The unsaturated soil mixture is poured in layers and rammed using a rammer (manually or mechanically) until the desired layer thickness is achieved. Since ramming occurs only on the top surface of a layer, the loss of energy from top to bottom of the layer creates a density variation gradient within the layer. With modern ramming (high compaction effort) tools, a dry density of 2.1 g/cc can be achieved for unstabilised rammed earth (USRE) (Gerard et al. 2015). Earlier studies reported that the dry density of in situ USRE walls generally

H. N. Abhilash (✉)
Indian Institute of Science, Bangalore, India
e-mail: abhi.hn387@gmail.com

J.-C. Morel
Coventry University, Coventry, UK
e-mail: ac0969@coventry.ac.uk

B. V. V. Reddy et al. (eds.), *Earthen Dwellings and Structures*, Springer Transactions in Civil and Environmental Engineering,
https://doi.org/10.1007/978-981-13-5883-8_18

varies between 1.75 and 2.0 g/cc (Gomes et al. 2014). Therefore, depending on the methodology of construction adopted, the dry density of the USRE varies. It should be noted that the compressive strength of the material is directly proportional to the dry density. Precise replication of in situ fabrication in the laboratory is quite essential for accurate investigation of mechanical properties of rammed earth.

The compressive strength and stress–strain characteristics of USRE for a soil is dependent on various parameters, some of the most important parameters are particle-size distribution (clay content), compaction energy, moulding water content and the moisture content at the time of testing. Dry density of USRE is directly related to the compaction energy and moulding water content. The process of manufacturing USRE specimens in the laboratory that resembles in situ method is quite debatable. The closest technique that resembles in situ manufacturing is proposed by Walker et al. (2005), directing to compact the soil until the ringing sound is heard upon the impact of hammer on the compacted layer surface. With this technique, uniform control of density in the individual layer might get affected due to the qualitative judgement system. The stress–strain relationships generated from adopting such qualitative specimen casting methods may not be reliable. The present study was focused on examining the strength and stress–strain characteristics of USRE under monotonic cyclic loading. The modulus and stiffness degradation due to cyclic loading was examined.

18.2 Material Properties

18.2.1 Particle-Size Distribution of Soils

The characteristics of the two soils used in the experiments are discussed here. The particle-size distribution (PSD) curves for the two soils designated as Soil-A and Soil-B were obtained are shown in Fig. 18.1. These soils are extensively used for earth constructions locally. The clay fraction of the soils was 17 and 20% for Soil-A and Soil-B, respectively. Also, there is a major concern on amount of clay present in the soil, high clay content in the soil will result in shrinkage and also impact the strength of the material (Reddy et al. 2007; Walker and Stace 1997). For cement-stabilised soil blocks, Reddy et al. (2007) and Walker and Stace (1997) propose the clay content to be in between 10 and 20%. In the case of unstabilised earth, from the existing literature, it is hard to conclude on the optimum clay content essential for greater mechanical properties. From the literature, the clay content in the soil suitable for earthen building material varies from 5 to 40% (Gomes et al. 2014; Maniatidis and Walker 2003). In general, the clay content of 10–30% is considered to be in acceptable limits (Ciancio et al. 2013). Therefore, the soils considered in this study have been accepted in their natural state, which is truly locally available building material.

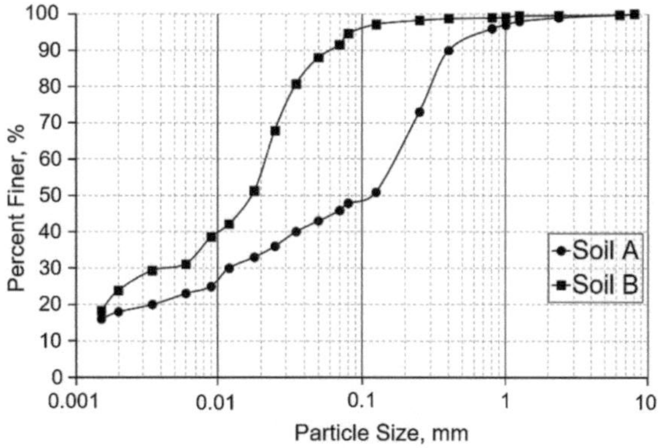

Fig. 18.1 Particle-size distribution curves for Soil-A and Soil-B

18.2.2 *Compaction Characteristics of the Soils*

The literature shows that the density of the material is widely accepted as an indicative factor of material strength and durability. The Proctor test is used to generate the maximum dry density (MDD) and the optimum water content (OWC). For both Soil-A and Soil-B, the MDD and OWC at Standard and Modified Proctor energy have been generated and presented in Fig. 18.2. Moulding water content is an important parameter to be known and controlled to maintain density and excessive shrinkage problems. The NZS-4297 (1998) recommends the variation of mixing water content not to exceed less than 3% or greater than 5% of the OWC. The MDD and OWC values obtained for both the soils at different compaction energies are given in Table 18.1. It can be seen that, lower the compaction energy higher the OWC, and higher the compaction energy lower the OWC. In case of Soil-A and Soil-B, for any compaction energy in-between the standard and modified proctor energy, the moulding water content should be in the range of 9–10.5% and 11–13.5%, respectively.

Table 18.1 Maximum dry density and optimum water content values

Soil	Soil-A		Soil-B	
Compaction energy (kN m/m^3)	600	2731	600	2731
Optimum water content (%)	10.5	9.0	13.5	11.0
Maximum dry density (g/cc)	1.93	2.07	1.75	1.97

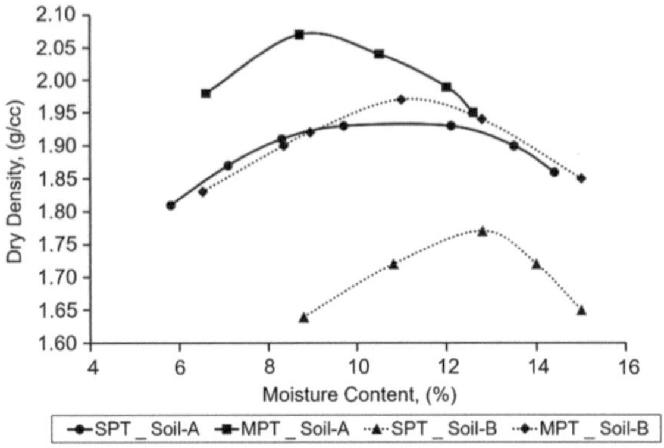

Fig. 18.2 Compaction curves for Soil-A and Soil-B

18.2.3 Sorption–Desorption Isotherms and Suction Value

The material isotherms will help in understanding the moisture ingress of the material at different relative humidities and temperatures. The moisture-carrying capacity of the specimens at given temperature and humidity can be estimated from these isotherm curves. The sorption and desorption isotherms of soil-A (Ferreira 2016) and B (shown in Fig. 18.3) were generated through dynamic vapour sorption (DVS) system at 23 °C. The maximum moisture-carrying capacity of Soil-A and Soil-B were found to be 4 and 4.8% at 97% RH. The ambient temperature and the relative humidity for this study are defined as 23 °C and 60% RH. The moisture-carrying capacity of Soil-A and -B at 23 °C and 60% RH were found to be 1.5 and 1.8%, respectively.

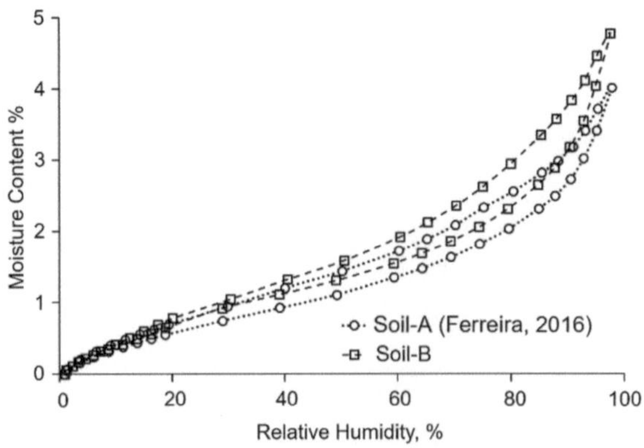

Fig. 18.3 Sorption and desorption isotherms of Soil-A and Soil-B

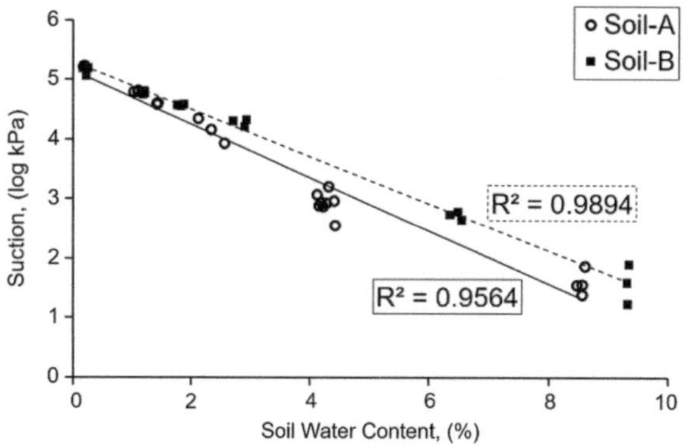

Fig. 18.4 Suction values for Soil-A and Soil-B with respect to the calibrated values

In unstabilised earthen building materials, clay, which acts as binder is vibrant to change in moisture content. Apart from clay, suction, which is also known as capillary suction plays a role in enhancing the compressive strength of unstabilised earth by increasing the apparent cohesion (Beckett 2011; Ciancio et al. 2013). As the drying takes place, the suction values of USRE will increase with reduction in moisture content (Gerard et al. 2015). Similarly, the suction values of the Soil-A and Soil-B (shown in Fig. 18.4) can be seen increasing with decrease in soil–water content. The suction values for both the soils were measured through non-contact filter paper (WHATMAN 42) technique, as per ASTM D 5298-03 (2003). The values presented in Fig. 18.4 are plotted with respect to the calibration curves established in the laboratory (Soudani 2016) generated according to ASTM D 5298-03 (2003).

18.3 Methodology and Experimental Programme

The USRE cylindrical specimens with aspect ratio 2 were subjected to cyclic compressive loading. The test was carried out on three sets of specimens having different dry densities for two soils that were cast by varying compaction energy using Proctor method. The influence of dry densities on compressive strength and stress–strain characteristics at three dry densities has been generated.

In this experimental programme, each set of USRE specimens has been subjected to cyclic loading at four pre-defined stress points, namely 0.01, 0.2, 0.4 and 0.75 MPa for Soil-A and five pre-defined stress points, namely 0.01, 0.2, 0.4, 0.7 and 1.0 MPa for Soil-B. In the case of cyclic loading, the specimen is loaded at a constant rate of displacement (10 μm/s) until the pre-defined stress point is reached, beyond which the specimen is unloaded before reaching higher pre-defined stress points. At each pre-defined stress, specimen is unloaded to a minimum load of 0.01 MPa, so that the specimen and the platen does not lose contact with each other. The axial strain and

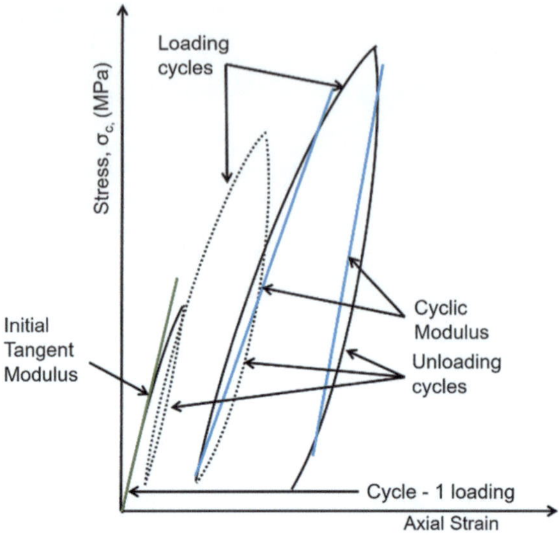

Fig. 18.5 Schematic representation of calculating initial tangent modulus and cyclic modulus

lateral deformation of the specimens were monitored using three extensometers and three LVDTs mounted at specimen's mid-height region. The modulus of elasticity for each loading and unloading cycle has been calculated by drawing a best-fit linear line as shown in Fig. 18.5. Modulus of elasticity calculated at first pre-defined stress point is defined as the initial tangent modulus (ITM). The average of the modulus calculated for each loading and unloading cycles from second pre-defined stress point till the final loading cycle is defined average cyclic modulus (ACM). To avoid damage to extensometer and LVDTs, they were demounted when the specimen reaches approximately 75% of the failure load.

18.4 Specimen Manufacturing and Conditioning

The Proctor method was chosen to manufacture the USRE specimens, and the main reason to choose Proctor method is due to its close resemblance with in situ ramming process of the rammed earth. The cylindrical specimens with dimensions 160 mm in diameter and 320 mm in height (large cylinder) and 110 mm in diameter and 220 mm in height (small cylinder) were cast. Large cylinders were fabricated within a metallic mould with the help of pneumatic and manual ramming, whereas small cylinders were manually fabricated within an impermeable cylindrical cardboard formwork designed in the laboratory. In both pneumatic and manual methods, ramming of each layer was carried out in circular sequential method as directed by Proctor procedure. In each series a minimum of three specimens were cast and tested. Due to the manu-

Table 18.2 Specimen properties and manufacturing details

Soil	Series	Compaction energy (kN m/m^3)	Manufacturing water content (%)	Dry density (g/cc)
A	A1	600	9.6	1.89
	A2	990	10.8	1.95
	A3	2730	10.2	2.02
	A4		11.1	2.02
B	B1	600	10.6	1.79
	B2	970	9.9	1.84
	B3	2580	11.2	2.02

facturing defects two specimens in Series B1 failed to provide any valuable results. The average dry densities and manufacturing water content of each series is given in Table 18.2. In series A3 and A4, specimens with similar densities were manufactured with manufacturing water content on either side of the OWC.

The ambient climate condition for the experiments was defined as 23 °C and 60% relative humidity. The specimens were manufactured and allowed to air dry in a climate-controlled room at ambient condition. Specimens were allowed to dry for a period of 4–5 weeks, until the constant mass of the specimen was achieved, which is an indication of the state of moisture equilibrium within the specimen. Due to ramming process, the top surface of the specimen will have uneven surface, which was levelled using mortar 24 h before testing.

18.5 Results and Discussion

18.5.1 Density Gradient

The manufacturing process of rammed earth wall will naturally induce a density gradient within the compacted layer. The occurrence of density gradient and the phenomenon were theoretically explained by (Bui and Morel 2009). For experimentally investigating the variation in density within the layers, three USRE specimens in each soil compacted with different energies were considered. Each layer was then cut into two halves (top and bottom portion), which were again cut into three pieces each from top and bottom portion of a layer. The dry density of each piece is then calculated using Archimedes principle. The variation in dry densities within the layer with respect to the compaction energy adopted is shown in Table 18.3. It is evident that, higher the compaction energy higher the density. But the dry density variation with increase in compaction energy reduces to a negligible value (at modified Proctor energy). At in situ, the compaction energy achieved to manufacture rammed earth structure is around 1500–2000 kN m/m^3. With this kind of compaction energy, there

Table 18.3 Density gradient within the USRE layers

Soil		Soil-A			Soil-B		
Compaction energy, kN m/m^3		600	990	2730	600	970	2580
Dry density, g/cc	Top	2.13	2.11	2.15	1.95	2	2.06
	Bottom	1.9	2.02	2.13	1.70	1.84	2.01
Difference in dry density, g/cc		0.23	0.09	0.02	0.25	0.16	0.05
% variation from top to bottom		11	4	1	13	8	2

will have a density variation of less than 5%, which can be still considered as negligible. Therefore, rammed earth walls can be considered as a homogeneous wall in global scale.

18.5.2 *Compressive Strength*

The variation of compressive strength with respect to dry density is shown in Fig. 18.6. The average unconfined compressive strength of Soil-A was found to vary between 2.3 and 3.2 MPa for dry density between 1.89 and 2.02 g/cc. Whereas, for Soil-B, it varied between 2.17 and 4.58 MPa for dry density between 1.79 and 2.02 g/cc. For Soil-A, A4 series specimens demonstrate lower compressive strength than A3 series, which has similar dry density in the range of 2.0–2.05 g/cc. The lower strength could be due to manufacturing water content, which was on the wet side of OWC. This suggests the compressive strength of USRE manufactured on the wet side of OWC

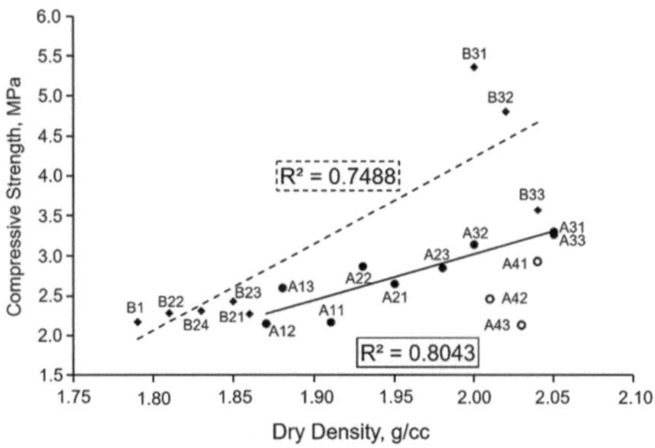

Fig. 18.6 Compressive strength of Soil-A and -B with respect to dry density

will provide lower strength than the dry side of OWC. As indicated by earlier works, the compressive strength of USRE increases with increase in dry density (Fig. 18.6). But it is to be noted that, the compressive strength of Soil-B at higher compaction energy is higher than the Soil-A specimens at similar dry density, this is attributed to higher clay content for Soil-B than Soil-A.

18.5.3 Stress–Strain Relationships

The stress–strain characteristics of both the soil specimens were generated through cyclic compression loading and unloading. Both soil specimens undergo irreversible strain at a very low compressive stress and increases with increase in stress. Even though the clay content in Soil-A is less than Soil-B, Soil-A specimens experience higher irreversible strain.

The USRE cylindrical specimens of Soil-A showed more shrinkage cracks, which interfered in the calculation of stress–strain characteristics. Through expert mason, it was learnt that the USRE walls constructed with Soil-A normally experiences heavy shrinkage cracks. Hence, a decision was made to test the specimens with shrinkage cracks. The stress–strain curves of Soil-A are not presented in this work due to its non-repeatable behaviour. The non-repeatable behaviour of stress–strain curves may be mainly due to shrinkage cracks present in the specimens. It is worth noting that the stress–strain curves generated through cylindrical specimens cored from compressed earth block manufactured with Soil-A produced repeatable behaviour (Champiré et al. 2016). The cylindrical specimens extracted by Champiré et al. (2016) were statically compacted and had the dimension of 64.4 mm in diameter × 140 mm in height without layered phenomena. Even though the non-repeatable stress–strain curves were encountered, the compressive strength reported by Champiré et al. (2016) and the values presented in this study are in close agreement with each other. The Soil-A specimens manufactured using Proctor process is consistent with in situ properties.

The compressive stress–strain curves of Soil-B, which was also manufactured using the Proctor method provides repeatable and accurate results. The stress–strain curves generated for Soil-B specimens are presented in Fig. 18.7. The stress–strain curves presented are limited to around 70% of the peak compressive stress. The initial tangent modulus (ITM) and average cyclic modulus (ACM) calculated are presented in Fig. 18.8. In agreement with compressive strength, ITM and ACM increase with increase in dry density. The difference between the values of ITM and ACM are negligible at low dry densities when compared to values at higher dry density. This indicates increase in damage caused to the material structure at repeated as well as at higher loading conditions.

For understanding the stiffness degradation of the specimen, the mean cyclic modulus of loading and unloading curves at each pre-stress interval is plotted in Fig. 18.9. It can be seen that the rate of stiffness degradation increases for the specimens with higher dry density and compressive strength. The compressive stress of 1.5 MPa is around 70% of the failure compressive strength for specimens in the series B1 and

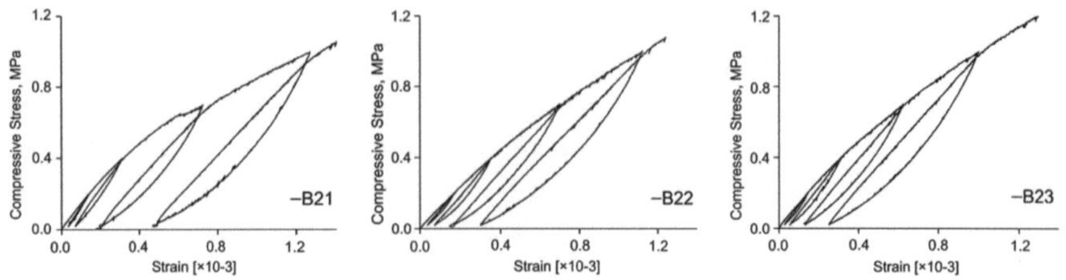

Fig. 18.7 Compressive stress–strain curves of Soil-B2 series

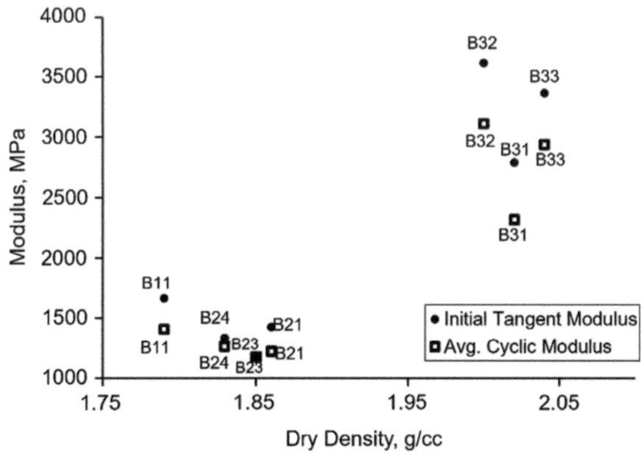

Fig. 18.8 Initial tangent modulus and average cyclic modulus variations with respect to dry density

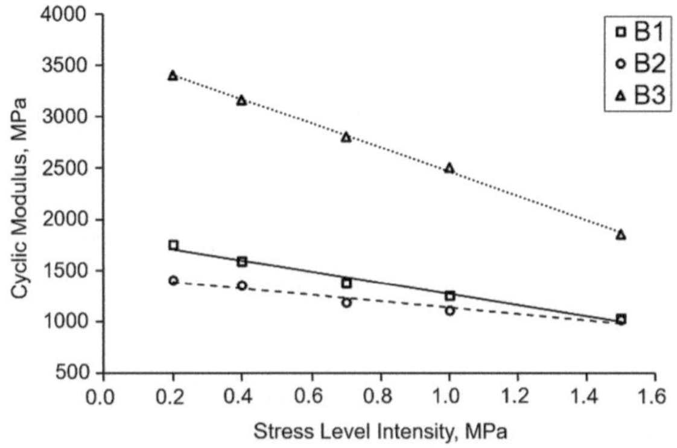

Fig. 18.9 Stiffness degradation with respect to pre-defined compressive stress points

B2, whereas it is around 30% for series B3. Assuming stiffness degradation varies linearly, the cyclic modulus of B3 series at 70% of failure compressive stress should be around 250 MPa, which indicates heavy damage at higher stress level. The change in modulus at each incremental pre-stress indicates the irreversible strain the material had undergone.

The difference in the Poisson's ratio measured for specimens with different dry densities is negligible. The average Poisson's ratio calculated for Soil-B specimens was 0.23, which is in agreement with the Poisson's ratio of USRE reported in earlier studies.

18.6 Conclusion

The main focus of the study was to generate strength and stress–strain characteristics of USRE using two local soils, where the specimens were cast using Proctor method. The following conclusion can be drawn from the results discussed.

- Density and compaction energy as well as density and modulus of USRE are linearly related.
- There is a variation in the density across the depth of the compacted layers, and it varies between 1 and 11%. The difference in density across the layer thickness is negligible at higher compaction (modified Proctor energy).
- The stiffness is sensitive to the stress level. The cyclic modulus decreases with increase the stress level.
- The Proctor compaction method adopted for manufacturing USRE specimens provide more homogeneous specimens across the layers with control on dry density.

Acknowledgements Authors would like to thank the Indo-French Centre for the Promotion of Advance Research (IFCPRA/CEFIPRA) for their support and aiding first author scholarship under the project CEFIPRA project:4608-1, and the support of French Research National Agency (ANR) through the "Villes et Bâtiments Durable" programme (project PRIMATERRA no. ANR-12-VBDU-0001). Authors would also like to thank Mr. Stephane Cointet for providing technical aid in the laboratory works.

References

ASTM D 5298-03 (2003) Standard test method for measurement of soil potential (suction) using filter paper, vol 11

Beckett C (2011) The role of material structure in compacted earthen building materials: implications for design and construction. Durham University

Bruno AW, Gallipoli D, Perlot C, Mendes J (2017) Mechanical behaviour of hypercompacted earth for building construction. Mater Struct [Mater Constr] 50(2). https://doi.org/10.1617/s11527-017-1027-5

Bui Q-B, Morel JC (2009) Assessing the anisotropy of rammed earth. Constr Build Mater 23(9):3005–3011. https://doi.org/10.1016/j.conbuildmat.2009.04.011

Champiré F, Fabbri A, Morel J, Wong H, McGregor F (2016) Impact of relative humidity on the mechanical behavior of compacted earth as a building material. Constr Build Mater 110(in revision):70–78. https://doi.org/10.1016/j.conbuildmat.2016.01.027

Ciancio D, Jaquin P, Walker P (2013) Advances on the assessment of soil suitability for rammed earth. Constr Build Mater 42:40–47. https://doi.org/10.1016/j.conbuildmat.2012.12.049

Ferreira JP (2016) Influence of temperature on the sorption-desorption curves of earth- based materials and consequences on their hygrothermal behaviour. Faculdade de Ciencias Tecnologia da Universidade NOVA de Lisboa

Gerard P, Mahdad M, Robert McCormack A, François B (2015) A unified failure criterion for unstabilized rammed earth materials upon varying relative humidity conditions. Constr Build Mater 95(2015):437–447. https://doi.org/10.1016/j.conbuildmat.2015.07.100

Gomes MI, Gonçalves TD, Faria P (2014) Unstabilized rammed earth: characterization of material collected from old constructions in South Portugal and comparison to normative requirements. Inter J Archit Herit 8(2):185–212. https://doi.org/10.1080/15583058.2012.683133

Maniatidis V, Walker P (2003) A review of rammed earth construction. University of Bath. Retrieved from http://staff.bath.ac.uk/abspw/rammedearth/review.pdf

NZS-4297 (1998) Engineering design of earth buildings. Standard New Zealand

Reddy BVV, Lal R, Rao KN (2007) Optimum soil grading for the soil-cement blocks. J Mater Civ Eng 19(2):139–148

Rigassi V, Craterre E (1995) Compressed earth blocks : manual of production. Deutsches Zentrum für Entwicklungstechnologien—GATE, vol I

Soudani L (2016) Modelling and experimental validation of the hygrothermal performances of earth as a building material. Université de Lyon, ENTPE

Walker P, Keable R, Martin J, Maniatidis V (2005) Rammed earth design and construction guidelines. BRE Bookshop, WATFORD

Walker P, Stace T (1997) Properties of some cement stabilised compressed earth blocks and mortars. Mater Struct [Mater Constr] 30(203):545–551. https://doi.org/10.1007/BF02486398

Chapter 19
Studies on Strength Development of Geopolymer Stabilised Soil-LPC (Lime-Pozzolana-Cement) Mortars

P. T. Jitha, B. Sunil Kumar and S. Raghunath

19.1 Introduction

The advantage of structural masonry is time-tested and attractive in terms of aesthetics and economics. Masonry units and mortars are the important structural components in masonry and are known to significantly influence the strength and performance of masonry. In the current scenario, in India, cement mortar is being extensively used as mortar in almost all masonry application. However, the use of ordinary Portland cement (OPC) alone as a binder in mortar is known to be inefficient both from workability and performance aspects. Hence, masonry codes encourage the use of lime and its products in mortar. They are also compatible with a wide variety of masonry units. Of late, there have been extensive studies in trying to reduce quantity of OPC in mortar by way of using industrial by-products and other cementitious materials. There have been extensive studies on use of lime, lime-surkhi, ground granulated blast furnace slag (GGBS), fly ash etc. as cementitious materials in mortar. The studies have been extended by using alkaline activated materials as binders, also known as geopolymers.

While workability and water retentivity are important parameters to study the properties of fresh mortar, compressive strength, tensile strength, shear strength and water absorption are some of the important parameters to be studied after hardening

P. T. Jitha (✉) · S. Raghunath
Department of Civil Engineering, BMS College of Engineering, Bangalore, India
e-mail: jithapt.cv17@bmsce.ac.in

S. Raghunath
e-mail: raghu.civ@bmsce.ac.in

B. S. Kumar
Cushman & Wakefield India Private Limited, Bangalore, India
e-mail: Sunilkumar.B@cushwake.com

B. V. V. Reddy et al. (eds.), *Earthen Dwellings and Structures*,
Springer Transactions in Civil and Environmental Engineering,
https://doi.org/10.1007/978-981-13-5883-8_19

of mortar. The following section briefly discusses some of the studies carried out on lime and geopolymer-based mortars.

In a very recent study by Jyothi (2017), an attempt has been made to stabilise soil by the combination of geopolymerisation and pozzolanic reaction. It is highlighted that LPC can be used along with alkaline solution instead of industrial by-products to produce feasible alternative to conventional masonry units. The formation of C-S-H gel as secondary reaction product is reported.

It has also been summarised that;

- The presence of calcium hydroxide ($Ca(OH)_2$) along with clay mineral like metakaolin leads to formation of C-A-S-H and N-A-S-H gels. The addition of alkali activators influences this gel formation and leads to the formation of geopolymeric network and C-S-H gel.
- The amount of polymerisation and hydration is dependent on the concentration of alkaline activators and percentage of $Ca(OH)_2$.

Quite a few researchers (Provis and Deventer 2009; Nair et al. 2006; Alonso and Palomo 2001; Jagadish and Yogananda 1984) have all reported that there is a scope to find low energy intensive and cost-effective sustainable alternative building material such that binders/blocks with LPC along with alkaline solution and other industrial by-products.

19.2 Objectives of the Study

Generally conventional geopolymer mortar is produced by thermal activation of alkali-activated products. In the present study, an attempt has been made to activate geopolymer-based mortar by combination curing process which includes solar curing without moisture loss. This is to ensure thermal activation for geopolymeric action and hydration for LPC. Following are the main objectives of the present study.

- To study the strength gain of soil-based LPC-industrial by-products-geopolymer mortar.
- To study the applicability of above mentioned mortar in load bearing masonry.

19.3 Details of the Study

19.3.1 Ingredients and Physical Properties

Soil—The soil used in the study is procured locally from two different sources, and the physical characterisation of the same is done as per the IS 1498-1970 classification as shown in Table 19.1. The soil is found to be a low expansive one, and the clay type is

Table 19.1 Physical properties of soil

S. No.	Parameters tested	Soil	
		Source 1	Source 2
1	Specific gravity	2.7	2.7
2	Liquid limit	34	43.73
3	Plastic limit	16.63	27.65
4	Shrinkage limit	14.84	22.23
5	Plasticity index	17.73	16.08
6	Free swell ratio	1.05	1.02
7	Sand content (%)	14	17.09
8	Silt content (%)	36	41.45
9	Clay content (%)	50	41.46
10	Type of soil based on IS classification	CH—fine grained, highly compressible clay	CI—clay of intermittent plasticity

mixture of swelling and non-swelling, i.e. mixture of Kaolinitic and Montmorillonitic minerals (Prakash and Sridharan 2004).

The lime-pozzolana-cement is produced in lab condition in the custom-made kiln. The process of manufacturing can be explained as following: soil and calcium carbonate ($CaCO_3$) in 1:1 proportion is used in the production of lime pozzolana cement. Soil is dry sieved through 600 μm sieve. Then, the finer soil and $CaCO_3$ are dry mixed first, after which required quantity of water is added and all the constituents are thoroughly mixed using pan mixer. This mix is then spread on an acrylic sheet to a thickness of 10 mm, cut to suitable size (100 × 100 mm) to make briquettes. It was allowed to dry in ambient condition and stored. These briquettes are burnt in a purpose-built kiln in lab at a temperature of 600–900 °C for 3–4 h (Jyothi 2017). The burnt briquettes are then powdered to a size less than 90 μm and stored in air tight containers.

The physical characteristics of LPC as per IS 4098-1983 detailed in Table 19.2.

The sand used as one of the ingredients in the mortar is locally procured which is confirming to the zone II as per IS 383-1970.

The industrial by-product, fly ash is from Raichur thermal power plant and is conforming to the code IS 3812:1982 requirements.

The ultra-fine slag is procured from a commercial supplier and the fineness is observed to be 100% when the wet sieve analysis using 45 μm is done. It consists of fine ground slag with significant amount of free lime.

Alkaline solutions used for the investigation are sodium hydroxide (NaOH) and sodium silicate (Na_2SiO_3). Earlier studies by Jyothi (2017), concentrated on 8 M geopolymer solution and for the present study, sodium hydroxide–sodium silicate combination is used in the ratio of 1:2.5 and 1:1 of 2 M and 4 M solutions and optimised to 2 M, 1:1 proportion.

Table 19.2 Physical properties of the LPC

S. No.	Parameters tested	Values	Requirements
1	Specific gravity	2.58	–
2	Fineness by air permeability test	844.6 kg/m^2	Minimum 250 kg/m^2
3	Initial Setting time	4 h	Minimum 2 h
4	Final setting time	24 h	Maximum 48 h
5	Normal consistency	80%	–
6	Soundness test	6–3 mm	<10 mm
7	7 days Average wet compressive strength of mortar cubes of 50 mm size (1LPC:3Sand)	0.73 MPa	Minimum 0.3 MPa for LP7 grade LPC
8	28 days Average wet compressive strength of mortar cubes of 50 mm size (1LPC:3Sand)	1.25 MPa	Minimum 0.7 MPa for LP7 grade LPC

Two times molecular weight of NaOH (40 g), i.e. 80 g of NaOH is dissolved in 1 l of distilled water to obtain 2 M solution. As the proportion of $NaOH:Na_2SiO_3$ is selected as 1:1, 1 l of Na_2SiO_3 is added to the NaOH solution to prepare the alkaline solution and thoroughly mixed. The solution was prepared one day prior to casting of specimen since the mixing of sodium hydroxide and sodium silicate is exothermic and to bring it to room temperature.

19.3.2 Process

The preparation of test specimen includes mixing of ingredients, casting and curing.

The production process includes varying the proportion of the ingredients as whole or part and using low-energy intensive method. The mix proportion is shown in Table 19.3.

Table 19.3 Mixes along with their designations

GP Mix	Materials used (%)				
	Soil	Sand	LPC	Fly ash	Ultrafine slag
GP-I	80	–	10	10	–
GP-II	40	40	10	10	–
GP-III	45	45	–	–	10
GP-IV	40	40	10	–	10
GP-V	90	–	5	–	5

Fig. 19.1 Mortar sample curing

The measured quantity of materials is dry mixed in the planetary mixer, subsequently added the alkaline solution for few minutes. The wet mix is filled into different moulds based on test requirements. The standard 70 × 70 × 70 mm steel cube moulds are used for compressive strength tests sample preparation.

Elevated curing or heat curing is done usually for geopolymer-based samples for the combination curing. Moisture trapped heat curing under sun is availed in this study. The specimens were kept in water tight plastic bags so as to ensure the prevention of moisture loss and kept in sun to avail day time ambient temperature.

The process of curing the test samples is shown in Fig. 19.1.

19.3.3 Experiment Details and Discussion

The listed experiments, as shown in Table 19.4 are carried out to check the usability and strength characterisation of mortar.

Wet Compressive strength: The results for all the 5 mixes are provided in Table 19.5. The comparison of the details is done with that of cement mortar of 1:6 (CM 1:6). It is clear that GP-III has outperformed all other mixes.

Flexural strength: Test specimens were cast for all the five mixes and were tested after 28 days on CBR machine. The flexural strength of GP-III is comparable with that of CM 1:6.

Tensile strength: Briquettes are cast by placing the mortar in the mould, and tensile strength is determined after 28 days. The tensile strength of GP-III and GP-IV is comparable/more than that of cement mortar.

Properties of bricks: The burnt clay bricks of size 230 × 105 × 75 mm which are procured locally are used for tensile adhesion test and triplet test. The average

Table 19.4 List of experiments with code specifications

S. No.	Test	Code specifications
1	Workability test (flow test)	IS 5512-1983
2	Initial and final setting times	IS 8142-1976
3	Wet compressive strength (7 and 28 days)	IS 2250-1981
4	Water absorption test	IS 2386 Part 3-1997
5	Flexural strength	ASTM C348-2014
6	Tensile strength	ASTM C307-03-2012
7	Dry and wet mortar densities	IS 2386 Part 3-1997
8	Tensile adhesion test (cross-couplet test)	ASTM C1072-13e1-2013
9	Triplet test	BS: 5628-1-2005

Fig. 19.2 Cross-couplet test tested sample

compressive strength and water absorption are determined as per IS codes, and the values observed are 7.5 Mpa and 18%, respectively.

Tensile adhesion Test (Cross-Couplet test): Tensile adhesion test can be determined by cross-couplet test, the test set-up (Lisaragga and Gavilan 2017) and custom-made steel brackets for testing are as displayed in Fig. 19.2. For the test in CBR testing equipment, loads are applied, at a load rating of 1.25 mm/min, till the failure of brick–mortar interface. In all the samples of GP-III mix, it was observed that failure did not occur at brick–mortar interface; instead, the brick had failed and it is displayed in Fig. 19.2.

Triplet test: The test programme consisted of the triplet made up of joint thickness of 10 mm, and three specimens were tested to evaluate the average shear bond strength. In all the samples of GP-III mix, it was observed that failure did not occur at brick–mortar interface; instead, the brick had failed as observed in Fig. 19.3. Keshava (2012), the study on the shear bond strength developed under varying normal stress levels for brick masonry triplets, solid concrete block masonry triplets and hollow concrete block masonry triplets have been done. The shear bond strength for brick masonry triplet is observed as 0.064 MPa against normal stress of 0.025 and 0.392 MPa for a normal stress of 0.5 MPa. The results obtained for GP-III is comparable with the values obtained in the studies by Keshava (2012).

Table 19.5 Test results of all the 6 combinations of the mortars

S. No.	Mix	Geopolymer Mix					CM 1:6
	Tests	GP-I	GP-II	GP-III	GP-IV	GP-V	
1	Wet Compressive Strength, MPa (7 days)	4.1	6.75	8.6	16.1	5.48	5.2
	Wet Compressive Strength, MPa (28 days)	5	6.82	9.7	17.09	5.68	9.42
2	Water absorption, % (7 days)	21.9	15.38	13.89	15.6	16.03	–
	Water absorption, %(28 days)	17.2	13.63	12.65	17.17	20.76	–
3	Flexural Strength, MPa	0.61	0.99	3.05	1.72	0.46	1.93
4	Tensile Strength, MPa	0.49	0.71	1.15	1.02	0.47	0.89
5	Dry Mortar Density, kg/m^3	1930	1998	2064	2083	1800	2100
6	Wet Mortar Density, kg/m^3	1970	2090	2100	2120	1835	2160
7	Tensile adhesion test (cross-couplet test), MPa	0.017	0.0624	>0.1037	0.02005	0.022	–
8	Shear strength from triplet test, MPa	0.008	0.01	>0.3	0.043	0.007	0.0227
9	Workability test (flow test), % of solution	35	30	32	30	37	45 (water)
10	Initial Setting time, min	5	12	22	14	5	30
11	Final setting time, min	17	25	39	27	16	600

Note The results for water absorption (7 and 28 days) and tensile adhesion test for CM 1:6 were not available

Test results: The mortar tests results are tabulated in Table 19.5. Test results are compared with CM 1:6 (Keshava 2012).

Some of the observations can be summarised as

- The wet compressive strength after 28 days of curing is more than 5 MPa for all the geopolymer binders. The least being 5 MPa for GP-I and the maximum being 17.09 MPa for GP-IV. This means that all of them have achieved target compressive strength desired for masonry.
- The flexural strength is minimum for GP-V (0.46 MPa) and maximum for GP-III (3.05 MPa). Flexural strength and the tensile strength of GP-III and GP-IV are significantly more than the commonly used CM 1:6.

Fig. 19.3 Triplet test

- Interestingly, GP-III specimens have the highest tensile adhesion value and shear strength. Indeed, for the masonry specimens made with this mix, it was observed that the failure did not occur at the brick–mortar interface. This high amount of bond strength would be very useful in masonry for resisting in-plane and out-of-plane lateral loads masonry.

19.4 Concluding Remarks

The major objective of the present research work was to develop geopolymer binders by making use of locally available soil and a wide range of pozzolanic materials like fly ash, ultra-fine slag and LPC. The focus has been to develop these binders for possible outcomes in masonry, wherein the compressive strength requirement is about 5 MPa. Thus, it was decided to completely optimise the use of alkaline materials. Based on the experimental investigations, the following broad sets of conclusions are highlighted.

1. Locally available soil with good amount of kaolinitic clay content can be used for making geopolymer binders
2. Addition of 10% pozzolanic materials or even less has resulted in relatively better strengths
3. A combination of LPC and fly ash has led to compressive strength of more than 10 MPa. Interestingly, a combination of LPC and ultra-fine slag have given compressive strength of more than 15.0 MPa
4. It is clearly evident from the results that 2 M NaOH solution of $NaOH:Na_2SiO_3$ of ratio 1:1 is marginally better than 2 M NaOH solution of $NaOH:Na_2SiO_3$ of ratio 1:2.5. It is clear that Na_2SiO_3 has a lesser role to play than NaOH. This is because reactive silica required for pozzolanic action and polymerisation is provided by the binders.

5. Bond strength results for GP-III mix are maximum, but compressive strength results are maximum for GP-IV mix
6. The quick setting of GP-I and GP-V has to be further investigated as it may be due to the quick pozzolanic reaction and polymerisation. The soil content is high in these two mixes compared to other mixes and it is clear that the reactivity of soil plays an important role in geopolymerisation.

Thus, it can be broadly stated that the binders developed from these studies can be conveniently used as an alternative to conventional cement mortar, without a need for pre-treatment and post thermal treatment of the locally available soil.

References

Alonso S, Palomo A (2001) Calorimetric study of alkaline activation of calcium hydroxide-metakaolin solid mixtures. Cem Concr Res 31:25–30

ASTM C1072-13e1-2013—Standard Test Methods for Measurement of Masonry Flexural Bond Strength. ASTM International, West Conshohocken, PA

ASTM C307-03-2012—Standard Test Method for Tensile Strength of Chemical-Resistant Mortar, Grouts, and Monolithic Surfacing. ASTM International, West Conshohocken, PA

ASTM C348-2014—Standard test method for Flexural Strength of Hydraulic Cement Mortars. ASTM International, West Conshohocken, PA

BS: 5628-1-2005—Code of practice for use of masonry. Structural use of unreinforced masonry, British Standards Institution

IS 1498:1970—Classification and identification of soils for general engineering purposes. Bureau of Indian Standards, New Delhi, India

IS 383:1970—Specification for Coarse and Fine Aggregates from Natural Sources for Concrete. Bureau of Indian Standards, New Delhi, India

IS 8142:1976—Method of test for determining setting time of concrete by penetration resistance. Bureau of Indian Standards, New Delhi, India

IS 3812:1982—Specification for Pulverized Fuel Ash. Bureau of Indian Standards, New Delhi, India

IS 4098:1983—Specification for lime-pozzolana mixture. Bureau of Indian Standards, New Delhi, India

IS 5512:1983—Specification for flow table for use in tests of hydraulic cements and pozzolanic materials. Bureau of Indian Standards, New Delhi, India

IS 2250:1981—Code of practice for preparation and use of masonry mortars. Bureau of Indian Standards, New Delhi, India

IS 2386 Part 3-1997—Methods of test for aggregates for concrete. Part 3: Specific gravity, density, voids, absorption and bulking. Bureau of Indian Standards, New Delhi, India

Jagadish KS, Yogananda MR (1984) Production of alternative cements for rural applications. In: Proceedings of regional seminar on farm machinery and rural industries, Chulalongkorn University

Jyothi TK (2017) Strength of lime-geopolymer based blocks using tank-bed soil and brick powder. Research Scholar, Department of Civil Engineering, B.M.S College of Engineering, Bengaluru, India

Keshava M (2012) Behaviour of masonry under axial, eccentric and lateral loading. PhD thesis, Department of Civil Engineering, B.M.S College of Engineering, Bengaluru, India

Lizaragga JF, Perez Gavilan JJ (2017) Parameter estimation for nonlinear analysis of multi-perforated concrete masonry walls. Constr Build Mater 141:353–365

Nair DG, Jagadish KS, Fraaij A (2006) Reactive pozzolanas from rice husk ash: an alternative to cement for rural housing. Cem Concr Res 36(6):1062–1071

Prakash K, Sridharan A (2004) Free swell ratio and clay mineralogy of fine-grained soils. Geotech Test J 27(2):220–225

Provis JL, van Deventer JSJ (2009) Geopolymers: structure, processing, properties and industrial applications. Woodhead Publishing Limited

Chapter 20
Innovations in Construction of Cement-Stabilized Rammed Earth Dwellings Post Bhuj-2001 Earthquake

Kiran Vaghela and Tejas Kotak

20.1 Introduction

Earthquake of magnitude 7.7 struck Kutch region of Gujarat in India on January 26, 2001 (called Bhuj earthquake), causing heavy damage to buildings and basic infrastructure resulting in death of about 20,000 people and causing injuries to hundreds of thousands of people. As a result, about 150,000 houses were damaged spreading across 900 villages and several small towns. It was a daunting task for the local government to provide houses to the affected people within a short time frame. Rescue and rehabilitation operations were initiated by the government in coordination with various NGOs and volunteers coming from across the globe. Hunnarshala Foundation based at Bhuj (India) is a not-for-profit organization embarked upon construction of more than 2000 rammed earth houses in and around Kutch region of Gujarat in India. The paper presents the innovations developed and the methodology adopted to construct 2000 cement-stabilized rammed earth (CSRE) houses within a span of about two years. The paper discusses the number of innovative techniques developed by Hunnarshala in executing the construction of 2000 CSRE dwellings.

20.2 Objectives of Earthquake Rehabilitation Programme

Construction of 2000 CSRE dwellings in a short span of time is a daunting task. Major anticipated hurdles were availability of construction materials and skilled manpower. The objectives for delivering the larger number of dwellings are as follows:

K. Vaghela · T. Kotak (✉)
Hunnarshala, Foundation for Building Technology and Innovation, Bhuj District, Gujarat, India
e-mail: tejas.hunnarshala@gmail.com

B. V. V. Reddy et al. (eds.), *Earthen Dwellings and Structures*,
Springer Transactions in Civil and Environmental Engineering,
https://doi.org/10.1007/978-981-13-5883-8_20

(a) Identifying construction technologies which maximize the use of local raw materials and local skills
(b) Minimum use of energy-intensive technologies
(c) Quicker delivery of durable dwellings respecting the traditional/vernacular house forms
(d) Capacity building through training programmes for unskilled local artisans
(e) Provide opportunities for income generation and livelihood options
(f) Incorporate earthquake-resistant features in the construction of new dwellings.

20.3 Building Typology and Types of Formwork

Two types of building typologies, (a) circular and (b) rectangular plan forms, can be seen in the dwellings in villages and peri-urban areas of Kutch region. Details of these building forms are as follows.

20.3.1 Circular Plan Forms

Since 1819, in the northern part of Kutch region, people have been living in traditional houses that are circular in shape with conical roofs. Generally, these types of circular dwellings are constructed using cob, adobe, wattle and daub walls, and thatch or tile roofs on wooden supporting structure. These types of traditional dwellings are termed as "Bhungas." Such dwellings have an inner clear diameter of about 4.5–5 m, with a 2.0-m wall. The conical roof has a steep slope with a rise of about 2.1 m from the eve height. Such houses have 20-m^2 built-up area. Figure 20.1 shows a traditional Bhunga house.

20.3.2 Rectangular Plan Form

Rectangular plan form was also adopted in the Kutch region of Gujarat state. These types of dwellings are slightly larger in plan area (40 m^2). Such dwellings have flat reinforced concrete slab as well as sloped roofs.

Fig. 20.1 Traditional "Bhunga" house

20.3.3 Types of Formwork Developed for Rammed Earth Work

In order to build circular and rectangular plan houses, two types of metal formwork were developed.

20.3.3.1 Circular Formwork

In order to generate circular house form, a circular formwork was developed to construct the traditional Bhunga houses. The criteria adopted to design the circular formwork are as follows.

(a) Simple and easy to dismantle and assemble
(b) Sustain repetitive use
(c) Should be lightweight and economical
(d) Should facilitate to build the 2-m-height wall in a short period of one day.

Figure 20.2 shows the details of the circular formwork assembly. The formwork has segmental stiffened plates attached to vertical stiffener frames, which control the wall thickness as well as verticality of the wall. The segmental plates are fastened through simple locking pins, and height of the plate is about 0.53 m.

Fig. 20.2 Circular formwork

Fig. 20.3 Rectangular formwork

20.3.3.2 Rectangular Formwork

Figure 20.3 shows the rectangular formwork, which has features similar to the circular formwork. The rectangular formwork has additional stiffened plates to facilitate the wall construction at corners and T-junctions. This formwork was dovetailed to build a house of 40-m^2 built-up area. A battery of such houses was built in several localities.

20.4 Construction Process and Details

20.4.1 Circular Bhunga Houses

The Bhunga houses are individual units and sometimes two houses are built close by, sharing some of the common facilities. Typical Physical dimensions of the house are as follows:

(a) Inner diameter: 4.57 m
(b) Wall height: 2.1 m
(c) Wall thickness: 0.23 m
(d) Conical tile roof with wooden under-structure

The house has three small windows (size: 0.45 × 0.45 m) positioned below sill band level. The Kutch region of Gujarat in India falls under earthquake zone V (IS 1893 - 2002), and hence, the house construction has to adhere to certain earthquake-resistant features following the guidelines of IS 4326 (1993) code. The code suggests sill, lintel and roof bands, and vertical reinforcement bars in the walls. Figure 20.4 shows the details of earthquake-resistant features adopted in the building.

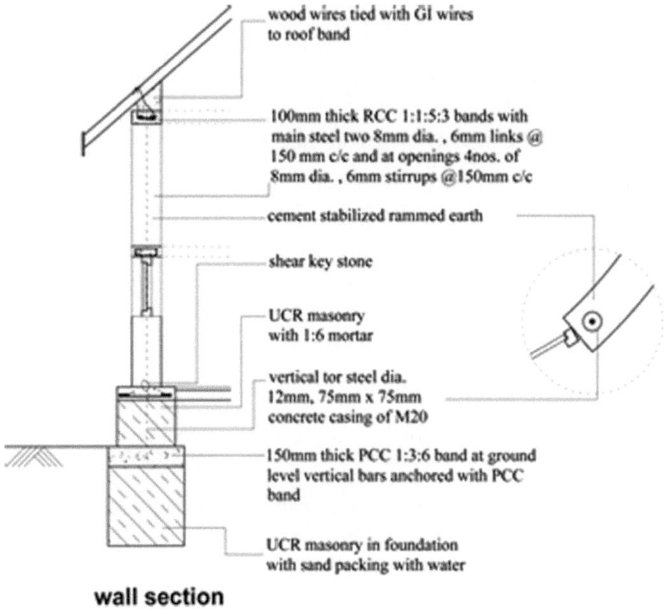

Fig. 20.4 Earthquake-resistant features in Bhunga house

20.4.1.1 Construction Process

Prior to commencement of constructing cement-stabilized rammed earth walls, it was ensured that the following activities were complete:

- Foundation and circular reinforced concrete (RC) plinth beam.
- Sieved soil, sand, and other materials were kept ready.

Walls of circular Bhunga house were constructed in a day utilizing the services of a group of 22 artisans with the following sequence of activities.

- Construction process commences at 8:00 A.M.
- A team of 6–8 artisans including one expert remove the formwork from the previous-day Bhunga construction and assemble the formwork for the new construction. Simultaneously, the other artisans start preparing the dry soil–sand–cement mixture.
- A team of 6 artisans prepare the wet soil–sand–cement mix and pour it into the formwork in layers. The soil is reconstituted to have 70% sand fraction, and the cement content was 8% (by mass).
- Immediately, the ramming of the processed mix commences and was carried out using manual rammers by a team of eight artisans. A loosely filled layer of about 120 mm is rammed into a compacted thickness of about 80 mm. A big rammer (150 mm^2 base, weighing 8.5 kg) and a small rammer (75 mm × 150 mm rectangular base weighing 7.0 kg) were used for the compaction. The compaction density was controlled in a narrow range of about 2000 kg/m^3.
- Two artisans simultaneously commence the preparation of steel for RC work in sill and lintel bands. This team completes the casting of sill concrete before lunchtime.

After a gap of one hour, the formwork for the upper 1 m level (above sill) was installed and ramming started again. The wall reaches 2.0 m level by 5:30 P.M., and then, the lintel band was cast. Curing of CSRE wall commences from the next day and continues for 28 days. Figure 20.5 shows the Bhunga house construction sequence and completed houses.

20.4.2 Rectangular Houses

These are bigger houses when compared to Bhunga houses. The construction technique is similar to that adopted for Bhunga house, but it takes 1.5 days to complete the wall construction which is 3 m height. After the walls are constructed and cured, the flat RC roof is cast. The CSRE walls have exposed surface (form finish) in the exterior side, and the interior surfaces have a coat of paint (Fig. 20.6).

Fig. 20.5 Construction sequence of Bhunga house and completed houses

Fig. 20.6 A typical rectangular building with CSRE walls

20.5 Utilization of Non-organic Solid Wastes for Rammed Earth

In some regions of the Kutch district, finding suitable soil for rammed earth construction was difficult. Non-organic solid wastes such as mine wastes were explored for rammed earth construction. There are factories processing China clay mineral in this region. These industries generate about 30,000 tonnes of solid waste every month. The solid waste basically contains about 70% sand and 30% silt and clay fractions. This is an ideal composition for the construction of CSRE. Since the clay present in the waste is kaolin, it does not pose a problem for constructing CSRE. China clay solid waste was blended with 20% sand and 8% cement and then used for rammed

Fig. 20.7 China clay CSRE wall

Fig. 20.8 200 CSRE houses with China clay waste

Fig. 20.9 Rammed earth sculpture

earth construction. Figure 20.7 shows CSRE wall with China clay waste. Figure 20.8 shows a cluster in which 200 houses were built with China clay solid waste.

20.6 Rammed Earth Sculptures

CSRE was also explored for constructing sculptures. Figure 20.9 shows two such sculptures. In these sculptures, 12% cement (by mass) was used. First rammed earth structure was constructed, and then, the sculpture work commenced after de-shuttering.

20.7 Training and Capacity Building for Quicker Delivery

Quick delivery of houses to disaster-affected people is of high priority after a natural calamity occurs, as majority of them have to stay in temporary shelters. Along with this, the implementing team has to ensure that the supply of raw material and construction quality is consistent. To achieve these, trained personnel are required.

To initiate the training process, a demonstration house was constructed by involving a team of 20 artisans. After the construction process was successfully mastered, these artisans were sent to the villages for constructing the houses. The team composition was as follows:

- **An experienced engineer and a rural expert**: trained to examine soil and recognize changes in soil quality by field tests. If they encountered any changes in the soil quality, it was immediately sent to the laboratory for further testing.

- **Supervisor**: The supervisors were specially trained in stabilized rammed earth technology. Their role was to monitor the construction quality of rammed earth construction as well as other construction activities of "*bhunga's*."
- **Building artisans**: Construction of one Bhunga in a day was possible with a team of 22 artisans. These artisans were specifically trained on construction practices for stabilized rammed earth walls. The training provided knowledge on preparing dry and wet soil mixes, efficiently erecting formworks, proper ramming techniques, testing rammed earth density, carefully de-shuttering the formwork, and curing.

These skilled artisans trained more teams during the process of construction, thus resulting in a multiplier effect. The whole operation of constructing large number of CSRE buildings was coordinated and managed by the Hunnarshala Foundation.

20.8 Concluding Remarks

It was an exciting experience in first learning about CSRE construction and then building the capacity to deliver 2000 rammed earth houses in span of two years to rehabilitate the earthquake-affected families. Involvement of local people in the processes such as development of formwork and rammed earth construction was generating income to local unemployed youth and improving the local economy. The unique feature was that unskilled workers could do construction with rammed earth easily after undergoing a specific training on rammed earth construction and the construction process was faster compared to other types of walling systems. Training of building artisans is a continuous ongoing process at Hunnarshala in Bhuj district.

Acknowledgements The whole exercise of CSRE house construction would not have been possible but for the technical inputs and initial training from Prof. K. S. Jagadish, Dr. M. R. Yogananda, and the R&D efforts at Indian Institute of Science Bangalore, India. The idea of converting this experience into a paper for the ISES-2018 conference was mooted and guided by Prof. B. V. Venkatarama Reddy. The authors would like to thank Prof. K. S. Jagadish, Dr. M. R. Yogananda, and Prof. B V. Venkatarama Reddy for their support and encouragement.

References

IS 1893-Part I (2002) Criteria for earthquake resistant design of structures. Bureau of Indian Standards, New Delhi, India

IS 4326 (1993) Earthquake resistant design and construction of buildings—code of practice. Bureau of Indian Standards, New Delhi, India

Chapter 21
Effect of Bamboo Fiber and C&D Waste on Moisture Content and Compressive Strength Relationship for Cement Stabilized Rammed Earth

K. Arpitha

21.1 Introduction

In the late nineties and early twenties, there was a surge in exploring alternate architecture and lifestyles all over India. Today many buildings were seen with rammed earth as a load-bearing wall but the strength and stability of these wall can be increased by adding stabilizer such as cement, lime, and other binder (Bryan 1998). Earthen materials are used for construction of load-bearing walls in various forms such as rammed earth wall, stabilized mud blocks, Cob wall, adobe masonry, and daub construction throughout the world. Earthen materials are energy efficient and a sustainable building material with low carbon footprint. High aesthetic quality, better thermal comfort, energy efficiency, environment friendly, and use of local material are some of the main attraction in favor of earthen material. Unstabilized soils tend to saturate and subjected to erosion but with the help of various soil stabilization techniques we can enhance the strength, quality, and efficiency of earthen material.

Unstabilized rammed earth wall is thicker in section, which will result in reduction of compressive stresses and increase in self weight. It is possible to construct a thin rammed earth wall by adjusting the stabilizer content, soil composition, density, and moisture content. Natural fibers as reinforcement in rammed earth will enhance their workability and strength. The disadvantage of using straw fiber and other natural fiber stabilization is that the compressive strength of soils decreases as the content increases (Minke 2000). But bamboo is a long growing grass in longitudinal direction and its fibers can be used to increase the shear strength and flexural strength of rammed earth material. To stabilize earthen material and to achieve required strength,

K. Arpitha (✉)
ACS College of Engineering, Bangalore, India
e-mail: arpithagowda.k@gmail.com

B. V. Reddy et al. (eds.), *Earthen Dwellings and Structures*, Springer Transactions in Civil and Environmental Engineering, https://doi.org/10.1007/978-981-13-5883-8_21

soil is generally reconstituted using sand, but it can also be replaced with C&D waste having gradation similar to sand. This will reduce the need of natural sand to reconstitute soil in CSRE, and recycling of building demolished waste is one of the main themes in Green rating system. Furthermore compared to sand, C&D aggregate has lower production cost and it is an eco-friendly way to recycle construction waste without exhausting the natural resource. The most economical way of increasing the strength of rammed earth will be with the use of bamboo and C&D waste, when compared to other alternative materials as additives. A brief review of the literature on the stabilized rammed earth is presented below.

Burroughs (2006), he inspected different soil sample for rammed earth under unconfined compressive strength by varying the amount of added additives by 0–6% per dry soil weight. In this study, authors carried out tests with a large number of 104 different soil samples by varying the percentage of stabilizer such as OPC, lime, and asphalt emulsion as stabilization material. Each indivisible soil samples were subjected to series of test to know their physical properties such as gradation, LL (liquid limit), PL (plastic limit), and linear shrinkage. By this one can identify the properties of soil as favorable or unfavorable for different stabilization technique. By conducting enormous tests, authors suggested that the soil with unconfined compressive strength (UCS) of over 2 Mpa is favorable for rammed earth construction for which soil should consist of more than 65% of sand, plasticity index (PI), and shrinkage limit of less than 15 and 6%. The soil samples over 2 Mpa unconfined compressive strength exhibit less strength when stabilized compare to unfavorable soils (<2 MPa), where unfavorable soils (<2 MPa) exhibited more strength when stabilized, but their compressive strength still did not meet the 2 MPa criteria.

Jiang et al. (2010) compared the strength of polypropylene fiber of different lengths and percentages, which were used to reinforce 14 different soil samples. He conducted test on 14 samples which were not rammed earth samples, but his study shows how polypropylene fibers can change the strength properties of soil. Polypropylene fiber was added at rate of 0.1, 0.2, 0.3, and 0.4% per dry weight of soil by varying length and evaluated the unconfined compressive strength and the shear strength using the direct shear test. This work shows fiber less than 0.6 inches or 1.5 cm length, and 0.3% per dry weight of soil increase the strength and stability of soil by increasing compressive strength and shear strength compare to longer polypropylene fibers. A part of his work shows effect of different sizes of aggregate in the fiber reinforced soil mixture. Where he evaluated that the angle of internal friction highest for 3.5 mm aggregate which, as it shows increasing in the aggregate size lowered the unconfined compressive strength (UCS). Therefore cohesion decreased as the aggregate size increased.

Reddy and Kumar (2011) conducted a study on full-scale rammed earth walls to examine the ultimate compressive strength. This study evaluates the relations between the strengths of small samples to full-scale walls of 3 different-sized specimens measuring 150 × 150 × 300 mm prism, 625 × 150 × 700 mm small wall, and 750 × 150 × 3000 mm wallet. The results shows that the prism had the highest average compressive strength compare to the small wall, and the wall of size 750 × 150 × 3000 mm exhibits the lowest strength. Slenderness ratio is the contributing factor toward the strength of each specimen, where walls were the most slender and this leads to buckling as the walls deflect laterally when loaded. The higher compressive strength of soil sample is evaluated on the wet-side the optimum moisture content curve, where 3 different samples are tested under fully saturated condition same as recommended since this approach is done in testing of masonry.

21.2 Methodology

The strength under compression and moisture content of CSRE specimens were examined for small prisms of size 15 cm × 15 cm × 30 cm. The specimens were examined while varying the moisture content and trying to evaluate the strength of bamboo fiber and C&D aggregate at dry and saturated condition (Fig. 21.1).

a. Bamboo fiber (12–8 months old) is mechanically extracted and cleaved into small pieces by the roller. The pin-roller looser was used to extract small pieces into coarse fibers before removing fat by the boiler at 90 °C for 10–12 h. Then, the fibers were put into the dehydrator and finally dried in the rotary dryer and cut manually for required length (*l*/*d*) (Fig. 21.2).
b. Construction and demolished debris used in this investigation were grained by the roller and kept in oven for 24 h to eliminate the excess moisture. All materials passing by the 4.75-mm sieve are used to reconstitute the soil.
c. First set of samples considered for this study includes reconstituted soil with sand, and 10% of cement with and without (1 and 3%) bamboo fiber.
d. Second set of samples considered for this study includes reconstituted soil with C&D waste, 10% of cement with and without (1 and 3%) bamboo fiber.
e. Three specimens were tested under compression for each case, and the mean values were used in interpreting the result (Figs. 21.2, 21.3, 21.4 and 21.5).

Fig. 21.1 Oven dried bamboo fiber, soil stabilized with cement and C&D

21.3 Material Used in the Investigation

For this study, we have considered a mix proportion of 1:5:5 (cement:soil:sand or C&D) where first sample soil is reconstituted with natural sand (1:1) and for second sample soil is reconstituted with C&D waste (1:1) (from retained material above 0.075 mm sieves to particle passing from 4.75-mm sieve) with and without bamboo fiber at a rate of 1 and 3%. A cement content of 10% (by weight) was used in this study. Reconstituted soil–sand mixture contains 20.1% clay and 67.5% sand material, and approximately same proposition is considered for C&D waste mixture too.

Soil: Finding a favorable soil is an important step in rammed earth construction. The mixture of stabilized rammed earth consists of over 90% of soil and 10% of stabilizer such as cement and lime. The poorly graded soil and soil sample with

Fig. 21.2 Bamboo fiber diameter 1.156 mm, 2.12 and 700 μ

Fig. 21.3 Grinded C&D debris

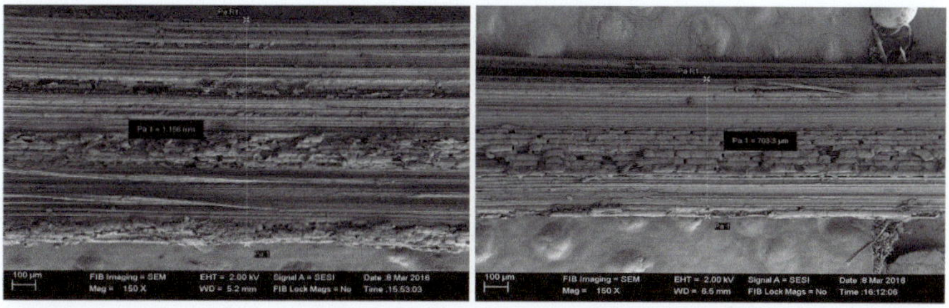

Fig. 21.4 SEM scoping image of bamboo fiber diameter 1.156 mm and 700 μ

Fig. 21.5 Three specimens under compression

higher percentage of clay is considered to be unfavorable soil which requires large quantities of stabilizer to increase its strength (Venkatarama Reddy et al. 2007). Several research papers such as Bryan 1988 and Burroughs 2006 offer guidance on selecting soil sample for rammed earth. The soil mixture is typically about 15–50% clay and silt, 60% sand, and 15% gravel (Burroughs 2008). To know the favorable or unfavorable soil condition, the sample is graded through a stack of sieves with progressively smaller openings (Fig. 21.6). The optimum moisture content of soil sample considered at 1 m depth is 11.9% with a specific gravity of 2.16 and dry density of 1.86 g/cc.

In this study, we carry out geotechnical investigation on to evaluate strength of soil at Nirmithi Kendra near Mandya, where they need to excavate and dump the excessive soil mass of 4–5 m depth to level the site with road way. This investigation includes drilling 150 mm diameter boreholes at two locations using manual auguring (Table 21.1).

C&D waste: Results in terms of strength, workability, and durability tests on certain percentage replacements of fine recycled aggregate were found to be encouraging for part replacement of natural aggregate (Fig. 21.7). The construction and demolition debris as shown in Fig. 21.3 were grinded to its minimum size and sieved in a set of sieve from 4.75 to 0.075 mm to known their particle size distribution (Fig. 21.6).

Table 21.1 Laboratory test results of soil sample

BH No.	Depth (m)	Bulk density (g/cc)	Water content (%)	Grain size distribution (%)					Atterberg limits (%)		
				Gravel	Sand			Silt and clay	Liquid limit	Plastic limit	Plasticity index
					Coarse	Medium	Fine				
1	1	1.78	21	5	6	19	17	53	40	28	12
	1.5			3	1	12	19	65	48	31	17
	2.5			–	1	6	20	73	47	30	17
	3.5			1	1	25	32	41	37	25	12
	4.5			–	–	48	8	44	36	25	11
	6			–	–	12	47	41	35	25	10
2	1	1.75	24	8	–	20	24	48	41	29	12
	1.5			–	1	6	25	68	42	27	15
	2.5			1	–	6	32	61	40	28	12
	3.5			–	–	34	21	45	42	28	14
	4.5			–	–	4	42	54	45	31	16
	6			–	–	3	44	53	44	29	15

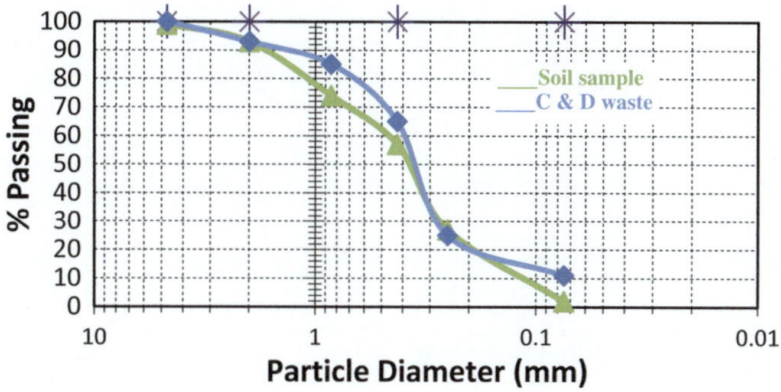

Fig. 21.6 Particle size distribution

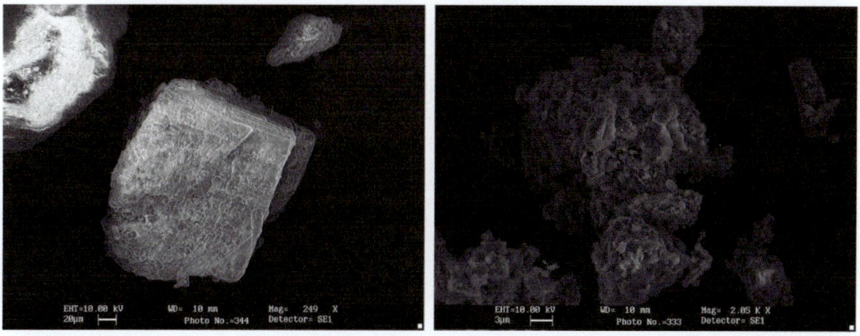

Fig. 21.7 SEM scoping image of natural sand before and after 28 days of curing in rammed earth

The specific gravity of recycled fine aggregate is 2.06 with a water absorption capacity of 12.68% (as it content more fine particle and air voids). SEM scoping image of C&D aggregate (Fig. 21.8) shows a small voids or micropores, which directly affect the density of specimen. As density of prism decrease, the strength decreases by affecting the optimum moisture content and water absorption capacity of sample (Figs. 21.9 and 21.10).

Bamboo: The mechanically extracted bamboo fibers of diameter 700 μ and (*l*/*d* = 20) 1.4 cm long were used (Fig. 21.7), where it helps the earthen material to gain maximum density and increase its durability. Compare to natural fiber such as straw fiber, bamboo fiber imparts more strength due to its cellulous micro structure (Kavitha et.al 2016). Bamboo fiber comes with a disadvantage of having high water intake similar to other natural fiber due to the presence of hydroxyl and other constituents, which leads to low and weak interfacial bonding between fibers and matrix that is usually relatively more hydrophobic as shown in Fig. 21.4.

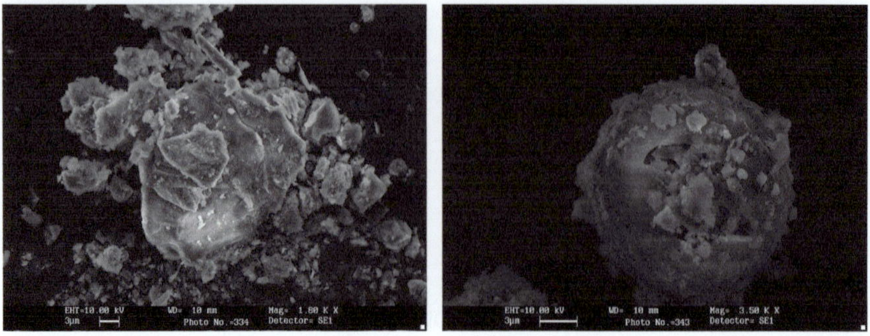

Fig. 21.8 SEM scoping image of C&D waste before and after 28 days of curing in rammed earth

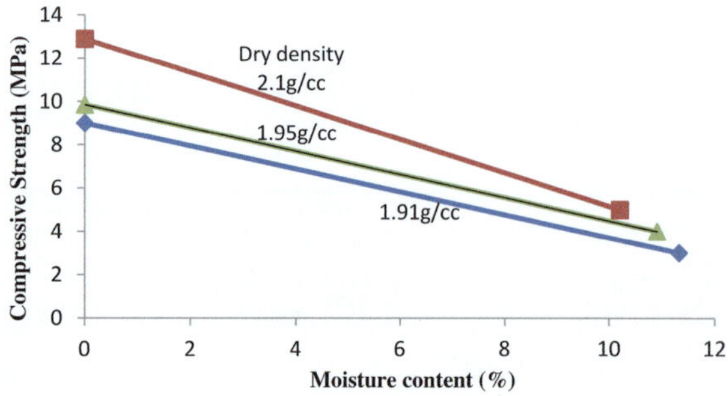

Fig. 21.9 Compressive strength and moisture content relationship for sample A at dry condition and saturated condition

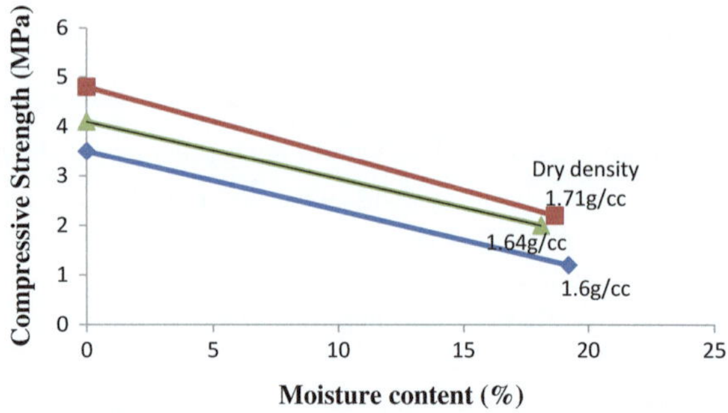

Fig. 21.10 Compressive strength and moisture content relationship for sample B at dry condition and saturated condition

21.4 Experimental Program

The standard proctor test (IS: 2720-Part 7) was performed on the all six samples, and the maximum dry density and optimum moisture content (OMC) of these samples were shown below (Table 21.2).

A prism of size 150 × 150 × 300 mm is prepared by compacting the cement stabilized wet soil mix in three layers of 100 mm each. A uniform mixture was prepared by adding required quantity of water and compacted with a metal rammer to ensure uniform dry density. Compacted specimens were removed

Table 21.2 Types of sample used in experiment

Case	Sample A	Material	Avg. dry density (g/cc)	OMC %	Sample B	Material	Avg. dry density (g/cc)	OMC %
Case 1	SA1	Soil + Sand + Cement	1.91	11.3	SB1	Soil + C&D + Cement	1.6	14.5
Case 2	SA2	Soil + Sand + Cement + Bamboo fiber (3%)	2.10	10.5	SB2	Soil + C&D + Cement + Bamboo fiber (3%)	1.71	13.9
Case 3	SA3	Soil + Sand + Cement + Bamboo fiber (1%)	1.95	11	SB3	Soil + C&D + Cement + Bamboo fiber (1%)	1.64	14.1

from the molds after 24 h from casting and cured under wet burlap for 28 days. The samples A and B were tested for compressive strength for two different moisture conditions. A 28-day cured sample was dried at room temperature for a week and kept in an oven to dry at 70 °C for 2 days to remove excess moisture in sample as moisture content of 0% (case 1), whereas soaking them in water for 2 days will lead to saturated condition (case 2). The prisms were tested under compression in a universal testing machine, and the ultimate compressive strength of the material is noted for all six cases (Fig. 21.5).

21.5 Result and Discussion

Moisture content and compressive strength relationship obtained for prism specimens under dry and saturated condition. As compressive strength of cement stabilized rammed earth is sensitive to the moisture content, as strength decreases when moisture content increases irrespective of the dry density. Figures 21.9 and 21.10 show variation of mean compressive strength with respect to moisture content respectively.

The fall in compressive strength for sample A as the moisture is varied from dry to saturated condition is 66.67, 59.4, and 61.2% for dry density of 1.91, 1.95, and 2.1 g/cc. The natural fiber reinforced CSRE prisms with bamboo gradually increase the strength but due to high water intake due to the presence of hydroxyl at saturated condition it fails achieve target strength.

Sample B, where soil is reconstituted with C&D waste at 1:1 proportion, shows less strength compare to soil reconstituted with natural sand. Adding bamboo fiber to the sample A and B shows increase in strength, but due to finesse and collapsed microstructure of C&D aggregate tend to reduce strength of prism. The high moisture content in the prism tend to decrease compressive strength as it is varied from dry to saturated condition. The variation of moisture from dry to saturated condition majorly affect the sample B with C&D waste as shown in result.

References

Bryan AJ (1998) Criteria for the suitability of soil for cement stabilization. Build Environ 23(4):309–319

Burroughs S (2006) Strength of compacted earth: linking soil properties to stabilizers. Build Res inf. 34(1):55–65

Burroughs S (2008) Soil property criteria for rammed earth stabilization. J Mater Civil Eng. 20(3):264. https://doi.org/10.1061/(asce)0899-1561

Jiang H, Cai Y, Liu J (2010) Engineering properties of soils reinforces by short discrete polypropylene fiber. J Mater Civ Eng. 22(12):1315–1322

Kavitha S, Felix Kala T (2016) Effectiveness of bamboo fiber as a strength enhancer in concrete. Int J Earth Sci Eng 09(03):192–196. ISSN 0974-5904

Minke G (2000) Earth construction hand book—the building material earth in modern architecture. WIT Press, Southampton, UK

Reddy BVV, Kumar PP (2011) Structural behaviour of story-high cement stabilized rammed-earth walls under compression. J Mater Civ Eng. 23(3):240–247

Venkatarama Reddy BV, Lal R et al (2007) Optimum soil grading for the soil-cement blocks. J Mater Civil Eng (ASCE) 19(2):139–148

Chapter 22
Strength and Elastic Properties of Tank-Bed Soil and Lime–Pozzolana-Based Geopolymer Units and Prisms

T. K. Jyothi, S. Raghunath, R. V. Ranganath and K. S. Jagadish

22.1 Introduction

Generally, for low-rise load-bearing masonry structures, the strength requirement is much less than that required in concrete structures. At present, cement-based production of masonry units, the solid and hollow concrete blocks, is relatively wide ranging. Cementitious materials like lime, fly ash (FA), and ground granulated blast furnace slag (GGBS) are used in production of stabilized mud blocks (SMBs) or compressed stabilized earth blocks (CSEBs) and stabilized adobe blocks. There have been extensive studies on these. Indeed, SMB houses are widely accepted in many parts of India, due to the evolution of block making machines as mentioned by Jagadish et al. (2007). In brief, one may state that there is a wide choice of masonry materials available in the current construction scenario. However, geopolymer (GP)-based masonry units/mortar is indeed in a nascent stage, although geopolymer concrete is relatively better understood. For the development of GP-based masonry units/binders, one can think of a variety of raw materials such as FA, GGBS, and red mud soil. However, these materials are industrial by-products, and as such they are not available naturally everywhere. Moreover, there is adequate amount of requirement of usage of

T. K. Jyothi (✉)
Department of Civil Engineering, Government Engineering College, Ramanagaram, India
e-mail: jyothi271016@gmail.com

S. Raghunath · R. V. Ranganath
Department of Civil Engineering, BMS College of Engineering, Bangalore, India
e-mail: raghu.civ@bmsce.ac.in

R. V. Ranganath
e-mail: rvranganath.civ@bmsce.ac.in

K. S. Jagadish
Department of Civil Engineering, Indian Institute of Science, Bangalore, India
e-mail: ksjagadish@gmail.com

B. V. V. Reddy et al. (eds.), *Earthen Dwellings and Structures*,
Springer Transactions in Civil and Environmental Engineering,
https://doi.org/10.1007/978-981-13-5883-8_22

these industrial by-products for the production of concrete and other value-added products. An under-utilized natural resource has been the ubiquitous soil. Another source of raw material for the production of GP-based units/binders is from debris of building which contains huge quantity of burnt bricks. The bricks from construction and demolition (C&D) waste can be potentially used for the production of GP-based units/binders. In a recent study by Varsha (2016), it has been stated that the total volume of masonry walls in a load-bearing masonry building would be about 45% of the total building materials volume. Thus, the demolition of old masonry buildings can potentially produce huge volumes of brick waste. Lime is another raw material which is amply available naturally. Limestones which may not be suitable for the production of ordinary Portland cement (OPC) has a potential of being utilized to produce lime–pozzolana cement at a small scale, thus being insulated by the capital-intensive costs of large-scale production. LPC is a binder which is ideally suited for masonry applications as stated by Rai (1998).

Masonry walls quite often need to be designed for lateral loads in addition to gravity loads. In such situations, the forces carried by the shear walls and cross-walls are to be obtained by apportioning the lateral loads in proportion to the relative stiffness of each wall. In such a situation, it is important to have knowledge of the elastic properties of masonry. Also, for carrying out stress analysis of masonry building for a combination of loads, it is essential to have knowledge of the stress–strain characteristics of masonry. Obtaining the elastic properties of masonry is indeed an arduous task since masonry, as a structural element, is quite complex. Nevertheless, the most commonly referred elastic property, i.e., the modulus of elasticity under compression, is found by conducting tests on masonry prisms. Usually, a stack-bonded masonry prism is evaluated under compressive load. An understanding of the stress–strain characteristics is crucial for evaluating certain design parameters like the 'stress reduction factor'.

It is instructive to have an idea of the ranges of strength and elastic properties of a few masonry types. The compressive strength of a wide variety of table-moulded bricks in India ranges from 4.0 to 18.0 MPa. Generally, the strength of table-moulded bricks from north India is on the higher side. The compressive strength of wire cut bricks ranges from 10.0 to 23.0 MPa. Of late, there has been a wide variety of masonry units being produced (like solid concrete block, cellular concrete block, aerated concrete blocks, and geopolymer block). The compressive strength of such blocks ranges from 2.6 to 24.56 MPa as stated by Keshava (2012), Kishan and Radhakrishna (2013). Thus, the compressive strength of blocks being produced can be considered as moderate strength blocks catering to the requirement of the 2–4 storied, load-bearing masonry buildings. The masonry efficiency is generally found to be higher when the size of the masonry unit is more. Masonry prisms made with table-moulded bricks generally have lower masonry efficiency as reported by Raghunath (2003), Jagadish (2015). The cement-based masonry units and prisms made from cement/concrete blocks generally have a higher modulus of elasticity when compared to that of table-moulded brick reported by Keshava (2012). The information available on the strength and elastic properties of geopolymer masonry is limited.

22.2 Objectives

A combination of lime–geopolymer products can be thought of for masonry applications. Of course, silicates and oxides of sodium are not naturally available materials; they need to be manufactured. On the other hand, one can think of very low quantities of these for the production of GP-based units/binders (at concentrations of 7–8 M). It is rather well known that GP concrete is better suited for precast products than cast in situ products. With this as the background, the following objectives were set:

- To develop process for the production of geopolymer masonry units through the 'adobe' which is the guiding parameter for these studies.
- To study the strength and elastic properties of masonry prisms made with developed masonry units and binder.

22.3 Experimental Programme

In the present investigation, GP-based units were cast with similar dimensions of table-moulded bricks. The size of unit was 230 mm length, 105 mm breadth, and 75 mm height with frog on one side. As per Bureau of Indian Standard specification IS 3495 (part 1-4)-1992, the minimum wet compressive strength of bricks shall not be less than be 3.5 MPa. Based on a series of trials, the mix proportions of cubes which gave wet compressive strength greater than 3.5 MPa were selected for the production of masonry units. Table 22.1 provides the details of the symbols used for raw materials and curing. The following tests were carried out on TBS-BP-L-/LPC-based masonry in order to obtain the critical parameters required for design of masonry.

1. Compressive strength of masonry units
2. Water absorption of masonry units
3. Stress–strain characteristics of masonry units under compressive load
4. Stress–strain behaviour of stack-bonded masonry prisms under compressive load.

22.3.1 *Materials Used*

The raw materials used in this study were obtained from local resources. Soil with different clay contents was procured from a nearby source in Bangalore. Lime–pozzolana cement was produced by using industrial-grade calcium carbonate from a local vendor in Bangalore. The brick powder was procured by crushing table-moulded bricks that are commonly used in and around Bangalore. The bricks were crushed to less than 90μ using a pulverizer. Limestone was procured from Bijapur district of Karnataka, India. It was stored in airtight bins and was slaked as per the procedure given

Table 22.1 Details of symbols used for raw materials and curing

S. No.	Description	Symbol
1	Tank-bed soil (in general)	TBS
2	Low-clay content tank-bed soil	LC
3	Medium-clay content tank-bed soil	MC
4	High-clay content tank-bed soil	HC
5	Brick powder (BP)	B
6	Slaked lime	L
7	Lime–pozzolana cement (LPC)	LP
Note The subscript indicates the percentage of material considered for the mix		
8	Preheat treatment of soil for 3 h at 300 °C	P_1
9	Preheat treatment of soil for 6 h at 300 °C	P_2
10	Curing under ambient conditions for 28 days	A
11	Wet burlap curing for 28 days	W
12	Combination curing for 28 days	CC

in IS 1635-1992. Sodium hydroxide flakes and sodium silicate of industrial grade were used to make alkaline solution to produce geopolymer units/prisms. Table 22.2 gives the details of mix proportions and wet compressive strength of masonry units studied in the present investigation. Sodium hydroxide to sodium silicate ratio was kept at 1:2.5 for all the mixes. The earlier parametric studies tried to understand the influence of curing (thermal and hydration) revealed that a combination of both resulted in better strengths. Thus, combination curing refers to a process where both thermal activation and hydration were provided simultaneously. This was achieved by keeping the specimens in sealed transparent plastic containers and exposing them to solar heating. The sealed bag prevented the loss of moisture and exposure to sun provided the thermal input for geopolymerization.

22.3.2 Preparation of Stack-Bonded Masonry Prisms

The design of unreinforced masonry walls for compressive loads is carried out by two approaches as per IS 1905-1987; they are the (i) block/unit strength approach and (ii) prism strength approach. In both methods, the permissible compressive stress is based on the value of basic compressive stress which is multiplied by stress reduction factor (k_s), the area reduction factor (k_a), and the shape reduction factor (k_p).

The unit strength method is based on the corrected compressive strength of masonry units and the choice of mortar type. This eliminates the need to find out the compressive strength of masonry assemblage. This is generally considered to be conservative. The prism strength approach (outlined in Appendix B of IS 1905-1987)

Table 22.2 Mix proportions and wet compressive strength of bricks

S. No.	Designation	Material proportion				Alkaline solution (%) and molarity	Wet compressive strength (MPa)	Water absorption (%)
		TBS	BP	Lime	LPC			
1	LC_{50}-B_{50}-A	50	50	–	–	30 (7 M)	8.57	1.8
2	P_1LC_{50}-B_{50}-A	50	50	–	–		7.39	2.1
3	P_2LC_{50}-B_{50}-A	50	50	–	–		7.98	1.5
4	LC_{65}-B_{20}-L_{15}-W	65	20	15	–	18 (7 M)	7.95	11.1
5	P_1LC_{65}-B_{20}-L_{15}-W	65	20	15	–		6.85	7.2
6	P_2LC_{65}-B_{20}-L_{15}-W	65	20	15	–		7.42	8.5
7	MC_{50}-B_{50}-A	50	50	–	–	30 (7 M)	5.7	2.01
8	P_1MC_{50}-B_{50}-A	50	50	–	–		4.19	1.99
9	P_2MC_{50}-B_{50}-A	50	50	–	–		3.56	0.89
10	HC_{50}-B_{50}-A	50	50	–	–		7.97	2.05
11	P_1HC_{50}-B_{50}-A	50	50	–	–		8.44	1.87
12	P_2HC_{50}-B_{50}-A	50	50	–	–		7.27	2.14
13	HC_{80}-BP_{10}-LP_{10}-CC	80	10	–	10	30 (8 M)	5.88	0.2
14	HC_{63}-B_{27}-LP_{10}-CC	63	27	–	10		8.8	0.18

Note The quantity of solution was based on flow test

Table 22.3 Mix proportioning for masonry units to cast prism

S. No.	Nomenclature	Material proportions			
		TBS	BP	Lime	LPC
1	LC_{50}-B_{50}-A	50	50	–	–
2	LC_{65}-B_{20}-L_{15}-W	65	20	15	–
3	MC_{50}-B_{50}-A	50	50	–	–
4	HC_{50}-B_{50}-A	50	50	–	–
5	HC_{80}-B_{10}-LP_{10}-CC	80	10	–	10

is considered to be more rational since it takes into account the evaluation of masonry through the testing of assemblage of units and mortar in the form of prism. Thus, the interaction of unit and mortar is taken into account. Thus, it is imperative that the strength of the masonry units, mortar, and prisms needs to be evaluated to prescribe the permissible stresses.

A masonry prism is the assemblage of masonry units with mortar. Compressive strength of prisms was tested by casting prisms using different proportions of tank-bed soil, brick powder, lime, and LPC. It was decided to cast the masonry prisms by considering the relatively better masonry units. Thus, 5 types of prisms were prepared among the 14 types of masonry units tried (Table 22.3). In all the cases, soil without preheating in combination with BP gave good results. Hence, TBS and BP in the proportion 50:50 were selected to cast masonry prisms. For lime-based masonry prism, LC, BP, and lime in the proportion of 65:20:15 proportion were considered, and for LPC-based masonry prism, HC, BP, and LPC in the proportion of 80:10:10 proportion were selected.

The mortar selected for the assemblage of prism for the present work was 75TBS:25BP. The compressive strength of mortar cubes was conducted as per IS 2250-1981. The mortar was tested for its 7 days' wet compressive strength, and its value fell under the category of M_2 mortar as per IS 1905-1987. The amount of alkaline solution required was determined by conducting flow table test. The initial setting time of this mortar was 20 min, and the final setting time was 2 h. The average thickness of mortar in masonry prism was 10 mm. The selected proportions for masonry units to cast prisms have been illustrated in Table 22.3. The process of prism construction has been illustrated in Fig. 22.1.

22.3.3 *Test Procedure to Obtain Strength and Elastic Properties of Masonry Prism Under Compression*

Usually, compression tests on stack-bonded masonry prisms are carried out to determine its properties, namely compressive strength, masonry efficiency (ratio of masonry strength to unit strength), elastic properties. Lourenco (1996) has reported

Units for prism construction | Mortar preparation | Prepared mortar

Casting of prisms | curing of prisms

Fig. 22.1 Process of prism construction

that 'the compressive strength of masonry in the direction normal to the bed joints has been traditionally regarded as the sole relevant structural material property'. The test was conducted as per IS 1905-1987 specifications. The compressive strength of masonry prisms can be conducted by considering that prism should be of height at least 400 mm with *h*/*t* ratio ranging between 2.0 and 5.0. A demountable mechanical strain gauge of 200 mm gauge length was used to measure the strains along the axis of loading. The load at failure was noted to describe the prism strength.

22.4 Results and Discussions

The proportions of ingredients chosen for making masonry units were based on cube strength results. Compressive strengths of the 14 different types of units were all above 3.5 MPa ranging from 3.56 to 8.57 MPa. Five types of stack-bonded masonry prisms were made using TBS-BP-L/LPC units and geopolymer mortar. They indicated compressive strength in the range of 1.81–5.53 MPa with masonry efficiency ranging from 22.7 to 64.5%. Although modulus of elasticity was not needed for routine design of unreinforced load-bearing walls, especially when the working stress method of design is adopted (IS1905-1987), there was the need to specify the modulus for all other purposes such as stress analysis, limit state method of design, and reinforced wall design. The best-fit curves were plotted to obtain stress–strain rela-

Table 22.4 Prism strength, efficiency, and modulus

h/*t* ratio = 3.95							
S. No.	Prism type	Corrected compressive strength		Masonry efficiency (%)	Initial tangent modulus (MPa)	Chord modulus (MPa)	E_m/f_m
		Prism (MPa)	Brick (MPa)				
1	LC_{50}-B_{50}-A (230 × 105 × 415 mm)	5.53	8.57	64.5	6413.9	4075	736.9
2	LC_{65}-B_{20}-L_{15}-W (230 × 105 × 415 mm)	1.81	7.95	22.7	1122.4	1122.4	620.1
3	MC_{50}-B_{50}-A (230 × 105 × 415 mm)	3.14	5.7	55	1137.4	1137.4	362.2
4	HC_{50}-B_{50}-A (230 × 105 × 415 mm)	4.28	7.97	53.7	4141.9	4141.9	967.7
5	HC_{80}-B_{10}-LP_{10}-CC (230 × 105 × 415 mm)	2.68	5.88	45.5	3282.9	2946.2	1099.3

tionship. In addition to initial tangent modulus, the chord modulus of elasticity taken between 0.05 and 0.33 of the maximum compressive strength of each prism was determined. Initial tangent modulus, chord modulus, and masonry efficiency with corrected compressive strength of the prisms have been indicated in Table 22.4, and its stress–strain behaviour plot has been presented in Fig. 22.2.

The revised draft of IS: 1905, 2016, and the draft code of reinforced masonry suggest empirical relation or the modulus of elasticity of masonry as

$$E_m = 550 f_m$$

where

E_m Modulus of elasticity of masonry made of either clay or concrete
f_m Compressive strength of masonry prism.

Alternatively, the modulus of elasticity is defined by the chord modulus taken between 0.05 and 0.33 of the maximum compressive strength of each prism determined in accordance with appendix B of the code.

Table 22.4 provides the compressive strength, masonry efficiency, average initial tangent modulus, and the average chord modulus for the five different types of masonry prisms. As observed from Fig. 22.2, they ranged from 1122.4 to 6413.9 MPa.

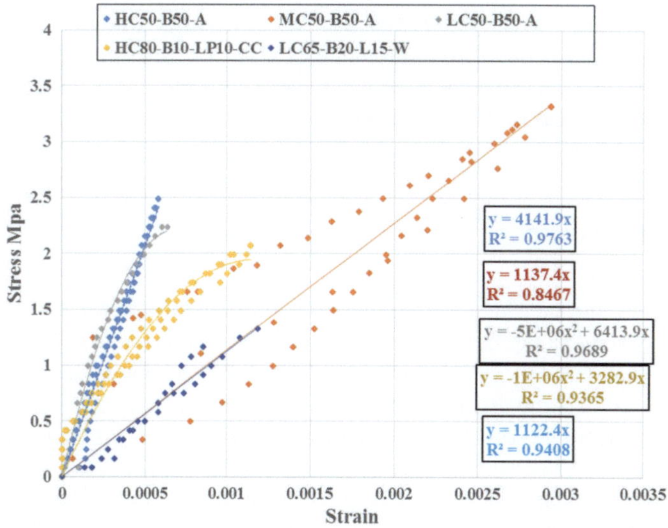

Fig. 22.2 Stress–strain characteristics in geopolymer prisms of different combinations of tank-bed soil, brick powder, lime, and LPC

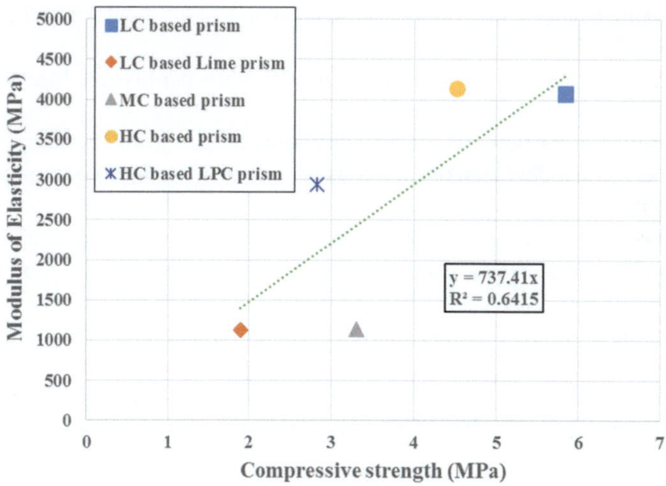

Fig. 22.3 Relationship between compressive strength and modulus of elasticity of masonry prisms

An attempt was made to establish a relation between the compressive strength and the modulus. Figure 22.3 gives a plot of compressive strength versus modulus of prism, and linear relationship was attempted.

The linear relation for brick and prism is:

$E_m = 456.21\, f_b$ for masonry units
$E_m = 737.41\, f_m$ for masonry prism.

Similar plots have been provided by Rai (2016) for bricks, mortar, and masonry. The scatter noticed in the results provided is typical of masonry. Even from the present study, the regression coefficient is rather low. There is the need to carry out further studies to establish better correlations.

22.5 Conclusions

The focus of the present study was to develop a process for the production of masonry units using a variety of ingredients that can be obtained or produced locally. The ingredients were tank-bed soil, brick powder, lime, and LPC. Adobe process was adopted for masonry units. Amongst the fourteen different mixes with different activation processes, five were shortlisted for evaluating them as masonry products. All the five of them appears to be better than the conventional table-moulded brick masonry. The prism strengths indicate that they can be possibly used for load-bearing applications.

Based on the experimental investigation, the following broad sets of conclusions have been highlighted:

- The proportions of ingredients chosen for making masonry units were based on cube strength results. Compressive strengths of the 14 different types of units were all above 3.5 MPa.
- The water absorption of all units was well within the permissible limit of 20% as per IS 1077-1992 (Reaffirmed 2007).
- The modulus of elasticity of the masonry units ranged from 1800 to 6000 MPa, which was much higher than the modulus of similar strength, table-moulded bricks of South India.
- From among the 14 different types of units, five were chosen for evaluating prism strength. The compressive strengths of the stack-bonded prisms were found to be in the range of 1.81–5.53 MPa. This was also much higher than that of the prisms made using table-moulded bricks of South India, with cement mortar. The modulus of elasticity of the masonry prisms ranged from 1122.4 to 6413.9 MPa
- The strength and elastic properties of TBS-BP-LPC masonry appeared to be adequate and comparable to that of other conventional masonry in South India.

References

IS1077-1992 (2007) Common Burnt Clay Building Bricks—Specification. Bureau of Indian Standards, New Delhi, India

IS: 1635-1992 (2003) Field slaking of building lime and preparation of putty—code of practice. Bureau of Indian Standards, New Delhi, India

IS: 1905-1987 (2002) Indian Standard Code of Practice for Structural Use of Unreinforced Masonry. Bureau of Indian Standards, New Delhi, India

IS: 1905 (2016 Draft) Indian Standard Code of Practice for Structural Use of Unreinforced Masonry. Bureau of Indian Standards, New Delhi, India
IS: 2250-1981 (2005) Code of practice for preparation and use of masonry mortars. Bureau of Indian Standards, New Delhi, India
IS 3495 (part1-4)-1992 (2002) Methods of tests of burnt clay building bricks. Bureau of Indian Standards, New Delhi, India, Reaffirmed 2002
Jagadish KS (2015) Structural masonry. I.K. International Publishing House Pvt. Ltd., Bengeluru, India
Jagadish KS, Venkataramareddy BV, Nanjunda Rao KS (2007) Alternative building materials and technologies. New Age International (P) Ltd., Publishers, Bengaluru, India
Keshava M (2012) Behaviour of masonry under axial, eccentric and lateral loading. PhD thesis, Department of Civil Engineering, BMS college of Engineering, Bengaluru, India
Kishan LJ, Radhakrishna (2013) Comparative study of cement concrete and geopolymer masonry blocks. Int J Res Eng Technol 361–365
Lourenco PB (1996) Computational strategies for masonry structures. TU Delft, Delft University of Technology
Raghunath S (2003) Static and dynamic behaviour of brick masonry with containment reinforcement. PhD thesis, Department of Civil Engineering, Indian Institute of Science, India
Rai M (1998) Building materials in India, 50 years: a commemorative volume. Building Materials & Technology Promotion
Rai DC (2016) Masonry in compression. Department of Civil Engineering, Indian Institute of Technology, Kanpur
Varsha BN (2016) Private communications. Research Scholar, Department of Civil Engineering, B.M.S College of Engineering, Bengaluru, India

Chapter 23
Effectiveness of Polypropylene Fibers on Impact and Shrinkage Cracking Behavior of Adobe Mixes

Gerardo Araya-Letelier, Federico C. Antico, Jose Concha-Riedel, Andres Glade and María J. Wiener

23.1 Introduction

The earthen construction materials are characterized by their wide range of applications, availability, recyclability, thermal inertia, acoustic performance, and lower cost and environmental impacts compared to industrialized materials such as fired clay bricks (Cataldo-Born et al. 2016; Donkor and Obonyo 2015; Millogo et al. 2014; Minke 2000, 2006).

Although the widespread use and advantages of earthen materials, their performance is reduced compared to industrialized construction materials in terms of tensile and flexural strength, toughness, fracture toughness, water erosion resistance, and drying shrinkage cracking (Avrami et al. 2008; Minke 2000, 2006). To lessen some of these shortcomings, earthen materials can be reinforced with the incorporation of natural fibers such as straw, sisal, and wool (Aymerich et al. 2012; Galan-Marin et al.

G. Araya-Letelier (✉)
Pontificia Universidad Católica de Chile, Santiago, Chile
e-mail: gerardo.araya@uc.cl

F. C. Antico · A. Glade
Universidad Adolfo Ibáñez, Viña del Mar, Chile
e-mail: federico.antico@uai.cl

A. Glade
e-mail: aglade@alumnos.uai.cl

J. Concha-Riedel
Universidad Adolfo Ibáñez, Santiago, Chile
e-mail: jose.concha@uai.cl

M. J. Wiener
Purdue University, West Lafayette, USA
e-mail: wiener@purdue.edu

B. V. V. Reddy et al. (eds.), *Earthen Dwellings and Structures*,
Springer Transactions in Civil and Environmental Engineering,
https://doi.org/10.1007/978-981-13-5883-8_23

2010; Millogo et al. 2014; Quagliarini and Lenci 2010) and industrialized fibers such as polypropylene and glass fibers (Balkis 2017; Donkor and Obonyo 2015; Yilmaz 2009).

Among industrialized fibers, the use of polypropylene fibers as reinforcement of earthen materials has been mechanically characterized and results are promising. Yilmaz (2009) assessed the compression and split tensile performance of sand-clay mixtures reinforced with different dosages of micro-polypropylene (MPP) fibers finding that the addition of MPP fibers had a limited effect in terms of the final strength characteristics, but MPP fibers increased the span length of the peak deviator stress. Donkor and Obonyo (2015) studied the effect of macro-polypropylene fibers on the flexural and compressive strength and deformability of stabilized earth blocks finding that flexural strength and post crack performance were improved by the addition of fibers. Balkis (2017) evaluated the effect of waste marble dust and MPP fibers on the compressive and flexural strength of gypsum stabilized earthen materials finding optimum combinations of marble dust and MPP fibers where both compressive and flexural strengths are improved compared to plain gypsum stabilized earthen materials. Although these studies have contributed significantly to the investigation and improvement of polypropylene fiber-reinforced earthen materials, there are still some properties such as impact strength and drying shrinkage cracking of MPP fiber-reinforced earthen materials that have not been explored thoroughly.

The novelty of this research resides in addressing some of the relevant benefits that have not been studied exhaustively such as drying shrinkage cracking control and impact strength increment of adding MPP fibers to earthen materials. Moreover, this study introduces two simple experimental procedures to assess distributed and concentrated drying shrinkage cracking reduction generated by fiber reinforcement of earthen materials. Since earthen material is a generic term, this study refers to the mix between clayey soil, water and fibers as adobe mix since it might be used to produce adobe bricks. The objectives of this study are to evaluate the impacts of different dosages of MPP fibers on: (i) the drying shrinkage cracking performance of adobe mixes; and (ii) the impact strength of adobe mixes.

23.2 Materials and Methods

23.2.1 Materials

The soil used for this study was obtained from southern Santiago, Chile. This soil has been previously used and characterized by Araya-Letelier et al. (2018), obtaining a particle size distribution with a content of 11, 69, and 20% of clay, silt, and sand, respectively. Atterberg liquid and plastic limits were also studied, as well as the plasticity index and specific gravity of solids, obtaining a 29.1, 17.4, 11.7, and 2.51%, respectively. Specific gravity of the soil was determined following ASTM D854 (2000).

Table 23.1 Main properties of micro-polypropylene fibers

Length (mm)	Diameter (mm)	Aspect ratio	Specific gravity (20 °C) (g/cm^3)	Elongation at break (%)	Tensile strength (MPa)
12	0.031	387	1.16	60–140	310

Table 23.2 Adobe mix ID number and material proportion

Adobe mix ID	Oven-dry soil (kg)	Water (kg)	MPP fiber (%)	MPP fiber (kg)
0	1,000	307	0	0
0.25	1,000	307	0.25	2.5
0.5	1,000	307	0.5	5
1	1,000	307	1	10

Commercially available "Sika® Fiber P-12" micro-synthetic polypropylene fibers are used in this study. These MPP fibers are used in concrete and mortar to reduce plastic shrinkage cracking and spalling and to improve impact strength and abrasion resistance. The main MPP fibers' properties are given in Table 23.1.

23.2.2 Adobe Mix Proportions and Specimen Preparation

The workability of the adobe was assessed by hand mixing oven-dry clayey soil with portions of potable water until a homogeneous mix was obtained. The weight of water divided by the weight of the oven-dry clayey soil was defined as the water to soil ratio, choosing a value of 0.307.

This study used four different adobe mixes: a plain mix, and three mixes incorporating MPP fibers (0.25, 0.5, and 1% by weight of oven-dry clayey soil), where the fiber-reinforced adobe mixes were compared with that of plain adobe mix. Table 23.2 shows the adobe mix identification (ID) numbers (where the number indicates the dosage, in percentage, of MPP fibers) and the material proportions for each adobe mix.

To mimic real-life adobe manufacturing, the mixture of the materials, as well as the casting of the specimens, was executed manually. To prevent cluster formation, the MPP fibers were gradually added to the soil and mixed, prior to the water incorporation. Following the inclusion of MPP fibers, water was incorporated in four steps, and the mix was carefully homogenized before each step. The preparation of the different adobe mixes was performed in parallel, covering for two hours each one, after the mixing was finished, to promote material uniformity and equal water absorption. Table 23.3 indicates the different specimens cast for each adobe mix.

The casting of the RILEM beam specimens (i), and slab specimens (iii) was carried out in consecutive layers of approximately 20–30 mm, and each layer was compacted

Table 23.3 Types of specimens cast in this study

Specimen's name	Dimensions (mm)	Test	
		Type	Number of specimens per adobe mix
RILEM beam (i)	160 × 40 × 40, with a 5 × 3 notch at midspan	Impact strength at 7 days	3
		Impact strength at 28 days	3
Flat (ii)	180 × 5 (diameter and height)	Restrained drying shrinkage distributed cracking at 7 days	2
Slab (iii)	600 × 600 × 50, with 2 stress raisers	Restrained drying shrinkage concentrated cracking at 24 h	2

with a tamper to reduce the void content of the mixture. The flat specimens (ii) were cast in only one layer. The RILEM beam specimens (i) were demolded 48 h after casting and kept at 22 °C and 45% relative humidity (RH) for 28 days and rotated 90° to the adjacent side every seven days until testing. The RILEM beam specimens (i) were covered with a plastic bag during their first two days after casting to prevent moisture loss. The flat specimens (ii) were kept at 22 °C and 45% RH for seven days, but they were not demolded neither covered with a plastic bag to generate a restrained drying shrinkage condition, and the generated cracks were assessed at seven days after casting. The slab specimens (iii) were kept in their molds and subjected immediately to an accelerated drying shrinkage process discussed in Sect. 2.3 of this paper.

23.2.3 Experimental Testing

To quantitatively evaluate drying shrinkage distributed cracking, two flat specimens were cast for each adobe mix and conserved at laboratory environmental conditions (22 °C and 45% RH) for seven days. The cracks were measured using a 20 × 20 mm grid as guide, a crack width comparator, and a caliper as shown in Fig. 23.1, for each specimen. This procedure has been previously implemented by Araya-Letelier et al. (2018). The values calculated were crack width average (CWA), crack width reduction ratio (CWRR) in accordance with Eq. (1), and crack density ratio (CDR) considering the density of the cracked areas with respect to the total area exposed to drying shrinkage.

$$CWRR_{IDx} = \left(1 - \frac{CWA_{IDx}}{CWA_{ID0}}\right) \cdot 100 \tag{1}$$

where $CWRR_{IDx}$ is the crack width reduction ratio (expressed as percentage) of adobe mix IDx with respect to the adobe mix ID 0, CWA_{IDx} is the crack width average of adobe mix IDx and CWA_{ID0} is the crack width average of the adobe mix ID 0.

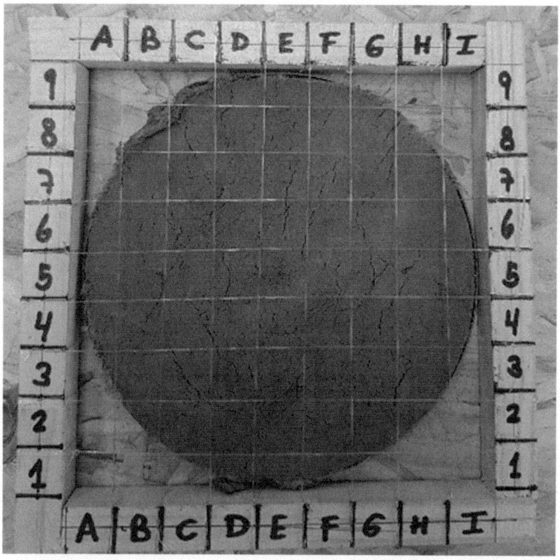

Fig. 23.1 Restrained drying shrinkage distributed cracking procedure

To complement the assessment of the influence of MPP fibers in reducing the restrained drying shrinkage cracking of adobe mixes, a restrained drying shrinkage concentrated cracking test is performed. The slab specimens (iii) were cast into the molds and placed immediately in a room with 42 °C and 20% RH during eight hours and then the specimens were kept at 22 °C and 45% RH for another 16 h. For each specimen, the values of CWA and CWRR were calculated 24 h after casting. The two restrained drying shrinkage cracking procedures presented in this study differ in terms of obtaining a random (distributed) cracking pattern, more representatives of the real working conditions of a material.

The impact strength of each adobe mix was evaluated at seven and 28 days after casting using an approach similar to previous studies (Araya-Letelier et al. 2017a, b, c, 2018), where a sphere projectile is thrown at the center of a specimen, which is supported by a steel frame as shown in Fig. 23.2.

To eliminate damage due to rebound, specimens (i) were attached with silicone to the steel frame, providing also equal support for each one. The total of specimens per adobe mix was six, recording for each one the number of blows required to fracture and collapse it. The total energy at collapse is given by Eq. (2).

$$E_c = n * m * g * h \tag{2}$$

where E_c is the total energy at the collapse, n is the number of blows of the projectile required to collapse the specimen, m is the mass of the projectile (0.047 kg), g is the gravitational constant (9.8 m/s^2), and h is the height of the fall (0.496 m). These

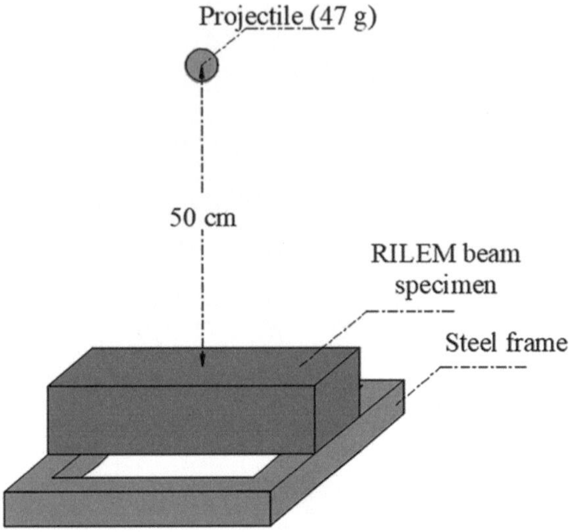

Fig. 23.2 Impact test setup

values were kept constant during the test and, therefore, each blow is equivalent to an impact energy of 0.22 J.

23.3 Results and Discussion

23.3.1 Restrained Drying Shrinkage Distributed Cracking

Figure 23.3 shows one specimen per type of adobe mix at seven days after casting, and it can be seen that a crack width reduction was observed as the MPP fiber dosage increased. Adobe mixes ID 0, ID 0.25, ID 0.5, and ID 1 showed crack widths up to 2.0, 0.85, 0.80, and 0.6 mm, respectively.

The estimated values of CWA and CWRR are shown in Fig. 23.4a in the left and right axis, respectively. It can be seen that values of CWA were reduced and, consequently, the resulting values of CWRR were increased with increasing dosages of MPP fibers. The CWA presented by adobe mix ID 0 (1.1 mm) can be reduced to values that vary between 0.50 mm (ID 0.25) and 0.19 mm (ID 1), which corresponds to CWRR values varying between 54% (ID 0.25) and 82% (ID 1). Overall, as fiber dosage increases the characteristic length between a piece of matrix material and fiber should be statistically reduced and, therefore, there are more chances to have fibers within the bulk that provide crack width control after the matrix develops macroscopic distributed cracks. In addition to reducing the values of CWA and CWRR, the incorporation of MPP fibers also reduces the CDR values as shown in

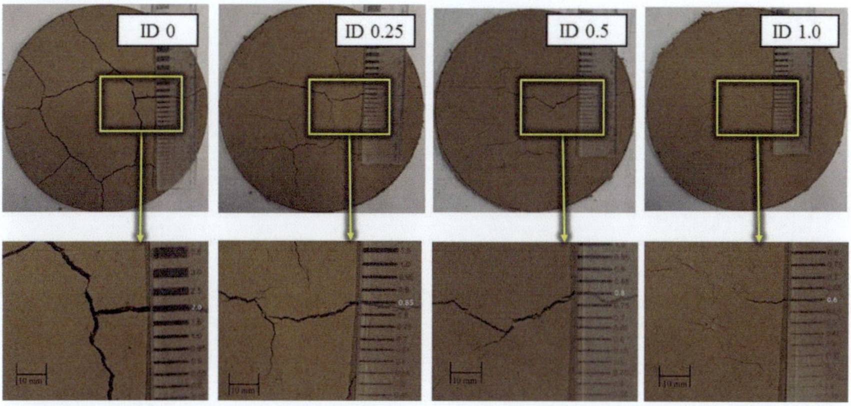

Fig. 23.3 Restrained drying shrinkage distributed cracking results

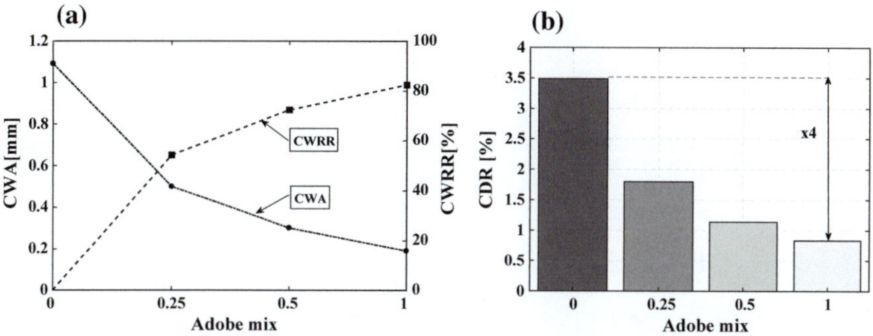

Fig. 23.4 Restrained drying shrinkage distributed cracking: **a** CWA and CWRR results, and **b** CDR results

Fig. 23.4b, where the CDR of the adobe mix ID 0 (3.5%) is four times larger than the CDR value of adobe mix ID 1 (0.84%). Therefore, the incorporation of increasing dosages of MPP fibers reduces both the maximum crack width and the density of cracks due to drying shrinkage when compared to plain adobe.

23.3.2 Restrained Drying Shrinkage Concentrated Cracking

Figure 23.5a presents one cracked slab specimen (iii) per type of adobe mix 24 h after casting and it can be seen that a crack width reduction was observed as the MPP fiber dosage increased. Adobe mixes ID 0, ID 0.25, ID 0.5, and ID 1 showed crack widths up to 6.1, 0.97, 0.88, and 0.71 mm, respectively. The resulting maximum and average crack widths of this test are larger than the corresponding results of the restrained drying shrinkage distributed cracking test for each adobe mix, which

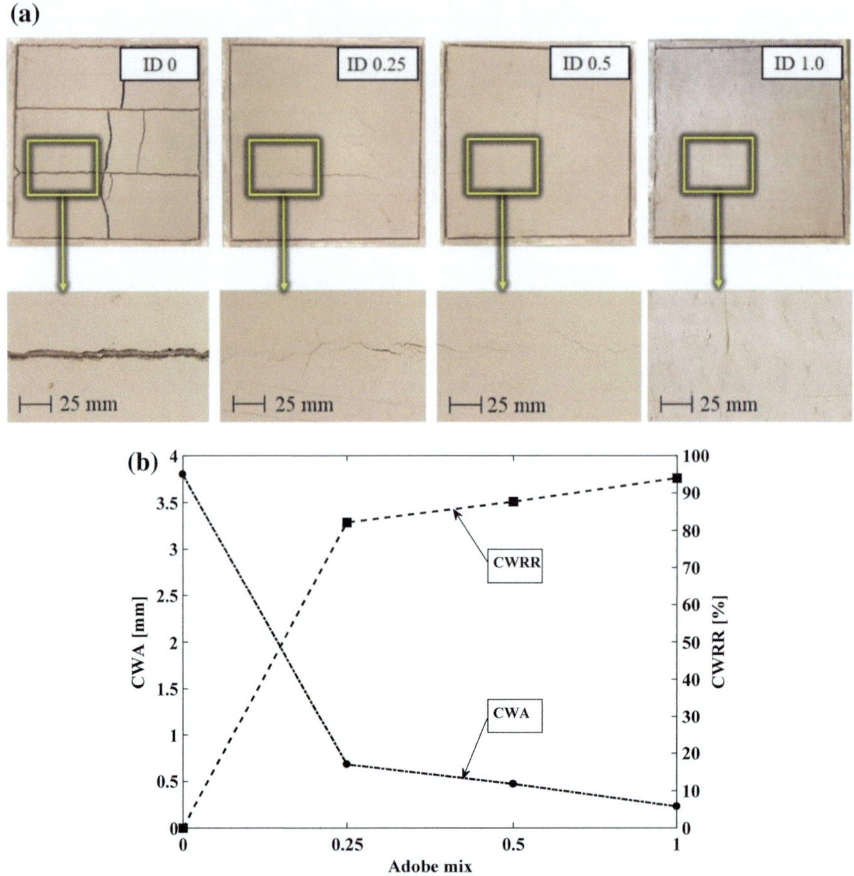

Fig. 23.5 Restrained drying shrinkage concentrated cracking: **a** each adobe mix at 24 h after casting, and **b** CWA and CWRR results

is reasonable due to the use of stress raisers and the exposure to a much more aggressive environmental conditions. However, the increment of the maximum and average crack widths is more significant for adobe mix ID 0 (plain adobe), where the maximum crack width increased from 2.0 mm (distributed cracking test) to 6.1 mm (concentrated cracking test). The MPP fiber-reinforced adobe mixes presented a more thermostable cracking performance compared to plain adobe, where the use of stress raisers and a more aggressive environment in the concentrated cracking test (42 °C and 20% RH) slightly increased the maximum and average crack widths compared to the distributed cracking test (22 °C and 45% RH).

Figure 23.5b shows the estimated values of CWA and CWRR in the left and right axis, respectively, and it can be observed that both, CWA and CWRR, are reduced with increasing dosages of MPP fibers. The CWA presented by adobe mix ID 0

(3.8 mm) can be reduced to values that fluctuate between 0.68 mm (ID 0.25) and 0.23 mm (ID 1), which corresponds to CWRR values oscillating between 82% (ID 0.25) and 94% (ID 1). Overall, as fiber dosage increases the characteristic length between a piece of matrix material and fiber should be statistically reduced and, therefore, there are more chances to fibers within the bulk that provide width control after the matrix develops macroscopic distributed cracks or fibers pullout from the matrix. (Antico et al. 2012; Mindess et al. 2002).

23.3.3 Impact Strength

It is known that manually compacted earthen materials have quasi-brittle behavior (Reman 2004) like other construction materials such as mortars (Araya-Letelier et al. 2017c). In the same way as fiber-reinforced mortars, fiber-reinforced adobe mixes are expected to improve fracture toughness with respect to unreinforced materials and the impact resistance is an effective test to measure the energy-absorbing capacity of fiber-reinforced quasi-brittle materials. At 7 days after casting (Fig. 23.6a), the mean values range from 0.16 (ID 0) to 4.43 J (ID 1), and standard deviation (SD) values range from 0.02 (ID 0) to 1.32 J (ID 1). At 28 after casting (Fig. 23.6b), the mean values range from 0.17 (ID 0) to 13.95 J (ID 1) and SD values range from 0.01 (ID 0) to 3.70 J (ID 1). It can be seen that the impact strength increased as the MPP fiber dosage increased at both ages, but this increment is more significant at 28 days (83 times) than a seven days (28 times). This might be the result of a stronger adobe matrix as well as a stronger bonding between the MPP fibers and the matrix that is developed at later ages. Even the smallest increment in impact energy at collapse, presented by adobe mix ID 0.25, is approximately eight times (at seven days) and 12 times (at 28 days) the required impact energy to collapse adobe mix ID 0. In terms

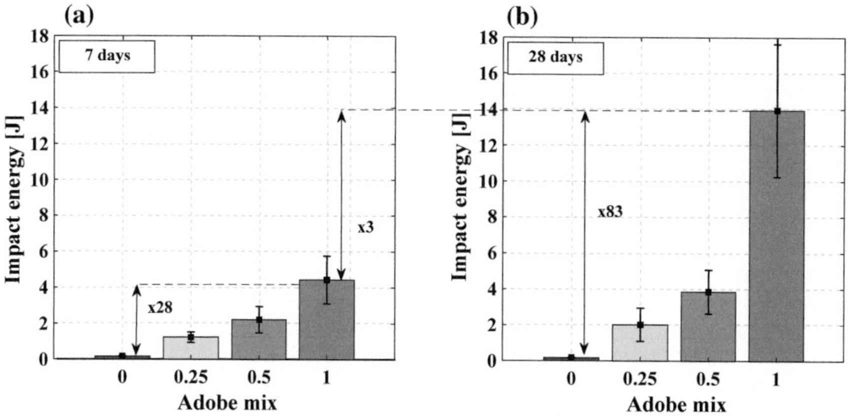

Fig. 23.6 Cumulative collapse impact energy: **a** at 7 days, and **b** at 28 days after casting

of the evolution of impact strength with time, there is a consistent increment from seven to 28 days for each adobe mix, ranging from 7% (ID 0) to 215% (ID 1).

23.4 Conclusions

The drying shrinkage crack widths reduced as fiber dosages increased since more fibers are located in the cross sections that are cracking. The CWRR ranged from 82% (ID 0.25) to 94% (ID 1.0). The CDR values range from 3.5% (ID 0) to 0.84% (ID 1). These results evidence that even small dosages of MPP fibers have a significant impact in mitigating the drying shrinkage cracks in adobe mixes.

The impact strength increased with the incorporation of higher dosages MPP fibers. For adobe mix ID 1 at seven and 28 days after casting, the impact energy at collapse was 28 and 83 times, respectively, to that mix ID 0 (plain adobe). In terms of evolution of impact strength with time, there is a consistent increment from seven to 28 days for each adobe mix, ranging from 7% (ID 0) to 215% (ID 1). This can be attributed to the reduction moisture due to drying.

Acknowledgements The authors would like to thank Arnaldo Puebla, Wladimir Vergara, Mauro Ortiz, Gian Piero Canevari, Matías Riveros, Cristobal Vargas, and Sabine Kunze, for the help provided for the sample preparation, and Sika S.A. Chile, for providing the MPP fibers used in this study.

References

Antico FC, Zavattieri PD, Hector LG Jr, Mance A, Rodgers WR, Okonski DA (2012) Adhesion of nickel–titanium shape memory alloy wires to thermoplastic materials: theory and experiments. Smart Mater Struct 21(3):035022

Araya-Letelier G, Antico FC, Parra P, Carrasco M (2017a) Fiber-reinforced mortar incorporating pig hair. Adv Eng Forum 21:219–225

Araya-Letelier G, Antico FC, Urzua J, Bravo R (2017b) Physical-mechanical characterization of fiber-reinforced mortar incorporating pig hair. In: Proceedings of the 2nd international conference on bio-based building materials and 1st conference on ecological valorisation of granular and fibrous materials. Clermont-Ferrand

Araya-Letelier G, Antico FC, Carrasco M, Rojas P, García-Herrera CM (2017c) Effectiveness of new natural fibers on damage-mechanical performance of mortar. Constr Build Mater 152:672–682

Araya-Letelier G, Concha-Riedel J, Antico FC, Valdés C, Cáceres G (2018) Influence of natural fiber dosage and length on adobe mixes damage-mechanical behavior. Constr Build Mater 174:645–655

ASTM D854 (2000) Standard test methods for specific gravity of soil solids by water pycnometer. ASTM International, West Conshohocken, PA

Avrami EC, Guillaud H, Hardy M (eds) (2008) Terra literature review: an overview of research in earthen architecture conservation. Getty Conservation Institute, Los Angeles

Aymerich F, Fenu L, Meloni P (2012) Effect of reinforcing wool fibres on fracture and energy absorption properties of an earthen material. Constr Build Mater 27(1):66–72

Balkis A (2017) The effects of waste marble dust and polypropylene fiber contents on mechanical properties of gypsum stabilized earthen. Constr Build Mater 134:556–562
Cataldo-Born M, Araya-Letelier G, and Pabón C (2016) Obstacles and motivations for earthbag social housing in Chile: energy, environment, economic and codes implications. Revista de La Construcción 15(3):17–26
Donkor P, Obonyo E (2015) Earthen construction materials: assessing the feasibility of improving strength and deformability of compressed earth blocks using polypropylene fibers. Mater Des 83:813–819
Galán-Marín C, Rivera-Gómez C, Petric-Gray J (2010) Effect of animal fibres reinforcement on stabilized earth mechanical properties. J Biobased Mater Bioenergy 4(2):121–128
Millogo Y, Morel JC, Aubert JE, Ghavami K (2014) Experimental analysis of pressed adobe blocks reinforced with Hibiscus cannabinus fibers. Constr Build Mater 52:71–78
Mindess S, Young JF, and Darwin D (2002) Concrete, 2nd edn. Prentice Hall
Minke G (2000) Earth construction handbook: the building material earth in modern architecture. WIT Press, Southampton
Minke G (2006) Building with Earth: design and technology of a sustainable architecture. Walter de Gruyter
Quagliarini E, Lenci S (2010) The influence of natural stabilizers and natural fibres on the mechanical properties of ancient Roman adobe bricks. J C Herit 11(3):309–314
Reman O (2004) Increasing the strength of soil for adobe construction. Archit Sci Rev 47(4):373–386
Yilmaz Y (2009) Experimental investigation of the strength properties of sand–clay mixtures reinforced with randomly distributed discrete polypropylene fibers. Geosynth Int 16(5):354–363

Chapter 24
Influence of Jute Fibers to Improve Flexural Toughness, Impact Resistance and Drying Shrinkage Cracking in Adobe Mixes

Jose Concha-Riedel, Gerardo Araya-Letelier, Federico C. Antico, Ursula Reidel and Andres Glade

24.1 Introduction

Revival of interest in the use of earthen construction materials can be seen over the last few decades due to the advantages such as thermal comfort, greenness and low environmental impacts (Cataldo-Born et al. 2016; Donkor and Obonyo 2015; Millogo et al. 2014; Minke 2006). However, earthen materials also have disadvantages such as low mechanical strength, poor erosion resistance and drying shrinkage cracking (Avrami et al. 2008; Minke 2000, 2006). The incorporation of industrialized fibers, such as polypropylene fibers (Donkor and Obonyo 2015; Yilmaz 2009), and of natural fibers, such as straw and even wool (Aymerich et al. 2012; Quagliarini and Lenci 2010), has been studied to mitigate some of the disadvantages of earthen materials. Among natural fibers, there have been recent studies addressing the use of jute fibers. Islam et al. (2008) studied the impact of natural fibers, including jute fibers, on strength and toughness of earthen blocks as well as shrinkage cracking of mortars

---

J. Concha-Riedel
Universidad Adolfo Ibañez, Santiago, Chile
e-mail: jose.concha@uai.cl

G. Araya-Letelier (✉)
Pontificia Universidad Católica de Chile, Santiago, Chile
e-mail: gerardo.araya@uc.cl

F. C. Antico · A. Glade
Universidad Adolfo Ibañez, Viña del Mar, Chile
e-mail: federico.antico@uai.cl

A. Glade
e-mail: aglade@alumnos.uai.cl

U. Reidel
Sika S.A. Chile, Santiago, Chile
e-mail: reidel.ursula@cl.sika.com

B. V. V. Reddy et al. (eds.), *Earthen Dwellings and Structures*,
Springer Transactions in Civil and Environmental Engineering,
https://doi.org/10.1007/978-981-13-5883-8_24

Table 24.1 Morphological, physical and mechanical properties of jute fiber after Daniel et al. (2002)

Diameter (mm)	Specific gravity	Modulus of elasticity (MPa)	Tensile strength (MPa)	Elongation at break (%)
0.102–0.203	1.02–1.04	25,990–31,992	250–350	1.5–1.9

and jute was found to be the most effective fiber, improving strength and toughness as well as preventing shrinkage cracking. Güllü and Khudir (2014) studied the effect of natural fibers, including jute fibers, on the unconfined compressive strength of fine-grained soil subjected to freeze-thaw cycles, and it was found that the incorporation of jute fiber increased both non-freeze-thaw and freeze-thaw unconfined compressive strengths compared to unreinforced soil. Saleem et al. (2016) assessed the effect of jute fiber on the compressive strength of earth bricks finding increments of compressive strength due to the incorporation of jute fiber. Although the contribution of the studies addressing the reinforcement of earthen materials with jute has been significant, there are still some properties that have not been studied intensely and quantitatively, such as impact resistance.

The novelty of this research resides in the incorporation of jute fiber into earthen materials such as adobe mixes, addressing quantitatively the impact of the incorporation of jute into flexural toughness, impact resistance and drying shrinkage cracking. This study has the following objectives: to evaluate the influence of different lengths and dosages of jute fiber on (i) flexural strength, (ii) flexural toughness, (iii) drying shrinkage performance and (iv) impact resistance in adobe mixes.

24.2 Materials and Methods

24.2.1 Materials

The present study uses a clayey soil from a district in Santiago, Chile. Araya-Letelier et al. (2018) conducted a characterization of the same soil, obtaining an 11%, 69% and 20% of clay, silt and sand, respectively. The liquid and plastic Atterberg limits, plasticity index and specific gravity of solids were also characterized, and their values are 29.1%, 17.4%, 11.7% and 2.51% respectively.

Jute is used as a natural reinforcing fiber for adobe mixes, and Table 24.1 summarizes some of the most relevant morphological, physical and mechanical properties of this vegetal fiber reported by Daniel et al. (2002).

The jute fiber used in this study was processed to obtain three different lengths (7, 15 and 30 mm) to perform a sensitivity analysis on the impact of different lengths and dosages of fibers on the performance of adobe mixes. Considering the range of diameters (0.1–0.2 mm) reported by Daniel et al. (2002) and the three different

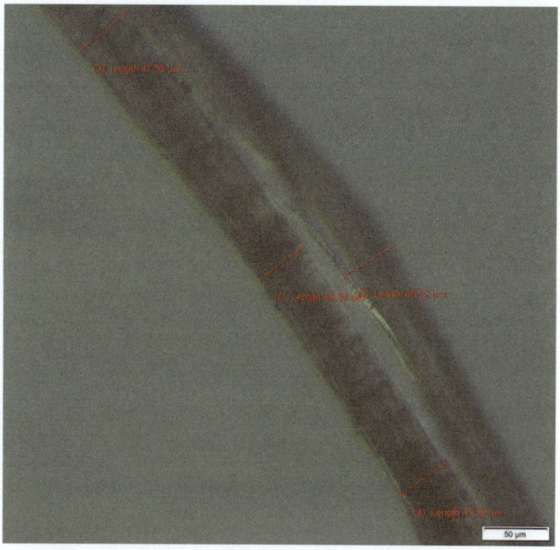

Fig. 24.1 Microscopy of jute fibers

lengths used in this study, the ranges of aspect ratios to be used are 35–70, 75–150 and 150–300 for fiber lengths of 7, 15 and 30 mm, respectively. Figure 24.1 shows a microscopy of the jute fibers.

24.2.2 Adobe Mix Proportions and Specimen Preparation

A water-to-soil ratio (i.e., the weight of water to the weight of dry soil) of 0.307 was determined based on the required workability of the mixes. Seven different adobe mixes were cast for this investigation, six of them incorporating fiber and one plain mix, to make comparisons between them. The six jute-reinforced mixes had two dosages (0.5 and 2% by mass of soil) and three different fiber lengths (7, 15 and 30 mm). The different adobe mix ID numbers and fiber dosages are presented in Table 24.2, where the identification number (ID) indicates first the dosage and second the length of the fiber.

The adobe mixes and the specimens were carried out manually. The fibers were mixed initially with the dry soil, until all clusters were disassembled. After all the fibers were included into the soil, water was added into the mixture in four equal portions, mixing after each incorporation. Finally, the adobe mixes were covered with a plastic bag to prevent water evaporation. The following specimens were cast for each adobe mix: (i) six notched RILEM beam specimens (160 mm × 40 mm × 40 mm, according to ISO/R679 (1968)) for impact strength test, (ii) six RILEM beam specimens for flexural strength, toughness indices and residual strength factors and (iii) two cylindrical specimens for drying shrinkage cracking. After 48 h, the

Table 24.2 Adobe mix ID number and material proportion

Adobe mix ID	Fiber (%)	Fiber length (mm)
0–0	0	0
0.5–7	0.5	7
0.5–15	0.5	15
0.5–30	0.5	30
2.0–7	2.0	7
2.0–15	2.0	15
2.0–30	2.0	30

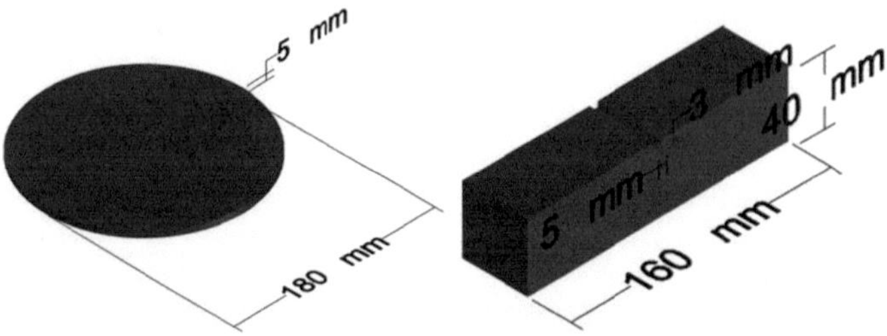

Fig. 24.2 Cylindrical specimen (right) and RILEM beam specimen (left) after Araya-Letelier et al. (2018)

specimens were demolded and kept at laboratory conditions (22 °C and 45% relative humidity) for 28 days until testing, rotating to the adjacent side 90° every 7 days. A fine coat of glue and sand was applied in the molds of the cylindrical specimens for ease of de-molding. The specimens for shrinkage crack testing were kept at laboratory conditions (22 °C and 45% relative humidity) for seven days in the molds, and cracking was measured afterward. Sketches in Fig. 24.2 show dimensions of the specimens.

24.2.3 Testing

Flexural strength of each adobe mix was assessed at 28 days after casting using a three-point bending configuration over six RILEM beam specimens with a span of 130 mm between supports and using a displacement control protocol with a rate of 1 mm/minute.

To assess the energy absorption capacity of each adobe mix, flexural toughness indices were calculated according to ASTM C1018 (1997) procedure. The flexural toughness indices were calculated as the area under the load-deflection curve up to a specific deflection value divided by the area under the load-deflection curve up to the deflection where the first crack was observed (δ). Specifically, flexural toughness indices I_5, I_{10} and I_{20} are the values obtained using deflections of 3δ, 5.5δ and

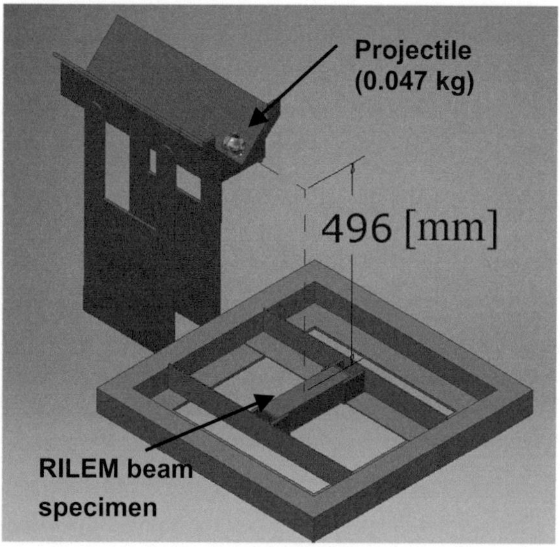

Fig. 24.3 Impact test setup after Araya-Letelier et al. (2018)

10.5δ, which are stated in ASTM C1018 (1997), respectively. Moreover, the residual strength factors, representing the average post-crack load retained over a deflection interval as a percentage of the load at the first crack, were estimated directly from the flexural toughness indices using Eqs. (24.1) and (24.2) (for further information, see ASTM C1018 (1997)).

$$R_{5,10} = 20(I_{10} - I_5) \quad (24.1)$$

$$R_{10,20} = 10(I_{20} - I_{10}) \quad (24.2)$$

where $R_{5,10}$ and $R_{10,20}$ are the residual strength factors between the intervals of 5.5δ and 3δ, and 10.5δ and 5.5δ, respectively. Mean and standard deviation of toughness indices and average residual strength factors for each adobe mix were obtained from the individual flexural load-displacement curves at 28 days after casting.

To measure the drying shrinkage cracking, widths and lengths were measured using a crack width comparator and a caliper, guided by a 20 × 20 mm grid, which was drawn on the surface of specimens. Crack width reduction ratio (CWRR) was calculated with respect to plain adobe, as well as crack width average (CWA) for all dosages.

To assess the impact energy absorption of the adobe mix after an age of 28 days, a projectile was thrown to the center of a RILEM beam specimen with a notch in the middle section as presented in Fig. 24.3. To avoid damage due to rebound, the specimens were glued with silicone to the steel frame. The impact energy was calculated using Eq. (24.3).

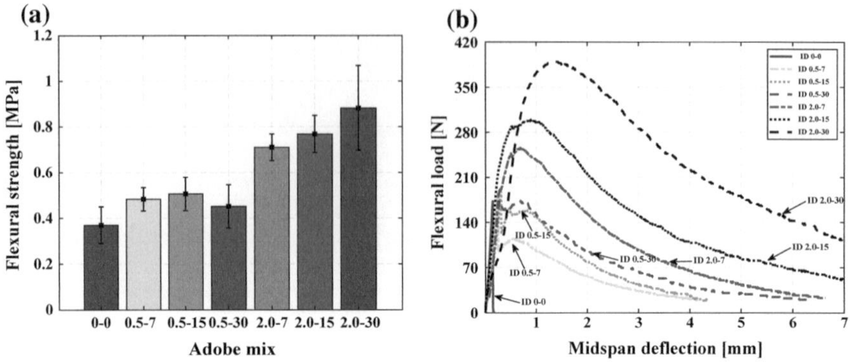

Fig. 24.4 **a** Flexural strength results and **b** compressive strength results

$$E_C = n * m * g * h \tag{24.3}$$

where E_C stands for the total energy at failure of the specimen, n is the required number of blows to cause failure to the specimen, m is the mass of the projectile (0.047 kg), g is the gravitational constant (9.8 m/s^2), and h is the height of the fall of the projectile (0.496 m).

24.3 Results and Discussion

24.3.1 Flexural Strength, Toughness Indices and Residual Strength Factors

The results for flexural strength are presented in Fig. 24.4a, along with the error bars, considering one standard deviation above and below the mean value. Plain adobe had the worst flexural strength (0.37 MPa), whereas adobe mix ID 2.0–30 had the strongest flexural strength (0.88 MPa). Adobe mix ID 0.5–7 presented the less scattered data (SD = 0.05 MPa), whereas the most scattered data was obtained for adobe mix ID 2.0–30 (SD = 0.19 MPa). Analyzing the error bars, if the SD value of mix ID 0–0 is added to the mean, and the SD value of mix ID 2.0–30 is subtracted from the mean, the resulting value is still larger for ID 2.0–30. The latter might indicate that the length and dosage of jute fiber in adobe mixes do increase its flexural strength. The increase in flexural strength when jute is incorporated might be explained by the adequate mixing of the fibers and the satisfactory bonding between long fibers and the matrix.

Typical load-deflection curves are presented in Fig. 24.4b for each adobe mix, tested as mentioned in Sect. 2.3. Figure 24.4b shows that initially, the flexural stress increases linearly at the same stress-deflection rate until a visible crack is observed

Table 24.3 Toughness indices and residual strength factors at 28 days after casting

Adobe mix ID	I_5		I_{10}		I_{10}		$R_{5,10}$	$R_{10,20}$
	Avg[1]	SD[2]	Avg[1]	SD[2]	Avg[1]	SD[2]	Avg[1]	Avg[1]
0–0	1.16	0.14	1.16	0.14	1.16	0.14	0.0	0.0
0.5–7	6.28	5.47	9.85	5.78	13.64	4.30	71.38	37.90
0.5–15	4.38	0.80	7.98	1.93	12.45	3.81	71.99	44.68
0.5–30	3.76	0.97	5.84	2.74	8.24	5.82	41.45	24.03
2.0–7	3.50	0.24	5.10	0.59	6.23	1.08	32.19	11.26
2.0–15	3.54	0.73	5.32	1.62	6.71	2.61	35.53	13.88
2.0–30	3.10	0.32	4.29	0.63	4.82	1.10	23.67	5.32

[1]Average; [2]standard deviation

growing from the lower surface of the sample (under tension). The later occurs regardless of the fiber dosage, which indicates that up to the peak load, the flexural response of the adobe mixes depends on the mechanical behavior of the matrix mainly, which is considered as brittle. After reaching the peak load, adobe mix ID 0 presents a brittle failure mode, whereas fiber-reinforced adobe mixes present an enhanced post-peak performance.

Table 24.3 presents average and SD values of I_5, I_{10} and I_{20} and the average values of residual strength factor for each adobe mix. It can be seen that all the values of toughness indices (I_5, I_{10}, and I_{20}) for adobe mix ID 0-0 are approximately 1.0, which means that this adobe mix has a brittle mode of failure where its flexural strength is followed by a sudden drop of the mechanical load joined by unstable macroscopic crack formation. In contrast, all the fiber-reinforced adobe mixes exhibit a load recovery after the first crack, joined by an increment of the mid-span deflection before failure, and toughness indices as well as residual strength factors increase mainly due to the presence of jute fibers.

24.3.2 *Restrained Drying Shrinkage Distributed Cracking*

The results of the drying shrinkage cracking can be seen in Fig. 24.5, where one of the two shrinkage specimens is shown. As can be seen, cracking is consistently reduced as the fiber length and dosage is increased. The CWA values of mix ID 0-0 and ID 2.0–30 were 1.2 and 0.2 mm, respectively.

CWA and CWRR results for each adobe mix are shown in Fig. 24.6 in the left side axis and right side axis, respectively. As was qualitatively observed in Fig. 24.5, the average width of the cracks, as well as the reduction ratio, is sensitive to the dosage and length of the fiber. ID 0–0 presented average values of crack width of 0.84 mm, whereas results for adobe mixes with jute fiber oscillated between 0.26 mm (ID 0.5–7) and 0.13 mm (ID 2.0–30). The corresponding CWRR values were 68 and

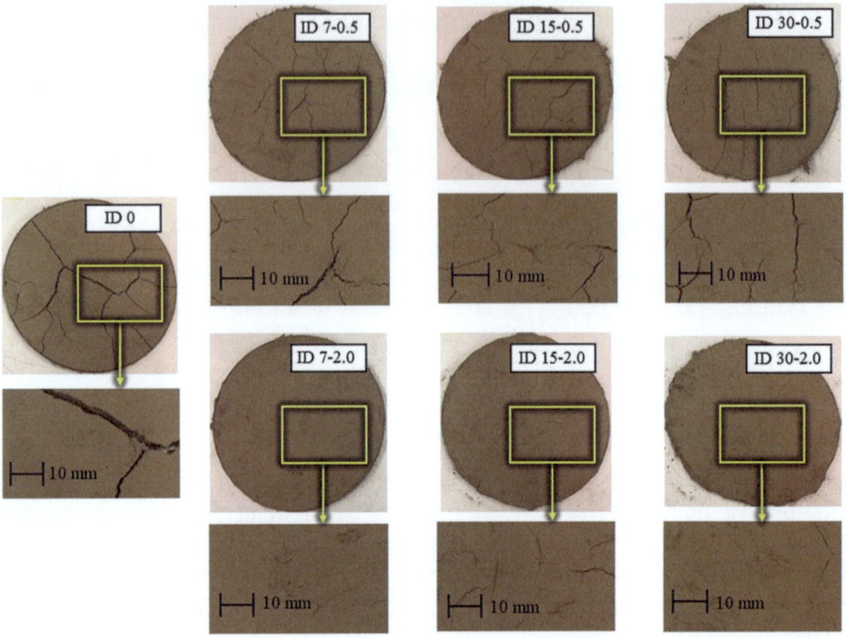

Fig. 24.5 Restrained drying shrinkage cracking results

Fig. 24.6 CWA and CWRR results

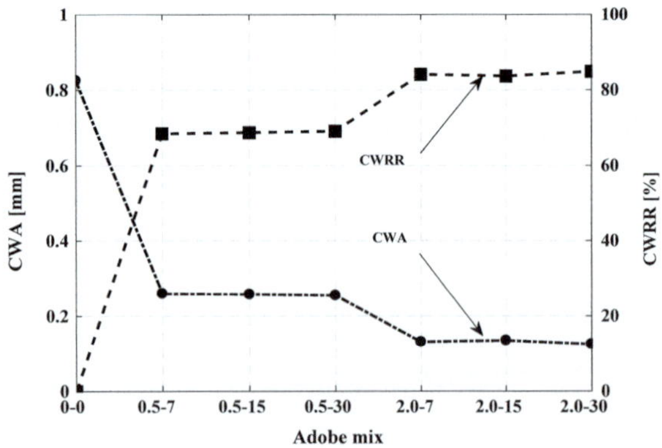

86%, respectively. As can be observed in Fig. 24.6, the CWRR results might indicate a higher sensitivity for fiber dosage than fiber length, where the increment for length is two percentage points and approximately 15 percentage points for increment of dosage. Antico et al. (2012) studied the behavior of material matrixes and the failure mechanisms for fibers, which is consistent with the effects observed by naked eye in this experiment. As the fiber dosages were increased, the characteristic length between the fiber and the matrix is reduced, increasing the chances that fibers are scattered throughout the matrix.

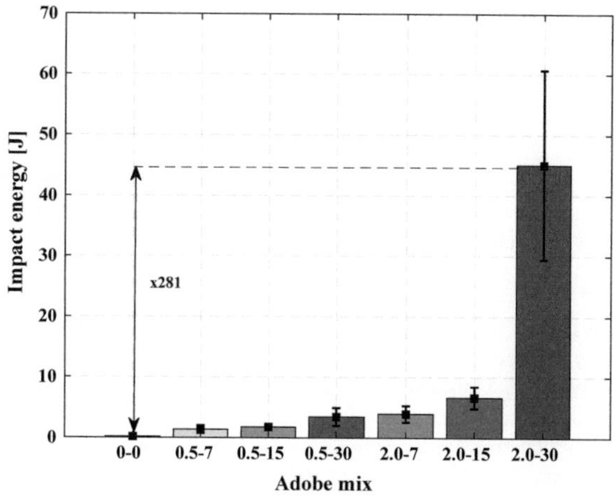

Fig. 24.7 Cumulative collapse impact energy

24.3.3 *Impact Strength*

Mean results of the impact test energy absorption are shown in Fig. 24.7, as well as error bars corresponding to one standard deviation above and below the mean. It can be observed that as the fiber length and dosage is increased, the energy absorption also increases. Adobe mix ID 0.5–15 and ID 2.0–30 show average results of 2.17 and 38 J (16 and 281 times the energy required to collapse plain adobe), respectively. As was studied by Araya-Letelier et al. (2017), quasi-brittle materials like mortars are expected to enhance its toughness when natural fibers are incorporated into the matrix. Like mortars, reinforced adobe could behave as a quasi-brittle material, and likewise, as fibers are incorporated it is expected that its energy-absorbing capacity is improved.

24.4 Conclusions

The results show that there is an improvement in the flexure strength of jute fiber-reinforced adobe mixes when compared with those of plain adobe specimens. The jute fiber reinforcement also resulted in an increase of flexural toughness. Incorporation of jute fibers reduced crack widths. The energy absorption due to impact was considerably enhanced by the incorporation of jute fibers. The impact energy absorption capacity of adobe mixes with jute fibers is sensitive to the length as well as fiber content.

Acknowledgements The authors would like to thank Arnaldo Puebla and Sabine Kunze, for the help provided for the sample preparation, and Sika S.A. Chile, for the use of their facilities for part of the experimental work presented.

References

Antico FC, Zavattieri PD, Hector LG Jr, Mance A, Rodgers WR, Okonski DA (2012) Adhesion of nickel–titanium shape memory alloy wires to thermoplastic materials: theory and experiments. Smart Mater Struct 21(3):035022

Araya-Letelier G, Antico FC, Carrasco M, Rojas P, García-Herrera CM (2017) Effectiveness of new natural fibers on damage-mechanical performance of mortar. Constr Build Mater 152:672–682

Araya-Letelier G, Concha-Riedel J, Antico FC, Valdés C, Cáceres G (2018) Influence of natural fiber dosage and length on adobe mixes damage-mechanical behavior. Constr Build Mater 174:645–655

ASTM C1018 (1997) Standard test method for flexural toughness and first-crack strength of fiber-reinforced concrete (using beam with third-point loading) (withdrawn 2006)

Avrami EC, Guillaud H, Hardy M (eds) (2008) Terra literature review: an overview of research in earthen architecture conservation. Getty Conservation Institute, Los Angeles

Aymerich F, Fenu L, Meloni P (2012) Effect of reinforcing wool fibres on fracture and energy absorption properties of an earthen material. Constr Build Mater 27(1):66–72

Cataldo-Born M, Araya-Letelier G, Pabón C (2016) Obstacles and motivations for earthbag social housing in Chile: energy, environment, economic and codes implications. Revista de La Construcción 15(3):17–26

Daniel JI, Ahmad SH, Arockiasamy M, Ball HP et al (2002) State-of-the-art report on fiber reinforced concrete. Reported by ACI Committee 544. Farmington Hills, MI

Donkor P, Obonyo E (2015) Earthen construction materials: assessing the feasibility of improving strength and deformability of compressed earth blocks using polypropylene fibers. Mater Des 83:813–819

Güllü H, Khudir A (2014) Effect of freeze–thaw cycles on unconfined compressive strength of fine-grained soil treated with jute fiber, steel fiber and lime. Cold Reg Sci Technol 106:55–65

Islam MS, Iwashita K, Rahman MM (2008) Performance of natural fiber reinforced soil in earthen houses. In: Proceedings of the sixth Asian young geotechnical engineers conference, Bangalore, India, Paper No. CP, vol 22

ISO/R679 (1968) Méthode d'essais mécaniques des ciments. Resistance á la compression et á la, flexion du mortier plastique (Méthode RILEM-CEMBUREAU)

Millogo Y, Morel JC, Aubert JE, Ghavami K (2014) Experimental analysis of Pressed Adobe Blocks reinforced with *Hibiscus cannabinus* fibers. Constr Build Mater 52:71–78

Minke G (2000) Earth construction handbook: the building material earth in modern architecture. WIT Press, Southampton

Minke G (2006) Building with Earth: design and technology of a sustainable architecture. Walter de Gruyter

Quagliarini E, Lenci S (2010) The influence of natural stabilizers and natural fibres on the mechanical properties of ancient Roman adobe bricks. J Cult Heritage 11(3):309–314

Saleem MA, Abbas S, Haider M (2016) Jute fiber reinforced compressed earth bricks (FR-CEB)—a sustainable solution. Pak J Eng Appl Sci 19:83–90

Yilmaz Y (2009) Experimental investigation of the strength properties of sand–clay mixtures reinforced with randomly distributed discrete polypropylene fibers. Geosynthetics Int 16(5):354–363

Chapter 25
The Effect of Incorporating Recycled Materials on the Load–Deformation Behaviour of Earth for Buildings

Kristopher J. Dick, J. Pieniuta, K. Arnold, P. Logan and Timothy J. Krahn

25.1 Introduction

The disposal of expanded polystyrene (EPS) is a great concern today as most EPS is ending up in landfills instead of being utilized in sustainable ways. Most EPS, consisting of up to 98% air, is finding its way to landfills because it is not economical to transport and thus not economically viable to repurpose sustainably (Babu et al. 2006). This landfill-bound material poses numerous hazards including the potential to harm wildlife if ingested, deterioration of soil fertility if mixed into arable soil and the possibility it will remain in landfills for centuries due to its inability to decompose (Saikia and de Brito 2012). By utilizing this waste in innovative ways, landfills can expect to have to bury less waste annually and several environmental concerns will also be addressed.

Many researchers have been investigating different means of using different forms of waste EPS to produce sustainable building materials as EPS can undergo large compressive deformations and absorb energy (Krundaeva et al. 2016). EPS earth is also relatively easy to fabricate on a construction site compared to other types of earth,

K. J. Dick (✉) · J. Pieniuta · K. Arnold · P. Logan
Biosystems Engineering, University of Manitoba, Winnipeg, Canada
e-mail: Kristopher.Dick@umanitoba.ca

J. Pieniuta
e-mail: Jennifer.Pieniuta@umanitoba.ca

K. Arnold
e-mail: arnoldk@myumanitoba.ca

P. Logan
e-mail: umlogan6@myumanitoba.ca

T. J. Krahn
Building Alternatives Inc., Codrington, ON, Canada
e-mail: built.alternatives@gmail.com

B. V. V. Reddy et al. (eds.), *Earthen Dwellings and Structures*, Springer Transactions in Civil and Environmental Engineering,
https://doi.org/10.1007/978-981-13-5883-8_25

whose fabrication process is often more complex. According to Babu et al. (2006), in addition to using earth containing EPS aggregate for building applications, similar mixes have been used for specialized applications including railway track beds, energy-absorbing material for the protection of military structures, and in floating marine structures.

Along with EPS, recyclable materials such as paper and cardboard can be utilized in construction materials as they demonstrate favourable binding properties in mixes (Demirbaş 1999). Although these materials do not pose the same challenges presented by EPS, the materials are still produced in vast amounts and can also be utilized in sustainable ways. In some developing countries, informal salvaging of recyclable materials is practised, so one can obtain a small income. If these individuals are aware of sustainable methods to utilize these materials, the recyclable material can also be used to improve their way of living in more ways than one (Demirbaş 1999).

The main objective of this research was to investigate the feasibility of using waste EPS, paper, polyethylene terephthalate (PET) and cardboard as aggregates in the production of earth cylinders. Many authors (Babu et al. 2006; Kan and Demirboğa 2009; Sayadi et al. 2016; Ramamurthy et al. 2009) have investigated the properties of EPS earth, but the authors are not aware of research on EPS earth with paper and cardboard.

25.2 Materials and Methods

25.2.1 Materials—Test Cylinders and Compressed Earth Blocks (CEB)

The test cylinders were made with a mixture of Type 10 General Use cement stabilizer, recycled cardboard and paper combined with soil. The earth material was classified as a loam soil textural class. Sieve analysis previously done on the soil indicated that it had an average of 45.11% sand, 35.33% silt and 19.56% clay (Griffith 2007). The density of the loose soil was 1.827 g/cm^3. The compressed earth blocks (CEB) were made with earth, chopped wheat straw and recycled EPS beads.

The recyclable cardboard and paper were collected from the University of Manitoba recycling depot. To prepare the paper and cardboard aggregate, a Powershred® H-8Cd cross-cut shredder was used (Fellowes, USA). This mechanical shredder cuts each full piece of paper or cardboard into 4 $\times$ 35 mm cross-cut particles. Given that variations from the specified size were observed, an average collected from randomly selecting ten pieces of shredded material was recorded. The average size of the paper and cardboard aggregate was 4 $\times$ 31 and 4 $\times$ 27 mm, respectively. Densities for the loose material were 0.045 and 0.087 g/cm^3 for the paper and cardboard, respectively.

The waste EPS material was broken into bead form by hand for this research. A drywall knife was used across the grain to shred the EPS. This method was effective in breaking the sheet of polystyrene into a uniform size but was time-consuming.

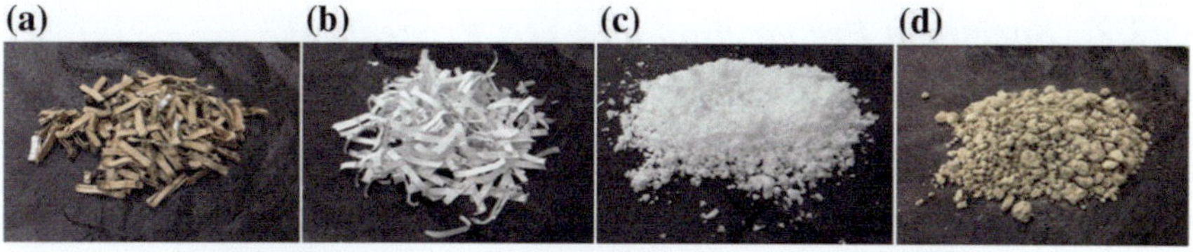

Fig. 25.1 Aggregate appearance before mixing, **a** cardboard, **b** paper **c** EPS and **d** earth

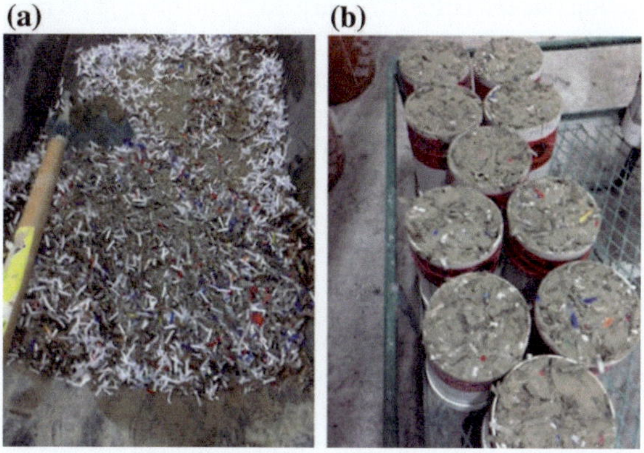

Fig. 25.2 **a** Plastic earth mix and **b** cylinders

The second method used a 75-mm wire cup brush and electric drill. This method was quicker and resulted in a final aggregate size suitable for mixing with a bulk density of 0.011 g/cm^3 with a nominal size of 5 mm. The final appearance of the aggregates, prior to mixing with water and cement, is shown in Fig. 25.1.

Using PET plastic as the variant in the mix design, the parts of paper, cardboard and earth were maintained consistently at a ratio of 1, 1 and 3, respectively. PET from recycled water and soft drink bottles was broken down with a paper shredder. Three separate mix designs were made with 1, 1.5 and 2 parts of plastic each. Every mix design used a cement binder at 10% of specimen weight. Five 100 mm × 200 mm cylinders were prepared for each mix design. Figure 25.2a, b shows the materials being mixed and the cylinders in their forms, respectively.

The compressed earth block (CEB) specimens were prepared using an EcoBrava® hydraulic press manufactured in Brazil. A total of 12 specimens, 125 mm wide, 250 mm long and 85 mm thick, were pressed. Earth, chopped wheat straw and EPS were combined in three combinations with a control specimen set with no EPS. The wheat straw was chopped to an average length of 12 mm. The earth and straw were kept at a constant volume, while the EPS was varied between 3:1 and 3:0.5 earth to EPS.

25.2.2 Specimen Preparation and Test Procedures

For the preparation of the earth cylinders, paper, cardboard and earth aggregates mixed into each mixture remained at a ratio of 1:1:3, respectively, on a volume basis, for all three mixes. The total mass of each part was also recorded, using a Ohaus Adventurer Pro Scale with a 0.1 g readability, to remain as consistent as possible. The volume of EPS increased by one part volume for each mix, therefore resulting in three mixes including one part EPS, two parts EPS and three parts EPS. A control mix containing 0 part EPS was also created. The amount of added cement was based on 10% of the total mass of all aggregates. Following the method discussed by Kazemi (1987), aggregates were mixed prior to adding cementitious material. All mixing was done in a mortar trough by hand with a hoe.

The water requirement for the mix depended on the volume of EPS added and, thus, was governed by the stability and consistency of the mixture. Potable water was added evenly to the mix on a mass basis. The wet mixture was worked for 2–4 min. If required, additional water was added, and remixing of the product was completed until an appropriate consistency was achieved based on the appearance and feel of the material. If a ball of the mixture could be dropped from shoulder height and only minimally break apart, the consistency was deemed acceptable. This form of mixing is quite subjective but is a practice found in various sources (Kan and Demirboğa 2009; Kazemi 1987; Ramamurthy et al. 2009). Table 25.1 summarizes the mix proportions used for the test cylinder specimens. A total of five cylinders were cast for each mix. The material was hand-compacted into 100 × 200 mm split cylindrical forms made from 100-mm-diameter split PVC forms.

The forms were removed 9 days following preparation and stored on shelving that provided air movement until day 20. It was of interest to investigate if heating would effect on compressive strength. The temperatures selected were based on melting points or transition temperature for EPS and PET. Two specimens from each mixture were then placed in an oven (Jelo Tech Lab Companion, Model OF-11E) and heated to determine whether the heat would modify the properties of the EPS and PET. Some studies suggest that by exposing EPS to heat treatment, service life can be maximized by improving strength properties and density (Kan and Demirboğa 2009). Currently, there is no literature describing how the properties of EPS are affected by heat once already mixed with other aggregates, cement and water. However, discussion on how EPS is impacted when exposed to heat treatment on its own is common. Based on the data collected by heating EPS alone, placing the EPS in an oven at a temperature of 130 °C for 15 min was found to be ideal. By heating the EPS at this specified temperature and duration, the EPS was expected to transform from a foamy state to a plastic state (Kan and Demirboğa 2009). Once cured, the objective was to heat specimens at 130 °C for 15 min. Once placed in the oven, it took approximately 30 min for the oven to equilibrate back to 130 °C. Specimens were left at this temperature for 15 min to allow for heat penetration into the specimen. Cylinders were left to cool at room temperature for 3 days prior to testing. To observe how this temperature would affect an unmodified piece of EPS, a sample size of 10 × 10 × 2.3 cm was cut,

Table 25.1 Mix proportions of aggregate and cement—EPS cylinder specimens

Mix #	Ratio[a] (EPS:P:CB:C)	EPS (kg)	Paper (kg)	Cardboard (kg)	Soil (kg)	Cement (kg)[b]	Water (kg)	% Recycled material[c]
Control	0:1:1:3	0	0.020	0.052	2.472	0.254	0.418	40
1	1:1:1:3	0.006	0.020	0.052	2.472	0.255	0.420	50
2	2:1:1:3	0.012	0.020	0.052	2.472	0.256	0.430	57
3	3:1:1:3	0.018	0.020	0.052	2.472	0.256	0.463	63

[a] *EPS* expanded polystyrene, *P* paper, *CB* cardboard, *C* soil
[b] Cement calculated based on 10% of the total mass
[c] Based on total volume of aggregates

with a density of 20.87 kg/m^3. After removing it from the heat, the EPS was hard, decreased in volume by approximately 90% and had a density of 208.3 kg/m^3. After curing the PET, two specimens selected for heat treatment were placed in an oven for 5 min at 260 °C corresponding to the melting point of PET. The purpose was to attempt to partially melt the PET to see if there would be any impact on compressive strength. After 20 days, the cylinders were tested under axial compression using a loading frame at the Alternative Village at the University of Manitoba. A load rate of 1.4 mm/min was used. Linear potentiometers were used to monitor the vertical and lateral movements of the specimen. A load cell was used to measure the applied axial compression force. A computer controlled data acquisition system was used to record the load–deformation measurements. Given that the resulting densities of the cylinders were in excess of 800 kg/m^3, the Standard Test Method for Compressive Strength of Cylindrical Concrete Specimens (ASTM C39/C39 M-16b) was used. For the specimens placed in the oven, the ASTM C39/C39 M-16b standard was also followed to determine the compressive strength.

The CEBs were tested in individually in compression. The mix constituents were combined by hand using a hoe and mortar-mixing pan. The dry ingredients were first combined, and then, water was added gradually until the desired consistency was achieved. A hydraulic pressure of approximately 10.4 MPa was used in the press to consolidate the material. The blocks were stored at room temperature misted, kept damp and stored under plastic. They were allowed to cure for 14 days. A hydraulic test machine with a load and linear potentiometer was used to load and record the response to compressive load. A load rate of 5 mm/min was used.

25.3 Results and Discussion

The aim of this study was to evaluate the effects of EPS and PET as aggregates on physical properties including density, lateral movement and compressive strength in both the unmodified and modified heat-treated states for test cylinders. The compressed earth block (CEB) specimens were not subjected to heat treatment and only contained EPS foam.

25.3.1 Density

The specimen bulk density was determined during each stage of the cylinder preparation process, including immediately following ramming, after 20 days of curing, and following exposure of heat. The densities of the EPS cylinders after ramming to the cured cylinders decreased by 10.7, 11.2, 10.6 and 11.3% for the control, mix 1, mix 2 and mix 3, respectively. Similarly, the density of the cured cylinders to the heat-treated cylinders decreased by 1.23, 1.51, 1.83 and 1.69% for the control, mix 1, mix 2 and mix 3, respectively. Table 25.2 provides a summary of the dry densities

Table 25.2 Average dry densities for each mixture after packing, curing and heat treatment

Mix #	After packing (kg/m^3)	After curing (kg/m^3)	After heat treatment (kg/m^3)
Control	2226.3	1988.5	1964.0
EPS specimens			
1	1950.6	1732.2	1706.1
2	1749.4	1563.5	1534.9
3	1570.5	1397.4	1373.8
PET specimens			
1	2206.0	2088.6	2042.7
2	2123.0	2010.0	2011.0
3	2127.1	2013.9	1932.3

Table 25.3 Dry compressive stress and density results compressed earth block with EPS

Specimen	Compressive stress at maximum load (MPa)			Average dry density (kg/m^3)
	Min	Max	Average	
Control	11.2	11.8	11.5	3289
Mix 1 high foam	2.06	2.76	2.31	2540
Mix 2 medium foam	3.57	4.99	4.32	2923
Mix 3 low foam	4.73	5.68	5.17	2930

for the specimen types. The PET specimens did not have as wide a range of density compared to EPS. The average dry density of the CEBs is presented in Table 25.3.

25.3.2 Compressive Strength

The compressive strength values were determined using the average maximum load for three non-heat-treated cylinders. The results indicate that a higher EPS content, with decreasing density, resulted in a lower compressive strength. Compared to the control, a 29.7, 52.4 and 69.8% decrease in compressive strength resulted for mix 1, mix 2 and mix 3, respectively. Similarly, in mixtures without heat treatment, an increase in EPS volume resulted in a decrease in density and compressive strength for the specimens exposed to heat. Compared to the control, on average, there was an 18.7, 42.1 and 61.4% decrease in compressive strength for mix 1, mix 2 and mix 3, respectively. Compared to the non-heat-treated specimens, there was a 2.81, 8.14 and 16.6% increase in compressive strength for mix 1, mix 2 and mix 3, respectively, and an 11.1% decrease for the control. Unlike the specimens not subjected to heat treatment, the samples containing EPS that were exposed to heat prior to

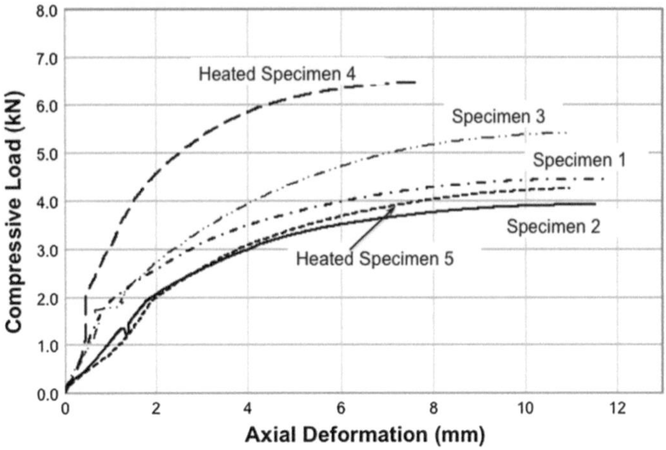

Fig. 25.3 Axial load versus deformation for EPS mix design 3 cylinders

testing demonstrated higher compressive strengths. It is surmised that heat treatment may have hardened the EPS resulting in a stiffer specimen. The control specimens illustrated a different behaviour, and it is likely that the heat dried the specimens to the point where the strength was compromised as a result. Figure 25.4 provides a graphical comparison of the compressive strengths, while Fig. 25.3 contains the load–deformation plots for EPS Mix 3. The results of compression tests on single CEBs are summarized in Table 25.3.

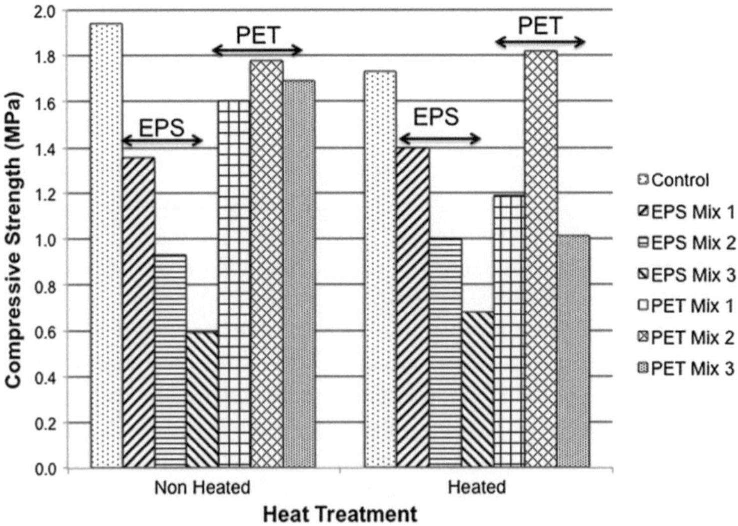

Fig. 25.4 Compressive dry strength comparison for cylinders

Fig. 25.5 CSEB compression test

During the compression tests, the CSEB that contained foam and straw produced a lot of crackling and popping noises when they were deformed as shown in Fig. 25.5. The outer edges of those bricks were also pushed further in a lateral direction when compared to the bricks with no foam or straw. The cause of that would indicate the straw and foam created many pores within the solid resulting in more deformation. The lighter density bricks resulted in a higher Poisson's ratio than the denser bricks that exhibited less lateral deformation

The combination of EPS with recycled aggregates has the potential to be used as a suitable building material. Qin et al. (2014) state that the compressive strength of adobe in rural areas of China resulted in a compressive strength ranging from 0.8 to 2.0 MPa. Cylinders tested containing 1 part and 2 parts of EPS resulted in compressive strengths within this range. Qin et al. (2014) also discussed that the strength of adobe, with the addition of 10% cement, resulted in compressive strengths within the range of 5.5–6.5 MPa, much higher than the compressive strengths observed in this study. Although compressive strengths were lower than those noted in cement stabilized adobe, earth blocks containing EPS still exhibited compressive strengths similar to adobe without cement. EPS earth blocks also show the ability to deform a substantial amount in compression before failing, a phenomenon not present in typical earth blocks. This observation is considered to be an important aspect of this study. The inclusion of EPS could allow for a structure to dissipate energy when subjected to lateral loads such as wind and seismic. Failure modes for the selected specimens are shown in Fig. 25.6. The EPS specimens did not display any clear shear behaviour at failure. The high EPS foam cylinders separated after compression. They bulged under load, and it is surmised that tension forces were induced within the specimen that contributed to a horizontal separation (Fig. 25.6a). The PET specimens did not display the same behaviour. The low PET content showed some shear failure (Fig. 25.6b) while the higher PET content tended to crush with vertical cracking (Fig. 25.6c).

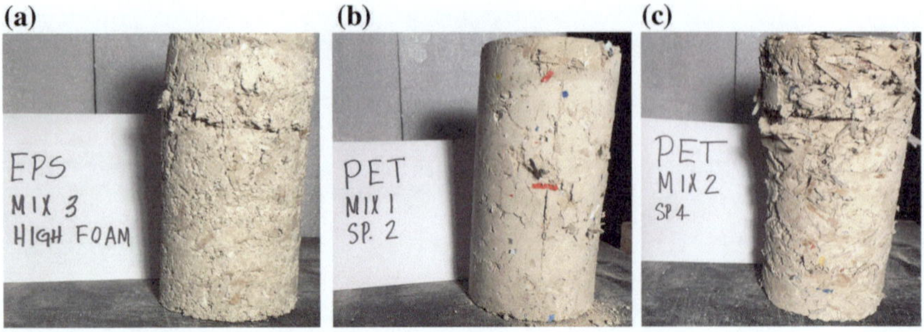

Fig. 25.6 Failure modes, **a** EPS mix 3, **b** PET mix 1 and **c** PET mix 2

25.3.3 Poisson's Ratio—Test Cylinders

It was postulated prior to testing that the incorporation of EPS might have an impact on the transverse deformation thus determining Poisson's ratio was of interest. Potentiometers were used to measure transverse expansion of the test cylinder in two orthogonal directions as well as the vertical directions. The ratio between the lateral and axial strain was used to obtain Poisson's ratio. As anticipated, bulging of the specimen during compression was clearly observed during testing. The specimens not heat-treated for mix 3 resulted in the highest Poisson's ratio for the EPS specimens. The high EPS content resulted in an average maximum axial deformation of approximately 12 mm. Specimens subjected to heat treatment resulted in lower Poisson's ratios for all mixes containing EPS. It is assumed that the transformation of the EPS from a foam state to a more plastic state created this difference. Mix 3 specimens did not exhibit shearing but rather separated horizontally when the load was relaxed. The ability to deform without shearing has the potential to be used as an energy-dissipating element in a structure. Lastly, based on average values for this limited sample, no apparent difference in Poisson's ratio for the control specimens occurred between the non-heat-treated specimens and heat-treated cylinder specimens. The PET specimens did not exhibit the same bulging behaviour without cracking. The ratios for the PET specimens were a result of specimen shearing and breaking apart as opposed to bulging out and were thus not considered to be valid at maximum load. Table 25.4 provides a summary of results for the EPS specimens.

Table 25.4 Average Poisson's ratio for control and EPS specimens at maximum load

Specimen type	Average Poisson's ratio	
	Non-heated	Heat treated
Mix 1	0.1561	0.0846
Mix 2	0.1923	0.0829
Mix 3	0.2974	0.0979

25.4 Conclusion

A series of tests were conducted on earth specimens that incorporated waste EPS and PET, recycled cardboard and paper. Compression tests were done on 100 mm diameter × 200 mm high cylinders and on single compressed stabilized earth block. Material properties including density, Poisson's ratio, stiffness and compressive strength were determined.

For cylinder test specimens, the axial deformation and Poisson's ratio increased with increasing amounts of EPS while stiffness decreased. The inclusion of PET in non-heated specimens had less impact on compressive strength when compared with EPS specimens. The effect of PET on Poisson's ratio was not significant when compared to control specimens. Based on load–deformation plots, the stiffness of the heat-treated specimens was greater than non-heat-treated specimens. Subjecting the cylinders to heat prior to testing resulted in higher stiffness values and compressive strengths, but decreased the average density, axial deformation and Poisson's ratio for EPS and PET mixtures. This is, however, not considered to be a sustainable practice nor practical for in situ production.

Based on the compressed stabilized earth block (CSEB) data, it is clear that the addition of EPS foam could be a viable option for improving structural characteristics with respect to energy dissipation. Results from this experiment show that it is possible to produce blocks with adequate compressive strength while at the same time reusing a material that is often discarded. The earth blocks are also lighter than those that contain no foam and have no drastic impact on the moisture content. From these data, mix 2 with 14–15% foam had an average maximum compressive strength of 4.15 MPa that shows the most promise from this test series.

The significant deformation of the high foam EPS mix is of particular interest to investigate the potential for use in energy dissipation in a structural system. While the compressive strength is low, if used in a framing system, then it is postulated that this block could tolerate racking loads while still remaining intact. Ongoing research is evaluating this behaviour.

Acknowledgements The authors would like to thank the Alternative Village at the University of Manitoba, Winnipeg, Canada, for supporting this research.

References

ASTM International (2016) Standard test method for compressive strength of cylindrical concrete specimens—C39/C39 M-16b. West Conshohocken, PA

Babu DS, Babu KG, Tiong-Huan W (2006) Effect of polystyrene aggregate size on strength and migration characteristics of lightweight concrete. Cement Concr Compos 28:520–527

Demirbaş A (1999) Physical properties of briquettes from waste paper and wheat straw mixtures. Energy Convers Manag 40:437–445

Griffith K (2007) Physical Properties of an Anola soil based cob re-inforced with chicken feathers. Unpublished Graduation project report, Dept. of Biosystems Engineering, University of Manitoba, Winnipeg, Manitoba, Canada

Kan A, Demirboğa R (2009) A new technique of processing for waste-expanded polystyrene foams as aggregates. J Mater Process Technol 209:2994–3000

Kazemi A (1987) Strength development in concrete blocks containing flyash. Unpublished M.Sc. thesis, Department of Civil Engineering, University of Manitoba, Winnipeg, MB

Krundaeva A, De Bruyne G, Gagliardi F, Van Parpegem W (2016) Dynamic compressive strength and crushing properties of expanded polystyrene foam for different strain rates and different temperatures. Polym Test 55:61–68

Qin L, Chen W, Li X (2014) Experimental research on compressive strength of adobe with cement. Appl Mech Mater 507:217–221

Ramamurthy K, Kunhanandan Nambiar EK, Indu Siva Ranjani G (2009) A classification of studies on properties of foam concrete. Cement Concr Compos 31:388–396

Saikia N, de Brito J (2012) Use of plastic waste as aggregate in cement mortar and concrete preparation: a review. Constr Build Mater 34:385–401

Sayadi AA, Tapia JV, Neitzert TR, Clifton GC (2016) Effects of expanded polystyrene (EPS) particles on fire resistance, thermal conductivity and compressive strength of foamed concrete. Constr Build Mater 112:716–724

Part II
Structural Performance and Durability

Chapter 26
Behaviour of Cement Stabilised Rammed Earth Walls Under Concentric and Eccentric Gravity Loading

B. V. Venkatarama Reddy, V. Suresh and K. S. Nanjunda Rao

26.1 Introduction

Monolithic rammed earth walls are constructed by the compaction of partially saturated soil and aggregate mixture in progressive layers in a rigid formwork. There are two types of rammed earth constructions: stabilised rammed earth and un-stabilised rammed earth. The stabilised rammed earth construction uses soil, aggregates and inorganic stabiliser additives (cement or lime). Cement stabilised rammed earth (CSRE) structures are popular and have been built since the last several decades in Australia, India, USA and many other countries (Verma and Mehra 1950; Easton 1982; Middleton and Schneider 1995; Hall 2002; Houben and Guillaud 2004; Walker et al. 2005; Reddy et al. 2014; Ciancio and Beckett 2015).

The compressive strength of the material is required for assessing the load-carrying capacity of the rammed earth walls. Apart from the strength of the material, compressive strength of the rammed earth wall is influenced by the load eccentricity and slenderness. Slender CSRE walls undergo lateral bending before collapse due to slenderness effects under concentric and eccentric gravity loads. The structural design of such walls should account for stress reduction due to the eccentricity of the loads and slenderness ratio. Therefore, it is essential to have the information on the characteristic compressive strength of the CSRE and the capacity reduction factors for determining the compressive strength of the rammed earth wall. There are

B. V. Venkatarama Reddy (✉) · V. Suresh · K. S. Nanjunda Rao
Department of Civil Engineering, Indian Institute of Science, Bangalore, India
e-mail: venkat@iisc.ac.in

V. Suresh
e-mail: pvksure@gmail.com

K. S. Nanjunda Rao
e-mail: ksn@iisc.ac.in

B. V. V. Reddy et al. (eds.), *Earthen Dwellings and Structures*,
Springer Transactions in Civil and Environmental Engineering,
https://doi.org/10.1007/978-981-13-5883-8_26

limited investigations on the structural behaviour of CSRE walls. Hence, the present investigation was focused on determination of the compressive strength of CSRE walls under concentric and eccentric gravity loading.

26.2 Earlier Studies

The literature on stabilised rammed earth can be grouped under three broad categories (a) characterising the properties of raw materials and production techniques, (b) assessing strength and durability characteristics (c) behaviour of rammed earth walls and the structural elements. There are many investigations on the first two categories and limited studies on the third category. A review of the papers on the compressive strength of CSRE walls is as follows.

Wallette, cube or cylindrical specimens are employed to assess the compressive strength of CSRE (Walker 2002; Hall and Djerbib 2004; Jayasinghe and Kamaladasa 2007; Reddy and Kumar 2009, 2011a; Reddy et al. 2016). The investigations of Reddy et al. (2016) showed that the compressive strength CSRE can be assessed by testing either a cylinder or prism having aspect ratio 2 (without using any correction factor), which can be used to determine the characteristic compressive strength of CSRE. Middleton and Schneider (1995) gave correction factors to account for the aspect ratio of the rammed earth for determining normalised compressive strength. But these corrections factors were based on the factors given in the Australian masonry code. The compressive strength of slender 3.0 m height (height to width ratio of 20) CSRE walls was experimentally determined by Reddy and Kumar (2011b). The tests were performed under concentric loads and found that there was considerable difference in compressive strength between the wallette/prism strength and 3.0 m height wall strength, because of the slenderness effect. Axial load capacity of square and rectangular CSRE columns was examined by Tripura and Singh (2015). CSRE columns of height (h) to thickness (t) of 6, 8 and 10 were tested in air dry condition. Strength reduced with the increase in h/t ratio and the capacity reduction factors derived were in tune with the reduction factors mentioned in different codes on masonry. In this study, the moisture content of the specimen during the test was not controlled and some of the masonry codes were based on working stress design and hence the capacity reduction factor estimated in such cases is not clear. The literature review reveals that there are no comprehensive studies on assessing compressive strength of CSRE walls, under concentric and eccentric loads.

26.3 Objectives, Scope and Experimental Programme

The main objective was to understand the compressive strength of CSRE walls and to understand the strength reduction capacity due to load eccentricity and slenderness effects. The scope of the investigations includes casting CSRE wall elements having different *h*/*t* ratios, testing the wall specimens (in dry condition) under concentric and eccentric gravity loading, capturing lateral deformations and analysing the results for capacity reduction. The *h*/*t* ratio of 10 and 14 was considered, and the load eccentricity was varied between 0 and 1/3 of the wall thickness. Totally three eccentricities were considered. The dry density and cement content of the specimens were 1800 kg/m^3 and 10% (by mass), respectively.

26.4 Materials Used in Experiments

CSRE wall specimens were cast using ordinary Portland cement (OPC) conforming to IS 12269 (1987). The 28-day compressive strength of OPC tested following the procedure outlined in IS 4031 (1988) was 69.2 MPa. The initial and final setting time for the cement was 148 and 312 min, respectively.

A local soil having 15.3% clay content was used in the preparation of the rammed earth walls and wallette specimens. Table 26.1 gives characteristics of the soil. The soil contains kaolinite clay mineral. Liquid limit and plasticity index values for the soil were 26.02 and 10.75%, respectively. The standard proctor optimum moisture content (OMC) and maximum dry density values for the soil with and without cement did not differ much.

Table 26.1 Characteristics of reconstituted soil

Soil property	Details
Textural composition (% by mass)	
Sand (4.75–0.075 mm)	73.7
Silt (0.075–0.002 mm)	11.0
Clay (<0.002 mm)	15.3
Atterberg Limits	
Liquid limit (%)	26.02
Plasticity index	10.75
Compaction characteristics	
(a) Without cement	
Standard Proctor OMC (%)	10.2
Maximum dry density (kg/m^3)	2039.0
(b) With 7% cement	
Standard Proctor OMC (%)	10.8
Maximum dry density (kg/m^3)	1990.0
Predominant clay mineral	Kaolinite

26.5 Casting CSRE Wall Specimens

The CSRE wall specimens of size 800 × 150 × 1500 mm and 800 × 150 × 2100 mm (width × thickness × height) were cast. The casting was carried out using a metal mould shown in Fig. 26.1. The mould assembly has two L-shaped split moulds joined at the corners using bolt-nut system. One of the L-shaped split moulds was of full specimen height and the other one was built-up (using bolts and nut systems) as the compaction proceeds. Also, Fig. 26.1 shows the partially cast mould assembly along with the casting process using a flat-headed rammer. After each layer was compacted, round-shaped dents were made on the compacted layer, as shown in Fig. 26.1, before commencing the compaction of the next layer. The compaction proceeds in two and three stages for 1.5 and 2.1 m height walls, respectively.

The soil and 10% cement (by mass) were mixed in dry condition in a mixer. The dry mix was spread into a thin layer and then required quantity of water (OMC) was sprayed onto it in stages as the mixing continued. It was ensured to spread the moisture uniformly in the cement-soil mix. The mass of the partially saturated soil-cement mix going into each compacted layer (of 100 mm thickness) was monitored such that the dry density of the specimen was maintained at 1800 kg/m^3. The specimens were cured under the wet burlap. After 28 days of curing, the specimens were air dried and then dried in a drying chamber (Fig. 26.2) at 45–50 °C until constant mass was obtained. Monitoring the mass of a small piece of CSRE specimen kept on top of the wall specimen in the drying chamber ensured constant mass. The dried specimens were used in testing for the compressive strength.

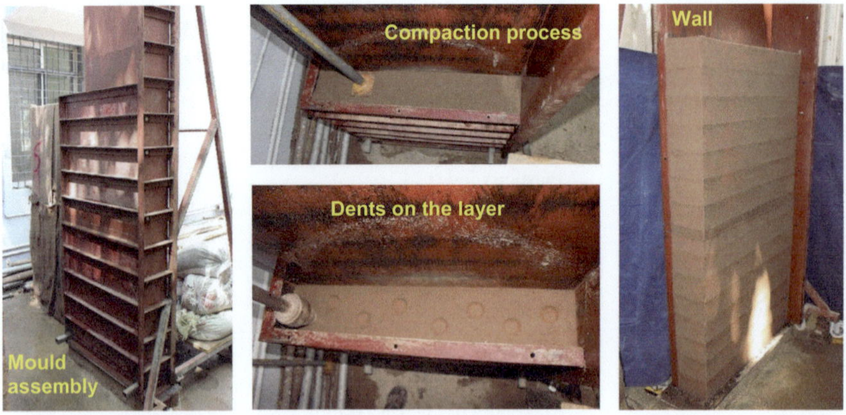

Fig. 26.1 Metal mould assembly, compaction process and CSRE wall specimen

Fig. 26.2 Drying set-up for CSRE walls

26.6 Testing Procedure for the CSRE Walls

The dried wall specimens were tested in a displacement controlled test rig. Figure 26.3 shows the wall testing set-up. The top and bottom of the wall had hinged end conditions supported on greased metal rollers. The lateral deformations of the walls were captured through LVDT's mounted as shown in the figure. The dry wall specimen after mounting in the loading test rig was subjected to uniform compression at a piston displacement rate of 6 μ/s until the specimen fails. The brittle failures were catastrophic and sudden. The tests revealed the failure load as well as the lateral deflections of the wall specimen.

Fig. 26.3 Compression test set-up for CSRE wall

26.7 Results and Discussion

The experimental results generated reveal (a) the influence of the wall slenderness on the compressive strength and (b) the reduction in wall compressive strength due to eccentric loading. The details of the results are discussed below.

26.7.1 Slenderness and Compressive Strength

A plot of compressive strength (for concentric loading) and *h*/*t* ratio (i.e. slenderness) is shown in Fig. 26.4. The values shown in the figure are the mean of three experiments. There is a linear relationship between the compressive strength and the wall slenderness. The compressive strength of the CSRE wall decreases by 14% as the *h*/*t* ratio was increased from 6 to 14. Reddy and Kumar (2011b) observed a reduction of 28% for the increase in *h*/*t* ratio from 5 to 20. There will be a considerable difference in the dry and wet compressive strength of CSRE (Reddy and Kumar 2011a; Reddy et al. 2016). In the present study, the CSRE specimens have nearly similar moisture content (~1.5%) during testing and hence, the comparison of compressive strength values is not affected by the moisture content. The CSRE strength is not affected by the *h*/*t* ratio in the range of 2–6 (Reddy et al. 2016). Typical failure patterns of the CSRE specimen are shown in Fig. 26.5. The figure shows progressive failure starting from crack initiation, deflected profile and the material failure. It is basically a material failure (shear failure) irrespective of the *h*/*t* ratio. Similar failures were observed by Reddy and Kumar (2011b) for the tests on slender ($h/t = 20$) CSRE walls.

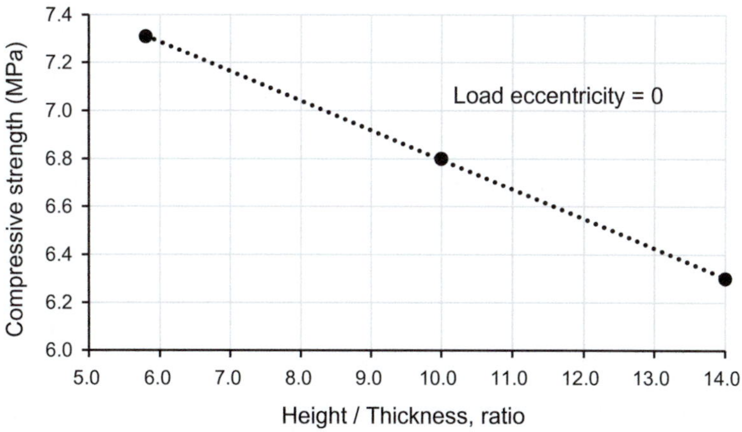

Fig. 26.4 Compressive strength versus height to thickness ratio

Fig. 26.5 Typical failure pattern for CSRE walls under zero eccentric compression

26.7.2 *Effect of Load Eccentricity on Compressive Strength of CSRE Walls*

The strength variation with load eccentricity for CSRE walls is plotted in Fig. 26.6. The values shown in the figure are the mean of three experiments, except for the cases of 1/3 eccentricity. In the case of 1/3 eccentricity, and for *h*/*t* ratio of 10 and 14, the number of specimens tested was one and two, respectively. The walls have similar moisture content at the time of testing (~1.5%). The results show that the compressive strength of the CSRE wall is sensitive to the load eccentricity. The strength decreases with the increase in eccentricity ratio. There is a linear relationship between strength

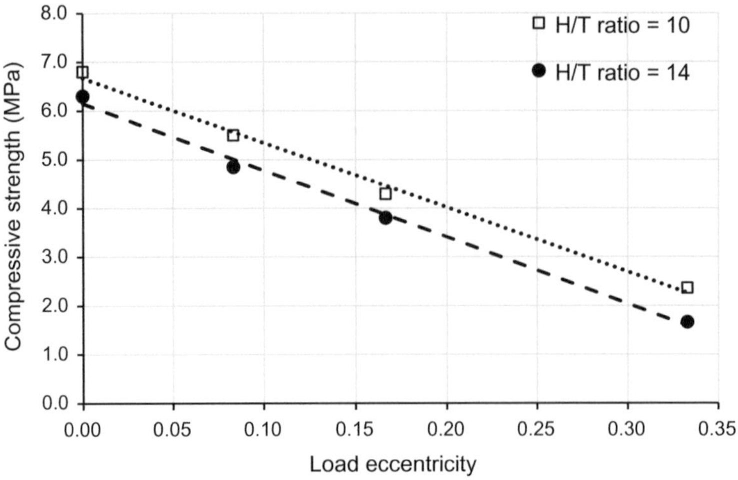

Fig. 26.6 Effect of load eccentricity on compressive strength of CSRE walls

and eccentricity ratio. The strength reduces by 65 and 75% as the load eccentricity increases from 0 to 1/3 for the walls with *h*/*t* ratio of 10 and 14, respectively. If the eccentricity ratio crosses 1/3, a portion of the wall will be under tension (Chapman and Slatford 1957).

26.7.3 Load Deflection Curves

Figure 26.7 shows the lateral deflection profiles for the CSRE wall with *h*/*t* ratio of 10. The profiles for 90% of the vertical load are shown in this figure, for different load eccentricities. The lateral deflection for the wall increases as the load eccentricity increases. For example, 1.5 m height wall ($h/t = 10$) shows a mid-height lateral deflection (at 90% failure load) as 2.69, 3.46, 4.29 and 4.77 mm, for the load eccentricity of 0, 1/12, 1/6 and 1/3, respectively. The mid-height lateral deflection doubles as the load eccentricity changes from 0 to 1/3.

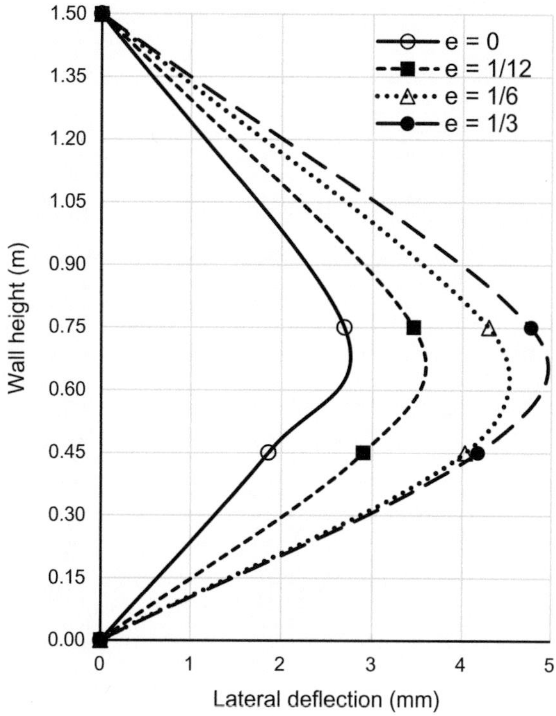

Fig. 26.7 Lateral deflections under concentric and eccentric loads at $0.9P_{ult}$ for 1.5 m wall

26.7.4 Failure Patterns Under Eccentric Loading

Typical failure patterns for eccentrically loaded walls are shown in Fig. 26.8. Due to considerable lateral deflection under eccentric loading, flexural tension crack develops across the wall width on one of the wall face and the opposite wall face experiences crushing of the material due to eccentric compression. The flexure tension crack is initiated at the interface of the compacted layers. Considerable lateral deflection can be seen before the wall collapses.

26.8 Concluding Remarks

Influence of slenderness and load eccentricity, on compressive strength of CSRE walls, was discussed. For the materials, specimens and test protocols presented herein, the results reveal that there is 14% reduction in strength as the wall slenderness increases from 10 to 14. The strength reduction is substantial (65–75%) with

Fig. 26.8 Crushing and flexure tension failures for CSRE walls under eccentric compression

the increase in load eccentricity from 0 to 1/3. The walls show material failure under concentric loads. In the case of eccentric loading, the walls show considerable lateral displacement and the failure is initiated by the development of tension crack growing at the interface of compacted layers.

Acknowledgements The authors would like to thank the IFCPAR/CEFIPRA for providing funds under the Indo-French collaborative project on "Developing design guidance for rammed earth construction". Some of the experimental facilities created under a project sponsored by the Department of Science and Technology (DST), Government of India were used in the experiments. The authors thank DST for this support.

References

Chapman JC, Slatford J (1957) The elastic buckling of brittle columns. Proceedings of ICE London 6:107–125

Ciancio D, Beckett C (2015) Rammed earth construction. In: Proceedings of 1st International conference on rammed earth construction. CRC Press, Perth, 10–13 Feb 2015

Easton David (1982) The rammed earth experience. Blue Mountain Press, Wilseyville, CA

Hall M (2002) Rammed earth: traditional methods, modern techniques, sustainable future. Build Engineer 77(11):22–24
Hall M, Djerbib Y (2004) Rammed earth sample production: context, recommendations and consistency. Constr Build Mater 18(4):281–286
Houben H, Guillaud H (2004) Earth construction—a comprehensive guide. London Intermediate Technology Publications, London
IS 12269 (1987) Specification for 53 grade ordinary Portland cement. Bureau of Indian standards, New Delhi, India
IS 4031 (part 7) & (part 5) 1988, Methods of physical tests for hydraulic cement, Bureau of Indian standards, New Delhi, India
Jayasinghe C, Kamaladasa N (2007) Compressive strength characteristics of cement stabilized rammed earth walls. Constr Build Mater 21(11):1971–1976
Middleton GF, Schneider LM (1995) Earth wall construction, 4th edn. Commonwealth Scientific and Industrial Research Organisation (Division of Building Construction and Engineering), North Ryde, NSW, Australia
Reddy BVV, Kumar PP (2011a) Cement stabilised rammed earth Part B: compressive strength and stress-strain characteristics. Mater Struct 44(3):695–707
Reddy BVV, Kumar PP (2011b) Structural behavior of story high cement-stabilized rammed-earth walls under compression. J Mater Civ Eng 23(3):240–247
Reddy BVV, Kumar PP (2009) Compressive strength and elastic properties of stabilised rammed earth and masonry. J Int Soc 22(2):39–46
Reddy BVV, Georg L, Sreeram VS (2014) Low embodied energy cement stabilized rammed earth building—a case study. Energy Build 68:541–546
Reddy BVV, Suresh V, Nanjunda Rao KS (2016) Characteristic compressive strength of cement stabilised rammed earth. J Mater Civ Eng 29(2):04016203
Tripura DD, Singh KD (2015) Axial load-capacity of rectangular cement stabilized rammed earth column. Eng Struct 99:402–412
Verma PL, Mehra SR (1950) Use of soil-cement in house construction in the Punjab. Indian Concr J 24:91–96
Walker P, Standards Australia (2002) The Australian earth building handbook, Standards Australia International, Sydney, Australia
Walker P, Keable R, Martin J, Maniatidis V (2005) Rammed earth design and construction guideline. BRE Bookshop, UK

Chapter 27
Alternative Methods in Numerical Modelling of Earth Masonry Under Seismic Loading

K. P. I. E. Ariyaratne, Chintha Jayasinghe, M. T. R. Jayasinghe and Pete Walker

27.1 Introduction

27.1.1 Background

Masonry buildings have been very popular as a low-rise structural type due to a series of advantages such as relatively low cost, availability of materials, thermal efficiency, sound insulation and adequate durability.

Masonry construction is dated back to the ancient era for the construction of dome structures, temples, load-bearing low-rise buildings, etc. Even at present, masonry construction is popular among the house builders. However, conventional masonry material manufacturing consumes a considerable amount of natural resources and the production process emits significant amounts of CO_2 to the atmosphere. This leads to a set of environmental issues which in turn warrants a need of alternative building materials such as earth-based masonry.

K. P. I. E. Ariyaratne (✉) · C. Jayasinghe · M. T. R. Jayasinghe
Department of Civil Engineering, University of Moratuwa, Moratuwa, Sri Lanka
e-mail: indunilerandi@yahoo.com

C. Jayasinghe
e-mail: chintha@uom.lk

M. T. R. Jayasinghe
e-mail: thishan@uom.lk

P. Walker
Department of Architecture and Civil Engineering, University of Bath, Bath, UK
e-mail: P.Walker@bath.ac.uk

B. V. V. Reddy et al. (eds.), *Earthen Dwellings and Structures*,
Springer Transactions in Civil and Environmental Engineering,
https://doi.org/10.1007/978-981-13-5883-8_27

With the aim of promoting sustainable construction technology, earth has been re-introduced as a raw material for the wall construction of low-to-medium-rise buildings. In the past two decades, several researchers have established the engineering properties of rammed earth (RE) and compressed stabilized earth blocks (CSEB). Most of the studies have given the emphasis on strength properties due to static loads. Performance of earth masonry under dynamic loading is yet to be established with detailed studies. Most of the dwellings are constructed using masonry materials. Since such masonry buildings have not been designed for seismic loads and those are highly vulnerable to seismic events, and hence, the severity of the damage is substantial which could even result in some fatalities.

Therefore, there is a necessity of investigating dynamic performance of earth masonry structures in regions of moderate seismicity to enhance the seismic capacity. The dynamic properties of structural elements made out of earth masonry such as rammed earth and compressed earth blocks (stabilized/un-stabilized) have been investigated rarely with experimental studies. Further to that, numerical studies with computer modelling have occasionally been carried out on earth masonry. Such numerical analysis will be made useful by validating with experimental results, so that more resource intensive and repetitive laboratory trials for varying types of masonry could be minimized.

In this paper, attention has been paid for analysing alternative numerical models using two different computer softwares. Further the experimental data have been used to validate the numerical results and find their limitations.

27.1.2 Aim and Methodology

The aim of this study is to develop numerical models using different computer software and then use currently available experimental results to validate and to find their limitations.

27.2 Numerical Studies Carried Out on Masonry Structures

Dolatshahi and Aref (2011) did a comprehensive numerical model with explicit computational procedures available in "ABAQUS" software because in the implicit analysis, several researchers found that numerical model cannot follow the masonry wall up to failure due to convergence issues. Therefore, the numerical model results were limited. From this method, bidirectional cyclic deformation of masonry walls

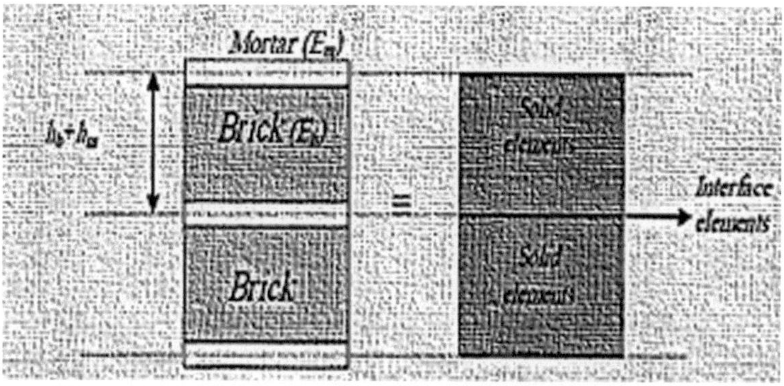

Fig. 27.1 Detailed model of brick and mortar (Dolatshahi and Aref (2011)

could be studied. The main objective of this study was to investigate the failure modes of un-reinforced masonry walls for various loading directions. Bricks and mortar were modelled as solid (C3D8R) and plane interface elements (COH3D8), respectively. Bricks were expanded by half the mortar dimension in both directions as in Fig. 27.1, and they were divided into two parts for capturing the exact behaviour and crack propagation of the wall.

For the joints, elastic and plastic behaviour were assumed. The failure modes for in-plane and out-of-plane loading were the diagonal crack and rocking mode, respectively. The numerical model was validated with past experimental studies.

Meillyta (2012) studied the performance of un-reinforced masonry walls with openings under horizontal loads by developing load–drift relationship. Finite element (FE) method (ABAQUS "software" with explicit solver), continuum elements and inelastic constitutive model (Drucker Prager) were chosen to numerically model the un-reinforced masonry wall. Interaction between the bricks was modelled using normal and tangential behaviour available in interaction module in ABAQUS. Static friction coefficient value of 5 and kinetic friction coefficient value in between 0.5 and 0.75 were used in modelling. The bottom face of the masonry was restrained for all translating degrees of freedoms. The model was validated using the load displacement results from an experimental un-reinforced masonry (URM) wall without openings.

Tarque et al. (2012) studied the non-linear seismic behaviour of adobe structures with numerical analysis since the experimental tests are costly and due to their limitations. The numerical model was validated with the experimental studies done at the Pontificia Universidad Católica del Perú by Blondet et al. (2006). The concrete damaged plasticity model was used in which the adobe was considered as an isotropic material. Further, the adobe masonry can be considered as a homogeneous material because adobe and mortar are basically made of mud. The material properties were obtained by Tarque (2011) in the plane cyclic test on adobe walls. The numerical model was subjected to an acceleration recorded at the base level in the experimental studies. The numerical model was validated fairly well with the experimental results under the crack pattern, failure mechanism and displacement response. The poor connection between the wooden beams of the roof and the adobe walls were simulated by reducing the element length to avoid the physical connection at the wall corners.

Betti et al. (2014) investigated the ability of estimating the seismic performance of un-reinforced buildings among different numerical models and analysis methods. The experimental model was a two-storey building tested on the shaking table under increasing natural ground motions. The first numerical model was built with the finite element method through a macro-model technique. The second numerical model was built using a macro-element approach. The main results of numerical and experimental studies have been compared. It has been concluded that FE model is able to predict damaged areas, initial collapse mechanism and collapse load.

The macro-element model is able to predict the collapse load accurately, but not the actual collapse mechanism. This method can be used only if the out-of-plane damage mechanisms are not initially activated. However, it can fairly estimate the fundamental dynamic response parameters by 6 degrees of freedom models. Therefore, paper suggested following both numerical and experimental approaches for the traditional, poor connected masonry buildings and locations like flexible floors where the global box behaviour cannot be assured.

Illampas et al. (2014) calibrated and validated a numerical model for adobe masonry building which is subjected to horizontal loading. For the experimental investigation, 1:2 scaled un-reinforced adobe masonry building was built and load was applied onto the rear wall using a hydraulic jack. The displacements were measured at the upper section of the rear wall, the façade and the side wall.

Experimental results concluded that adobe masonry structures subjected to horizontal loading are affected critically due to weak bonding between mortar joints and masonry units and lack of effective diaphragmatic function at roof level. Initiation of the damage was due to stress augmentation at the window corners and abutments of timber members. Further, the cracks formed were closed completely, leaving a damage indication with the removal of applied loads and they re-opened with re-applied loads. This observation reminds that an adequate inspection of earthen structures should be carried out after seismic events.

According to the experimental data, a 3D FE model was developed and calibrated with force–displacement response and failure mode. Isotropic damage plasticity constitutive law was adopted for numerical simulation. The FE analyses revealed that the global structural behaviour was affected by tensile response, and the structural behaviour of adobe masonry buildings subjected to horizontal loading is sufficiently accurate. For further investigation on the dynamic performance of adobe structures, the calibrated FE model was subjected to a time history analysis of a real earthquake.

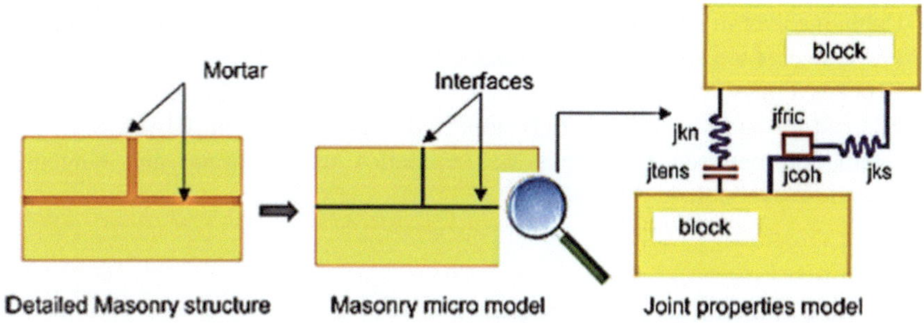

Fig. 27.2 Interface model (Çakti et al. 2016)

Ratnam (2014) has analysed dynamic behaviour of masonry wall panel of 1 m × 1.8 m with an opening made out of hollow cement stabilized soil interlocking blocks. Two masonry walls with and without reinforcement (sill and lintel band and vertical reinforcement) were tested to determine their in-plane cyclic performance. Lateral force was provided by a hydraulic actuator mounted horizontally at the height of the top surface of the wall. Lateral load and lateral displacement were measured. Further, a structural analysis program (SAP 2000) has been used to model reinforced and non-reinforced block walls and bamboo walls. It was concluded that lateral resistance and ductility of masonry walls were improved by the reinforcement.

Çakti et al. (2016) built a 1:10 scale model of the fifteenth-century Mustafa Pasha Mosque in Skopje and followed shake table tests. The experimental results of various dynamic excitations were used for the calibration of the discrete element model which represents masonry mosque and minaret by rigid blocks interacting via contact elements with tensile and shear bonds as in Fig. 27.2. Then, it was observed that numerical model can sufficiently simulate the time and frequency domain characteristics of low-level inputs and the damage regions. Generally, the discrete element approach can be used for the dynamic analysis of masonry structures which are relatively complex in laboratory conditions.

27.2.1 *Summary of Idealizations Used in Numerical Modelling of Masonry Structures*

- Micro- or macro-modelling technique was used.
- Bricks and mortar were modelled using C3D8R and COH3D8 elements in ABAQUS and shell elements in SAP.
- Bricks were expanded by half the mortar dimension in both directions.

- Explicit solver in ABAQUS was chosen as it is a computationally efficient and has low convergence problems over implicit method.
- Drucker Prager Plasticity Model or Concrete Damaged Plasticity Model was selected for quasi-brittle materials subjected to cyclic loads in ABAQUS.
- Interaction between the bricks was modelled using normal and tangential behaviour available in ABAQUS.

27.3 Experiments on Wall Panels

In this paper, experimental results of CSEB and RE wall panels which were subjected to time history and push over analysis, respectively, have been used for the comparison of two numerical applications.

27.3.1 Cement Stabilized Earth Blocks (CSEB)

Solid cement stabilized earth blocks which have been stabilized with 6% of cement were used as the main structural unit. Mortar joint thickness was 10 mm with cement: sand ratio of 1:6. According to the shake table capacities, wall panel dimensions were limited to 0.58 m × 0.58 m. Concrete layers were placed on the bottom and top of the walls for the confinement of the element. Two walls of the same size were subjected to moderate sized in-plane and out-of-plane seismic loading, and deflections were measured at the wall top, middle and bottom for each time increment of 0.00125 s.

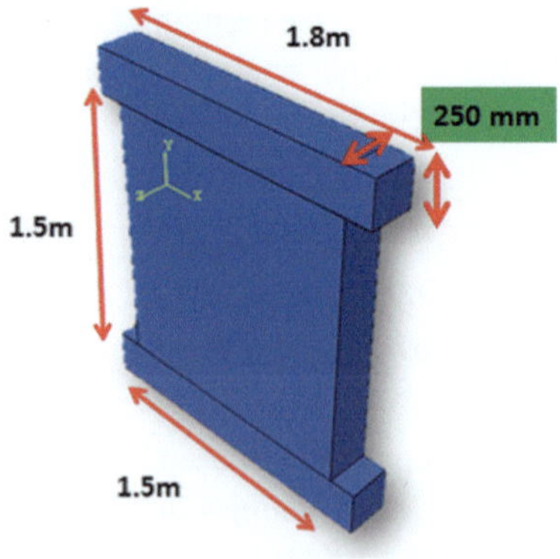

Fig. 27.3 Geometry of test specimens

27.3.2 *Rammed Earth (RE)*

Nabouch et al. (2016) constructed wall panels of size 1.5 m height × 1.5 m width × 0.25 m thickness by compacting earth layers using a pneumatic rammer. As shown in Fig. 27.3, top and bottom concrete layers were placed for the confinement of the element and to apply a horizontal load at the top of the wall. During the pushover test, vertical load of 0.3 MPa was applied on top of the beam to simulate the dead and live loads applied in a two storey building. The digital image correlation was performed to determine the displacement and crack propagation by comparing the images before and after the loading.

27.4 The Process of Numerical Modelling

The main steps of the dynamic analysis and the sequence of modelling the dynamic performance of earth walls using SAP and ABAQUS are shown in Figs. 27.4 and 27.5.

27.4.1 *SAP*

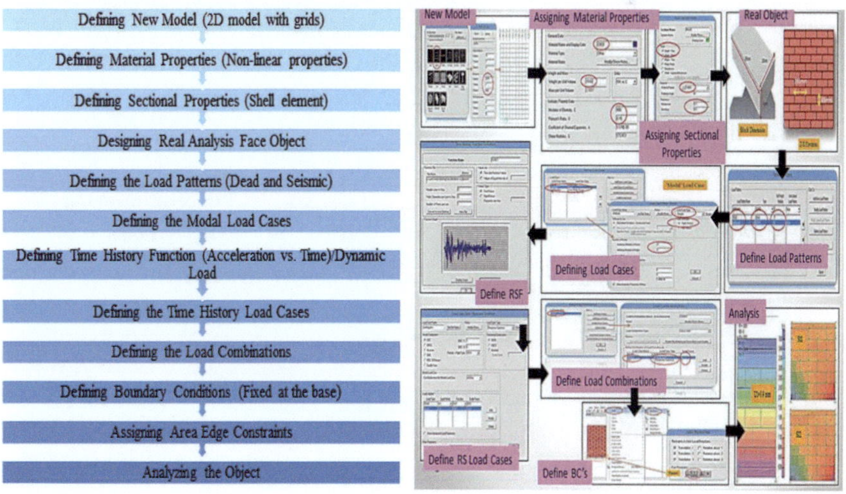

Fig. 27.4 Steps in SAP modelling

27.4.2 ABAQUS

The output of numerical analysis of CSEB and RE wall panels is presented in Table 27.1.

Table 27.2, illustrates the comparison of the results of numerical models of SAP and ABAQUS and other numerical modelling aspects.

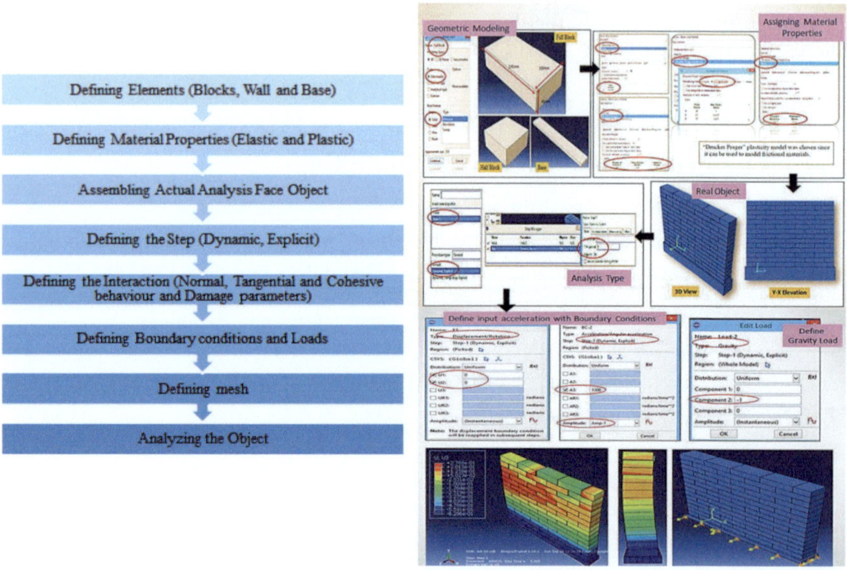

Fig. 27.5 Steps in ABAQUS modelling

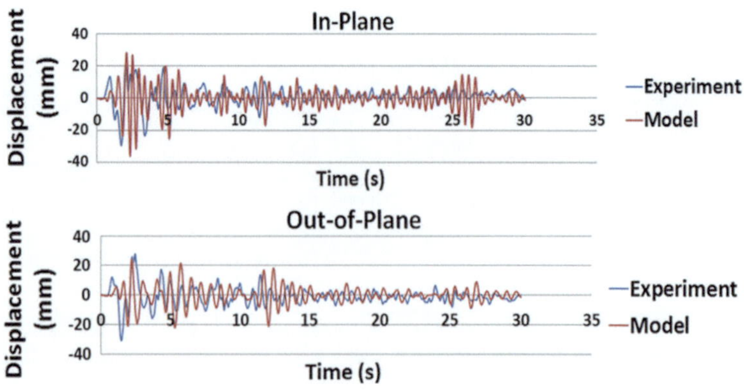

Fig. 27.6 Displacement versus time

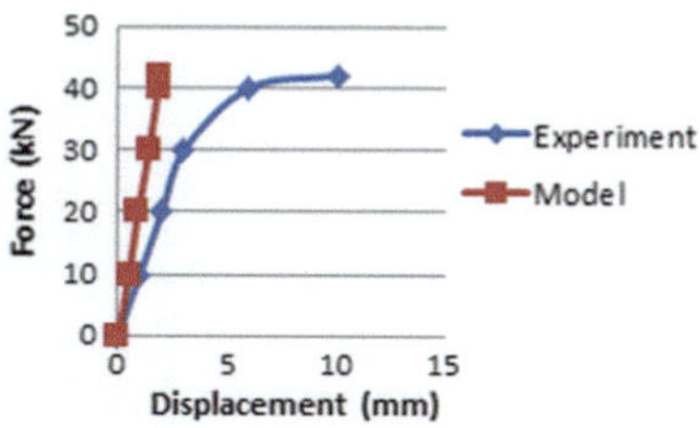

Fig. 27.7 Force versus displacement

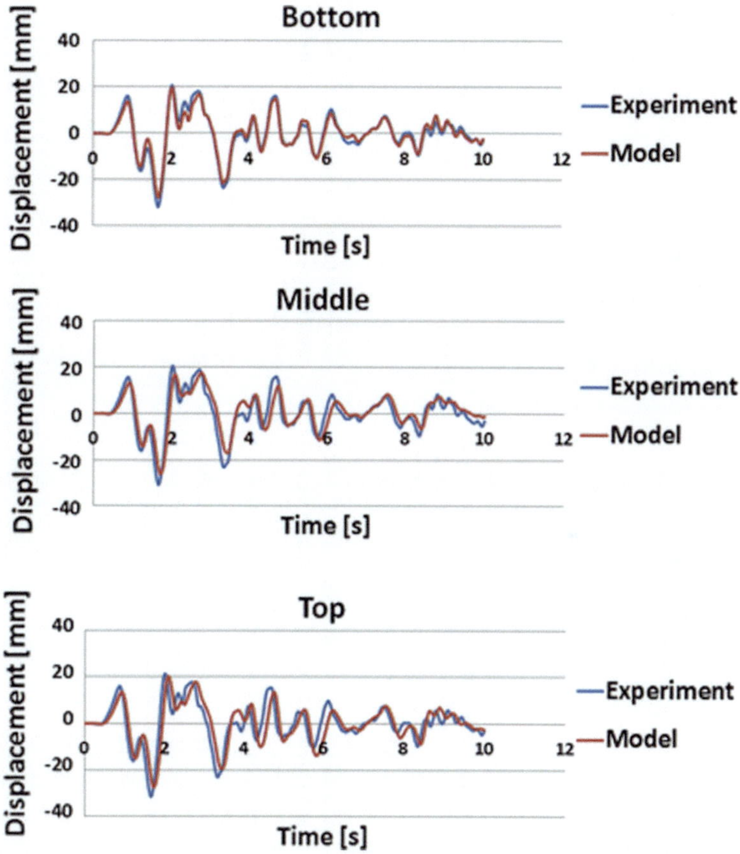

Fig. 27.8 Displacement versus time

Table 27.1 Numerical modelling results

	CSEB wall panel	RE wall panel
SAP	See Fig. 27.6	See Fig. 27.7
	Numerical displacement results in the middle and the top of the walls under the in-plane loading marginally coincide with the experimental results and for the out-of-plane loading, numerical results at the bottom and the middle of the walls also marginally coincide with the experimental results. The comparison graphs at the middle of the wall for in-plane and out-of-plane loading are shown in Fig. 27.6	For the top of the wall, numerical and experimental variation of the force versus displacement curve deviates with each other by more than 50%. The two curves for the experimental and numerical analysis are shown in Fig. 27.7
ABAQUS	Wall specimen was modelled using the solid element, concrete damaged plasticity model, dynamic explicit solver, tangential, normal, cohesive and damage interaction and applied the north–south component of the El-Centro earthquake in-plane and out-of-plane Numerical displacement results at the bottom, middle and the top of the walls under the in-plane and out-of-plane loading considerably coincide with the experimental results. The comparison graphs for out-of-plane loading are shown in Fig. 27.8	According to the above mentioned steps wall specimen was modelled using the solid element and pushover load was applied as point loads Under the in-plane loading, the experimental and numerical curves of force versus displacement converge to a greater extent as shown in Fig. 27.9

Table 27.2 Comparison of two computational models

Item considered	SAP	ABAQUS
Geometry modelling	Micro-macro 3D model	Micro-macro 3D model
Material properties	Anisotropic and Takeda hysteresis model	Elastic and concrete damage plasticity model
Assembling the actual element	Following grid pattern	Movement of elements
Defining the analysis type	Either modal or time history load case	Explicit solver
Interaction	Provide edge area constraints	Surface contact through tangential, normal, cohesive behaviour and damage
Defining the loads	Defining time history function	Defining displacement function and apply on the surface
Boundary condition	Fixed at the bottom (unless, huge distortions of the model)	Can apply displacement and acceleration constraints separately
Meshing the objects	Coarse mesh	Coarse mesh
Duration of analysis	Less than one minute	About half an hour
Displacements	Marginal convergence	Greater convergence

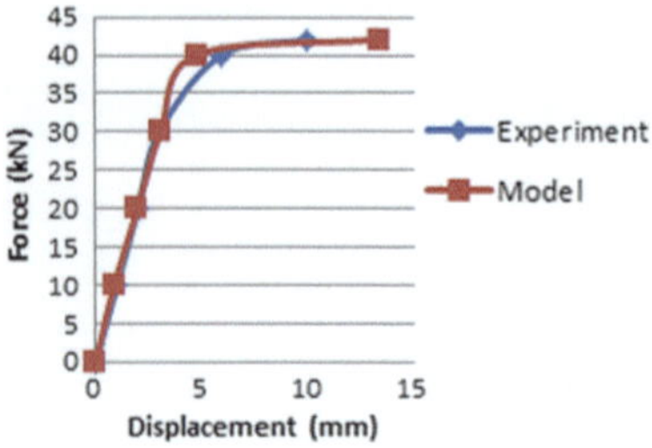

Fig. 27.9 Force versus displacement

27.5 Conclusion

Several attempts were made to assess the seismic performance of un-reinforced masonry with experimental and numerical studies using different computer softwares. The dynamic performance of earth masonry, in particular, was given a comparatively low attention by the past researchers.

Numerical analysis of dynamic behaviour of earth masonry will be given a higher prominence due to higher resource requirement needed for experimental studies. However, validation of numerical analysis with proper experimental programme presents more reliable results of the dynamic behaviour. This study has compared two different types of numerical modelling and proposed more reliable one.

In SAP, actual object was assembled according to the grid pattern using micro–macro-element approach. Masonry was assumed as anisotropic, and the Takeda hysteresis model was used. The bottom of the object should be fixed and edge area constraint must be applied to avoid model distortions.

In ABAQUS, actual object was assembled using 3D stress elements following micro–macro-element approach. Masonry was assumed as isotropic, and the concrete damage plasticity model was used. Surface interaction was applied through tangential, normal and cohesive behaviour and damage parameters. The dynamic explicit solver was used to analyse the object under displacement versus time function.

There are limitations in the SAP model to apply, true boundary and contact properties. Therefore, displacement values near the top and bottom of the wall do not coincide with experimental results. In the ABAQUS model, above limitations can be overcome, and hence, the results are considerably within the experimental values. Therefore, ABAQUS modelling of earth structures is good enough for evaluating their seismic performance compared to SAP models. But more sectional and material properties inherent to the actual structure should input in order to get accurate results and further solving time is much higher compared to SAP.

27.6 Future Work

In this paper, numerical comparisons were based upon the results of one type of experimental study. Therefore, better to confirm those observations with many experimental results by varying the parameters such as building height, wall thickness, scale of the structure, opening sizes, pre-compression load, block type, number of floors of the model and the interior structural arrangement of the model.

Acknowledgements Funding provided by University of Moratuwa and non-academic staff of Department of Civil Engineering, University of Moratuwa.

References

Betti M, Galano L, Vignoli A (2014) Comparative analysis on the seismic behaviour of unreinforced masonry buildings with flexible diaphragms. Eng Struct 61:195–208

Blondet M, Vargas J, Velásquez J, Tarque N (2006) Experimental study of synthetic mesh reinforcement of historical adobe buildings. In: Proceedings of structural analysis of historical constructions, New Delhi, India, pp 1–8

Çaktı E, Saygılı Ö, Lemos JV, Oliveira CS (2016) Discrete element modelling of a scaled masonry structure and its validation. Eng Struct 126:224–236

Dolatshahi KM, Aref AJ (2011) Three dimensional modeling of masonry structures and interaction of in-plane and out-of-plane deformation of masonry walls. In: Dolatshahi MK, Aref JA (eds) Engineering mechanics institute conference

Illampas R, Charmpis DC, Ioannou I (2014) Laboratory testing and finite element simulation of the structural response of an adobe masonry building under horizontal loading. Eng Struct 80:362–376

Meillyta (2012) Finite element modelling of unreinforced masonry (URM) wall with openings: studies in Australia. In: The proceedings of 2nd annual international conference Syiah Kuala University 2012 and 8th IMT-GT Uninet biosciences conference Banda Aceh

Nabouch R, Bui QB, Ple O, Perrotin P, Poinard C, Goldin T, Plassiard JP (2016) Seismic assessment of rammed earth walls using pushover tests. Proc Int Conf Sust Design, Eng Constr 145:1185–1192

Ratnam V (2014) Development of earthquake resistant designs, methodologies and construction technologies for masonry buildings in Sri Lanka. Research Project for Master of Engineering in Structural Engineering Design, Department of Civil Engineering, University of Moratuwa, Sri Lanka

Tarque N (2011) Numerical modelling of the seismic behaviour of adobe buildings. Ph.D. thesis, ROSE School, Istituto di Studi Superiori di Pavia IUSS, Pavia, Italy

Tarque N, Camata G, Spacone E, Blondet M, Varum H (2012) The use of continuum models for analyzing adobe structures. [a]Pontificia Universidad Católica del Perú, Peru. [b]Universita degli Studi 'Gabriele d'Annunzio' Chieti—Pescara, Italy. [c]Pontificia Universidad Católica del Perú, Peru. [d]Universidade deAveiro, Portugal

Chapter 28
Durability of Rammed Earth: A Comparative Study of Spray Erosion Testing and Natural Weathering

Inayath Kharoti, Pete Walker and Chintha Jayasinghe

28.1 Introduction

Lowering the emissions of greenhouse gases from various sources remains a major challenge for humanity. The production of building materials remains a significant contributor to such emissions. Finding alternatives, based on traditional vernacular materials, remains an active area for research and development. In this context, using earth as a modern construction material has regained popularity.

Natural earth construction methods have been used for thousands of years around the world and in many different climates. There are different types of earth buildings exist in various parts of the world (Minke 2000; Kamaladasa and Jayasinghe 2005; Reddy et al. 2014; Gray 2016; Kariyawasam and Jayasinghe 2016). The method of construction varies with the soil and the machinery available. In some countries, compressed stabilised earth blocks (CSEB) are popular, whilst in others rammed earth (RE) buildings have become more popular.

Rammed earth (RE) consists of moist soil compacted into a temporary formwork and rammed either manually or mechanically. RE walls can be left unplastered in dry climates though it needs to be protected in wet climates for durability (Gray 2016). Jayasinghe and Arandara (2007) have identified one of the key problems of unstabilised earth walls as is its susceptibility to water and the effects of driving rain over a period of time.

I. Kharoti
Department of Architecture and Civil Engineering, University of Bath, Bath, UK

P. Walker (✉)
BRE Centre for Innovative Construction Materials, University of Bath, Bath, UK
e-mail: p.walker@bath.ac.uk

C. Jayasinghe
Department of Civil Engineering, University of Moratuwa, Moratuwa, Sri Lanka
e-mail: chintha@uom.lk

B. V. V. Reddy et al. (eds.), *Earthen Dwellings and Structures*,
Springer Transactions in Civil and Environmental Engineering,
https://doi.org/10.1007/978-981-13-5883-8_28

Since concerns about the durability of earth buildings have been identified as one of the problems to hinder the popularity of earth building, various test methods have been developed to assess the performance under adverse weather conditions. Bui et al. (2009) monitored the natural weathering process of earth buildings whilst accelerated means of assessment was developed with the spray erosion apparatus, abrasion test and the drip test (Heathcote 2002). All those tests attempt to simulate natural weathering under different and accelerated conditions.

A strong and broadly consistent relationship between natural weathering and accelerated spray erosion testing is yet to be developed. The durability of rammed earth can be greatly improved using stabilisers such as cement. However, this increases carbon emissions associated with the material (Walker et al. 2005). One of the attractions of rammed earth walls is their aesthetic quality, and so they are often left without rendering or other protection on external surfaces. Whilst this reduces the cost of construction, it can leave the external surfaces exposed to weathering.

Out of the durability problems of natural rammed earth walls, loss of material due to erosion by wind driven rain has been identified as most significant (Heathcote 2002). There are several factors that contribute to the durability of earth walls. Some of those could be the particle size, composition of soil, compaction force, surface texture that varies with the construction method and the appearance needed. In order to popularise in the construction industry, rammed earth has to be established as a strong and durable material together with other desirable properties. Therefore, testing for durability has been given much attention. Accelerated spray erosion test (ASET) was developed in Australia (Middleton 1992) as a rapid means of assessing material suitability. However, the suitability of the ASET has to be assessed against actual weathering of unstabilised rammed earth walls subjected to water spray due to heavy rains. This paper covers an attempt made to compare the results of ASET with natural weathering.

28.2 Aims and Objectives

The work presented in this paper aimed to improve understanding in the weathering resistance (durability) of unstabilised rammed earth. Specific objectives were as follows:

- Evaluate the relative significance of different material factors that affect the durability of unstabilised rammed earth.
- Compare the natural weathering performance of rammed earth with accelerated erosion laboratory testing performance.

28.3 Methodology

In order to achieve the above objectives, the following methodology was adopted:

- The accelerated spray erosion test (ASET) was selected for the laboratory experiment on rammed earth specimens.
- The weathering of rammed earth specimens was assessed under full exposure to the natural environment with different weather conditions. This was carried out in the Building Research Park (BRP), Swindon, United Kingdom.
- An attempt was made to develop a correlation between the spray erosion test and weathering of rammed earth specimen exposed to the natural environment.

28.4 Experimental Programme

A detailed experimental programme was carried out with a selected rammed earth material to compare the laboratory testing on durability of rammed earth with that of natural weathering. The material was supplied by Claytec e.K., a German specialist supplier of materials for earth building. The rammed earth material was subjected to several laboratory testing, and two sets of rammed earth specimens were monitored for weathering in the laboratory as well as in the field. Figure 28.1 shows the apparatus used in the ASET. The rammed earth specimen is first mounted in the apparatus as shown, and then, a jet of water spray, delivered at 50 kPa pressure, is applied onto one face of the specimen for a period of one hour. The depth of erosion due to water spray is measured in 15-min time intervals.

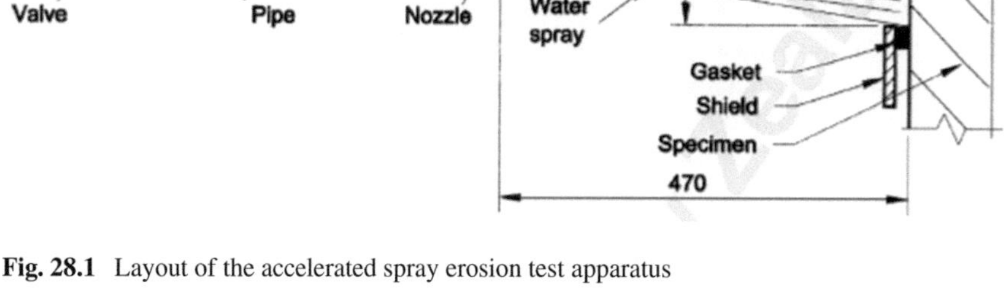

Fig. 28.1 Layout of the accelerated spray erosion test apparatus

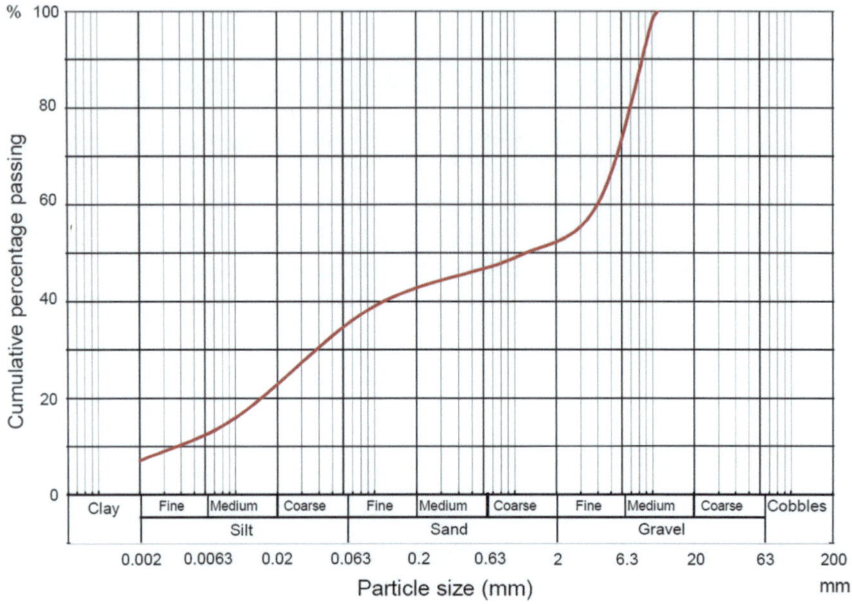

Fig. 28.2 Grain size distribution from wet sieve analysis

28.4.1 Soil Testing

Since the soil characteristics influence the selection of a material for rammed earth construction, the material chosen for the experimental programme was subjected to a set of laboratory testing to characterise the material.

The soil moisture content was determined in accordance with BS 1377-2, and the average moisture content of selected soil specimen was found to be in the range of 6.5%, which was maintained consistent throughout the experimental programme.

To determine the particle size distribution, wet sieve analysis and the hydrometer test were carried out (BS 1377-2). Figure 28.2 shows the particle size distribution curve. The fines content of the soil (below 63 μm) was around 35%, with just over 50% of the soil consisting of the particle size greater than 2 mm. The liquid limit and plastic limit of the selected soil were found to be 27 and 17%, respectively. ASTM D 1557-12 was used to determine Proctor density of the selected soil. The maximum dry density was found to be 2218 kg/m^3 at the optimum moisture content of 6.5%.

28.4.2 Preparation of Specimens

Two sets of specimens were prepared for the laboratory experiments and for the field study on natural weathering. The size of the specimen was maintained at 250 × 250

Fig. 28.3 Preparation of rammed earth specimens

Table 28.1 Variables introduced to the test specimens

Specimen No.	Maximum aggregate size (mm)	Moisture content (% dry mass)	Compaction	Surface finishes
1	14	6.5	25 blows per layer	Natural
2	14	6.5	25 blows per layer	Natural
3	14	6.5	25 blows per layer	Sol-gel applied
4	14	6.5	50 blows per layer	Natural
5	14	8.0	25 blows per layer	Natural
6	8	6.5	25 blows per layer	Natural
7	2	6.5	25 blows per layer	Natural

× 77 mm (thickness). The moisture content was maintained at 8% in the Specimen 5 to assess the effect of higher moisture content, whilst it was maintained at 6.5% in all other six specimens. The soil specimens were sieved to have the maximum aggregate size of 14 mm, and for some specimens, the maximum aggregate size was maintained at 8 mm. The soil was compacted, in five equal layers, into the metal U-shaped frame shutters. A 5 kg Proctor hammer was used to compact the soil in layers. Each layer was usually compacted with 25 blows. The average thickness of each compacted layer was maintained as 50 mm.

Specimen 4 was compacted by 50 blows for each layer to assess the effect of compaction on durability. Specimen 3 was applied with a sol-gel surface coating to find out the resistance to weathering with a finishing coat. Figure 28.3 shows the formwork and preparation of rammed earth specimens, including levelling of the top surface level. Table 28.1 presents the variables introduced to the seven specimens.

28.4.2.1 Specimens Cast for Natural Weathering

A metal frame was made out of steel that was painted to protect it from corrosion. As shown in Fig. 28.4, the steel frame was fabricated to hold the rammed earth specimens. The frame was designed to hold the rammed earth specimens at a sufficient height from the ground level to protect the test specimens from rainwater splashing. It was also covered at the top to prevent any rainwater seeping into the specimens.

Fig. 28.4 Specimens mounted in steel frame

C-section steel channels were welded together to form U-shaped casing in which to ram the soil specimens and to hold them during the natural weathering process.

28.4.2.2 Specimens Cast for Spray Erosion Test in the Laboratory

The sieved soil was rammed and the formwork was removed as shown in Fig. 28.5. The specimens were kept in the environmental chamber for one-week duration under controlled temperature of 19 °C and 50% relative humidity before testing. Figure 28.6 shows the ASET is being conducted.

Fig. 28.5 Casting of test specimen for the ASET

Fig. 28.6 Accelerated spray rig set up and in use

28.5 Results and Analysis

The results of the natural erosion and the laboratory spray erosion test are presented and compared below.

28.5.1 Results of the Natural Erosion

Natural erosion test was started in August 2017. After their preparation, as described above, the specimens were transported and set up in the steel frame at the Building Research Park, Swindon (UK). In this paper, their performance over their initial period of four weeks is reported, though this work is part of an ongoing study. Figure 28.7 presents the collection of photographic evidence for each specimen week by week. The test rig was placed facing south-west, in the direction of the prevailing wind. The site is located approximately 200 m above sea level. To measure natural weathering of the specimens, a 150–mm-diameter Perspex template, consisting of 81 holes spiralling towards the centre, was used. The changes in surface depth were measured using a digital dial gauge with a precision of 0.01 mm. Periodic measure-

Fig. 28.7 Photographs of natural weathering of specimens taken periodically

Table 28.2 Rate of natural weathering of test specimens

Specimen No.	Natural weathering rate (mm/year)
1	16.4
2	14.5
3	13.4
4	30.5
5	6.8
6	14.1
7	11.4

ments were taken for the depth of erosion. Over the test period, 90 mm of rainfall was measured, slightly higher than the average monthly rainfall.

The surfaces of the specimens visibly changed with increasing rainfall. As seen in Fig. 28.7, surface finer material was initially washed away exposing the coarser aggregates behind. The compaction layers become more evident with time as well. Measurement of erosion depth was complicated somewhat by the swelling of surface clays with the wetting. Consequently rather than showing loss of section, some specimens initially showed gain in section depth. Over the test period, the rate of erosion was equivalent to between 6.8 and 30.5 mm/year. Specimen 5 (higher compaction moisture content) showed least erosion, whilst Specimen 4 (higher compaction) showed most (Table 28.4). These findings are contrary to expectations, possible complicated by the clay swelling and limited time frame, with further measurements to establish longer term trends ongoing (Fig. 28.8).

The rate of natural weathering has been calculated in millimetres per year and presented in Table 28.2.

Fig. 28.8 A template fixed firmly onto specimen for measurements

28.5.2 Results of the Accelerated Spray Erosion Test (ASET)

The results of the accelerated spray erosion test are presented in Table 28.3. The measured erosion rates varied between 2.6 and 6.9 mm/min. The rate of erosion was evaluated by dividing the thickness of the specimen by the time taken for the full penetration to occur. At these rates, all specimens failed the test before 60-min exposure was reached.

Table 28.3 Results of the spray erosion test

Specimen No.	Rate of erosion (mm/min)	Weight loss (g)
1	6.9	2312
2	6.3	2536
3	5.7	1629
4	2.6	2029
5	2.9	990
6	4.4	1672
7	8.1	2201

Table 28.4 Comparison of natural weathering with laboratory testing

Specimen No.	Natural weathering		Accelerated erosion test	
	mm/year	Ranking	mm/min	Ranking
1	16.43	6	6.88	6
2	14.53	5	6.33	5
3	13.44	3	5.71	4
4	30.45	7	2.64	1
5	6.79	1	2.93	2
6	14.11	4	4.42	3
7	11.44	2	8.14	7

The ASET results show very little or no correlation to that of the natural weathering tests to date. For the accelerated spray testing, the specimens have been subjected to spraying through to full penetration, none of which lasted the full 60-min set out in the standard. Specimens 4 and 5 showed the highest resistance to the accelerated spray test lasted just over 29 and 26 min, respectively. Specimens 1 and 2 have the same aggregate size, compaction and moisture content. However, the Specimen 2 has shown a higher weight loss and lower erosion rate than that of Specimen 1. Specimen 3 which was applied with a sol-gel coating has shown relatively low erosion rate and weight loss. Specimen 4 with higher compaction has shown the lowest erosion rate although it was showing higher weight loss. The specimen with higher moisture content with the compaction has shown more or less the similar erosion rate with lower weight loss. Specimens 6 and 7 with relatively finer gravel sizes has shown higher erosion rate and higher weight loss. The seven specimens are compared on the performance based on spray erosion test and the field study. Specimens are assigned with a ranking and presented in Table 28.4.

Except for specimens 4 and 7, the relative ranking of the natural erosion and ASET is quite consistent.

28.6 Concluding Remarks

Since rammed earth (RE) is promoted as a sustainable building material for wall construction, it is important to ensure its durability. Acceleration spray erosion test (ASET) has been identified as a method of testing for the durability of rammed earth.

The results have not demonstrated a direct and accurate correlation between the ASET and actual weathering. When the rankings were considered, specimens 1, 2, 3, 5 and 6 have been assigned with either same or close ranking with respect to erosion due to water spray. This has shown that the ASET has a potential to assess the durability of RE, but there are important inconsistencies (specimens 4 and 7) too that need to be understood if the test is to be used more widely.

The higher compaction effort exhibited lower rate of erosion in the ASET. The specimen applied with sol-gel treatment has shown relatively lower erosion rate, paving the path for sol-gel to be used as a durability improvement measure. The effect of maximum particle size has shown mixed results, and further testing is needed for a conclusive findings.

With the useful trend patterns observed in this research, ASET can be recommended as a potential test for rapid durability assessment of unstabilised rammed earth. However, long-term testing is needed to establish a stronger correlation between the ASET and natural weathering. This work is ongoing.

References

ASTM International (2012) Standard test methods for laboratory compaction characteristics of soil using modified effort. ASTM

BS 1377 (1990) Methods of test for soils for civil engineering purposes—BS1377-2:1990. BSI

Bui Q, Morel JC, Reddy BVV, Ghayad W (2009) Durability of rammed earth walls exposed for 20 years to natural weathering. Build Environ 44(5):912–919

Gray J (2016) Earth and construction. [Online] Available at http://www.sustainablebuild.co.uk/constructionearth.html. Accessed August 2017

Heathcote KA (2002) An investigation into the erodibility of earth wall units, PhD, University of Technology Sydney

Jayasinghe C, Arandara K (2007) Identification of durability problems in earth buildings. The Institution of Engineers, Sri Lanka

Kariyawasam KKGKD, Jayasinghe C (2016) Cement stabilized rammed earth as a sustainable construction material. Constr Build Mater 105:519–527

Kamaladasa N, Jayasinghe C (2005) Development of an efficient construction Technique for Rammed Earth. Transactions, Institution of Engineers

Middleton G (1992) *Bulletin 5: Earth-wall construction*, 4th edn. CSIRO, Australia, North Ryde

Minke G (2000) Earth construction handbook. WIT Press, UK

Reddy BVV, Leuzinger G, Sreeram VS (2014) Low embodied energy cement stabilised rammed earth building. Energy Build 68:541–546

Walker P, Keable R, Maniatidis V, Martin J (2005) Rammed earth: Design and construction guidelines. IHS Press, BRE, Watford

Part III
Energy and Environmental Performance: Climatic Response and Thermal Performance

Chapter 29
Error Analysis on Thermal Conductivity Measurements of Cement-stabilized Soil Blocks

N. C. Balaji and Monto Mani

29.1 Introduction

The selection of materials in construction depends on their thermo-physical properties and thermal performance, because the materials regulate the indoor thermal comfort. The thermal properties of building materials are extensively available from various literatures representing various countries (Clarke et al. 1990). The basic thermo-physical properties such as density, specific heat capacity, and thermal conductivity of building materials have been extensively studied.

Cement-stabilized soil blocks (CSSBs) are modern alternative masonry units. CSSBs are also known as soil–cement blocks, stabilized soil/clay blocks, and compressed stabilized earth blocks. They are considered as low carbon and low embodied energy material. CSSBs do not require firing and can save up to 60–70% embodied energy compared with burnt clay bricks (Venkatarama Reddy 2009). The structural and mechanical characteristics of CSSBs have been extensively studied (production technique, density, soil–sand mix, soil–cement mix designs) and very well established in terms of varying clay content, cement content, and densities for durability and strength. Few studies have investigated the effects of CSSB constituents such as clay, cement content, density, and associated parameters (such as void ratio, porosity, and degree of compaction) on thermal conductivity of blocks. As the cement content increases, thermal conductivity of blocks affects pore size and pore structure (Horpibulsuk et al. 2011). Venkatarama Reddy and Gupta (2005) showed that the pore size of the blocks varies with the cement content and pore size decreases with an increase in cement content. The pores can affect the thermal conductivity

N. C. Balaji (✉)
Department of Civil Engineering, The National Institute of Engineering, Mysore, India
e-mail: balajinallaval@gmail.com

M. Mani
Centre for Sustainable Technologies, Indian Institute of Science, Bangalore, India
e-mail: monto.mani@gmail.com

B. V. V. Reddy et al. (eds.), *Earthen Dwellings and Structures*,
Springer Transactions in Civil and Environmental Engineering,
https://doi.org/10.1007/978-981-13-5883-8_29

of materials (Sugawara and Yoshizawa 1961). Bhattacharjee (1989) studied thermal conductivity variation for varying porosity and cement contents in soil–cement blocks. The thermal conductivity obtained for blocks with cement content varying from 4% to 10% and porosity ranging from 36 to 43% were 0.5009–0.7675 W/(m K). Generally, it was known that thermal conductivity is a function of density (Jakob 1949; Koenigsberger 1975), which is a function of porosity (Van Straaten 1967). Lesser the density means higher the air-filled pores in the material, contributing to lower thermal conductivity. However, the thermal conductivity of CSSBs has been extensively studied and results were reported (Bhattacharjee 1989; Adam and Jones 1995; Nagih and Ali 1995; Bahar et al. 2004; Khedari et al. 2005; Balaji et al. 2017). Further, due to scanty information in the literature on the thermal conductivity value variations, it is essential to understand the statistical inferences.

However cautiously planned and instrumented, no comparable experimental results are alike or give comparable/similar values. Some errors were associated with the experimentally measured values. This associated error can be statistically calculated by means of sample mean, coefficient of variation, and standard deviation values for particular experimental values. Errors involved in the testing materials are important to understand the reliable use of the test results obtained. Mosquera et al. (2013) performed repeatability and reproducibility tests on the adobe bricks from Amayuelas, Spain. The repeatability tests were conducted on the thermal conductivity of adobe bricks by a series of tests run on the same spot of the same sample of each type. Feng et al. (2015) showed the importance of the error analysis of measured data by analyzing the material errors, repeatability errors, between-lab errors, and reproducibility errors in determining the hygric properties of porous building materials. Salient observations from the earlier investigations are detailed error analysis that is required to understand the reliability of measured thermal properties of building materials.

The main objective of the study was to investigate the CSSBs thermal conductivity values error analysis through statistical analysis and also to study the significance of the material test results for errors such as repeatability and material errors. The errors due to test conditions, methods, and instrumentations used are associated with repeatability error, and an error caused by heterogeneity (composite nature) of materials is termed as a material error. The scope of the study is limited to measure the thermal conductivity of CSSBs with varying mix proportions by varying cement content (5, 8, 12, and 16%), clay content (10.5, 16, and 31.6%), and dry densities (1.7, 1.8, and 1.9 g/cc) in the study under room temperature, and calculate the associated measured errors of thermal conductivity values.

29.2 Experimental Program/Experimental Study

29.2.1 Selection of Building Materials

The CSSBs building material having low embodied energy (EE) was selected having EE in range of 0.45–0.85 MJ/kg (Hammond and Jones 2008; Sabapathy 2011; Hashemi et al. 2015). CSSBs with varying clay content, cement content, and densities were considered for the study. As such, thermal property data related to these materials are scanty and extensive data are required to understand the performance of CSSB as building envelopes.

29.2.2 Materials Used for CSSBs

Locally available natural red loamy soil comprising sand (48%), silt (28.6%), and clay (23.4%) was used for block making. This soil was reconstituted by blending river sand in three ratios so that the resulting soil–sand mixture comprised varying clay, silt, and sand fractions. The characteristics of clay mineral were altered; ordinary Portland cement conforming to IS 8112—1989 (BIS 1989) was used as a binder/stabilizer material.

29.2.3 The Manufacturing Process

CSSBs of dimensions 230 × 100 × 75 mm^3 for eight mix types with varying percentages of clay contents, cement contents, and dry density were studied (see Table 29.1). The blocks were cast by controlling the dry density of the blocks. The density of the material in this study is referred to as dry density and is determined during mix design and casting of the CSSBs. A total of 48 blocks consisting of six blocks in each category were cast.

29.2.4 Thermal Conductivity

Thermal conductivity (λ) of a material indicates the heat flux for a given temperature gradient. As per IS 3792—1978 (BIS 1978), thermal conductivity is "the quantity of heat in the 'steady-state' conditions flowing in unit time through a unit area of a slab of uniform material of infinite extent and of unit thickness, when unit difference of temperature is established between its face." In this study, the testing instrument QTM-500 (Fig. 29.2a), manufactured by Kyoto Electronics Manufacturing Co., Ltd. (KEM) Japan, was used to measure thermal conductivity, based on the transient hot

Table 29.1 Details of CSSBs having different clay, cement contents, and dry densities and their designation

S. No.	Material designation	Clay content (%)	Cement content (%)	Dry density (g/cc)
1	CSSB 31.6	31.6	8	1.7
2	CSSB 10.5	10.6	8	1.7
3	CSSB 5	16	5	1.7
4	CSSB 12	16	12	1.7
5	CSSB 16	16	16	1.7
6	CSSB 1.7	16	8	1.7
7	CSSB 1.8	16	8	1.8
8	CSSB 1.9	16	8	1.9

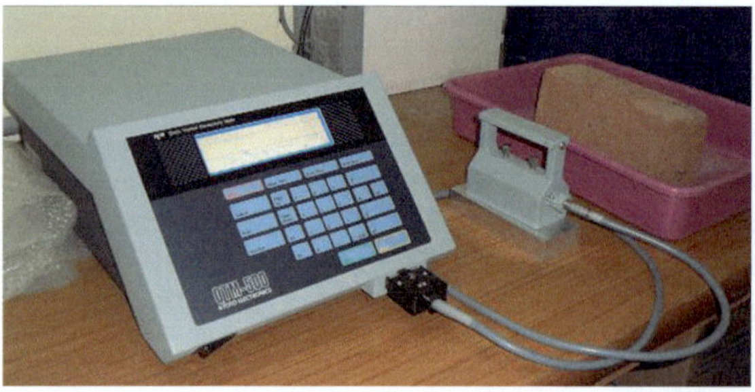

Fig. 29.1 Thermal conductivity testing instrument QTM-500

wire method. This method is a transient/non-steady-state process used to measure the thermal conductivity of the material under constant heat source and to measure the rise in temperature of the material. The thermal conductivity tests were carried out as per ASTM C1113-99 (ASTM 1999).

To ensure uniformity of test conditions, all the test specimens were conditioned (for ~72 h) and tested at 25 ± 1 °C room temperature and relative humidity in the range of 50–60% complying with ASTM C870-11 (ASTM 2011) standard. The materials have been assumed to be homogeneous and isotropic. The thermal conductivity tests at higher temperatures were also conducted using a QTM-500 instrument (see Fig. 29.1).

29.2.5 *Error Analysis—Test Repeatability and Material Errors*

In this study, the repeatability and material error involved in thermal conductivity of building material were analyzed. The materials considered have been illustrated in Table 29.1.

29.2.5.1 Basics of Error Analysis

Errors exist in all measurements, as no test is perfectly reliable (Feng et al. 2015) or repeatable. As per IS 15393 (part 1)—2003 and ISO 5725-1—1994, accuracy is termed as the closeness of agreement between a test result and the accepted reference value, and it is described as reliability of measured results. The two aspects of accuracy are trueness and precision. As per IS 15393 standard, trueness represents "*the closeness of agreement between the average value obtained from a large series of test results and an accepted reference value.*" It is expressed in terms of bias and precision representing "*the closeness of agreement between independent test results obtained under stipulated conditions,*" and it is expressed in terms of standard deviation. It is evident that the trueness is related to systematic errors ($e_{\text{systematic}}$) and precision defines random errors (e_{random}). Systematic errors in experimental observations are usually based on the measuring instruments and methods adopted. Random errors in experimental measurements are caused by unknown and unpredictable changes in the experiment. These changes may occur in the measuring instruments and/or in the environmental conditions (Hill 2015).

In this study, systematic error is an intrinsic behavioral attributed to instrumental error and it is corrected through instrument calibration using calibrated specimens. This error does not apply in the current study as the instrument has been regularly calibrated in between experimentation. Random errors may be due to the materials heterogeneity, material surface preparation (surface made even before testing, i.e., nearly even test surface), test conditions, materials temperature, and unsteady ambient conditions.

Besides the above-mentioned errors, there are other factors that influence accuracy of experiments according to the IS 15393. These include standard measurement method, accuracy maintained in the experiment (equipment user, type of equipment used, calibration of equipment), identical test specimens, short intervals of time (time elapsed between measurements), and participating laboratories. Repeatability is the precision obtained, under the "same" conditions, when independent test results are obtained with the same method, on identical test items, in the same laboratory, by the same operator, using the same equipment, and within short intervals of time; these are termed repeatability conditions. Repeatability leads to an estimate of the minimum value of precision (Pryseley 2010).

If the test material used and above conditions, methods, and instrumentations remain unchanged in replicate tests, these test conditions are defined as repeatability conditions (Feng et al. 2015) and the error will be expressed in relative standard deviation as repeatability error or $e_{\text{repeatability}}$. Another error termed material error is caused by heterogeneity (composite nature) of materials, and it is expressed in relative standard deviation as material error or e_{material}. This can be attributed to variation in how a specimen is made, stored, and prepared for the experiment and includes aging, material stability, and ambient exposed conditions.

29.2.5.2 The Calculation Methods for Repeatability and Material Errors ($\text{rs}_{\text{material}}$ and $\text{rs}_{\text{repeatability}}$)

In this study, two errors namely repeatability error and material error have been considered. Repeatability error is calculated for each experimental cycle, within the single sample, and by comparing the errors from duplicate samples to original material error. Errors such as $e_{\text{repeatability}}$ and e_{material} will be expressed in $\text{rs}_{\text{repeatability}}$ and $\text{rs}_{\text{material}}$, respectively.

For a single result

$$x_{i,j} \tag{29.1}$$

Average of one sample in replicate tests

$$\overline{x_{i,j}(j)} = \frac{1}{q}\sum_{j=1}^{q} x_{i,j} \tag{29.2}$$

Standard deviation of single sample

$$s_{x_{i,j}}(j) = \sqrt{\frac{\sum_{j=1}^{q}\left(x_{i,j} - \overline{x_{i,j}(j)}\right)^2}{q-1}} \tag{29.3}$$

Relative standard deviation of single sample

$$\text{rs}_{x_{i,j}}(j) = \frac{s_{x_{i,j}}(j)}{\overline{x_{i,j}(j)}} x100\% \tag{29.4}$$

Repeatability error

$$\text{rs}_{\text{repeatability}} = \overline{\text{rs}_{x_{i,j}}(j,i)} = \frac{1}{p}\sum_{i=1}^{p} \text{rs}_{x_{i,j}}(j) \tag{29.5}$$

For average of all results

$$\overline{x_{i,j}(i,j)} = \sum_{j=1}^{q}\sum_{i=1}^{p} x_{i,j} \tag{29.6}$$

Standard deviation of all results

$$s_{x_{i,j}}(i,j) = \sqrt{\frac{\sum_{j=1}^{q}\sum_{i=1}^{p}\left(x_{i,j} - \overline{x_{i,j}(i,j)}\right)^2}{p\cdot q - 1}} \tag{29.7}$$

Relative standard deviation of all results

$$\mathrm{rs}_{x_{i,j}}(i,j) = \frac{s_{x_{i,j}}(i,j)}{\overline{x_{i,j}(i,j)}} x100\% \tag{29.8}$$

Material error

$$\mathrm{rs}_{\mathrm{material}} = \sqrt{\left(rs_{x_{i,j}}(i,j)\right)^2 - \left(\mathrm{rs}_{\mathrm{repeatability}}\right)^2} \tag{29.9}$$

where p and q, respectively, represent the number of samples and tests; i and j, respectively, represent the test results of sample i in the test j; x is measured value; $\mathrm{rs}_{\mathrm{repeatability}}$ is relative standard deviation for repeatability errors; $\mathrm{rs}_{\mathrm{material}}$ is relative standard deviation for material errors; $s_{x_{i,j}}$ is the standard deviation.

Figure 29.2 illustrates the calculation process of the repeatability and material error of building materials in the study. Calculation process is adopted from Feng et al. (2015) study. Both the errors were calculated for all the materials considered in the study as shown in Table 29.2. The relative standard deviation of each sample over replicated tests has been calculated through Eqs. 2–4, and the $\mathrm{rs}_{\mathrm{repeatability}}$ is derived by taking average for all samples from a set using Eq. 5. By adopting Eqs. 6–8 for calculating the relative standard deviation of all results, $\mathrm{rs}_{\mathrm{material}}$ is calculated by eliminating the $\mathrm{rs}_{\mathrm{repeatability}}$ from samples according to Eq. 9.

29.3 Results and Discussions

For studying repeatability error tests, the number of sample (p) and tests (q) used is shown in Table 29.2. Repeatability conditions in thermal conductivity measurements were calculated through experimental results obtained from repeated/replicate tests on multiple specimens.

Test results show sound repeatability for all material, except for CSSB 31.6, CSSB 10.5, and CSSB 5 materials (see Table 29.2). The repeatability error and material error are calculated as described in above Sect. 29.2.5.2. The relative standard deviation of each sample over replicate tests has been found to be greater than relative standard deviation of all results (i.e., $\overline{rs_{x_{i,j}}(j,i)} \geq rs_{x_{i,j}}(i,j)$), showing that the error caused

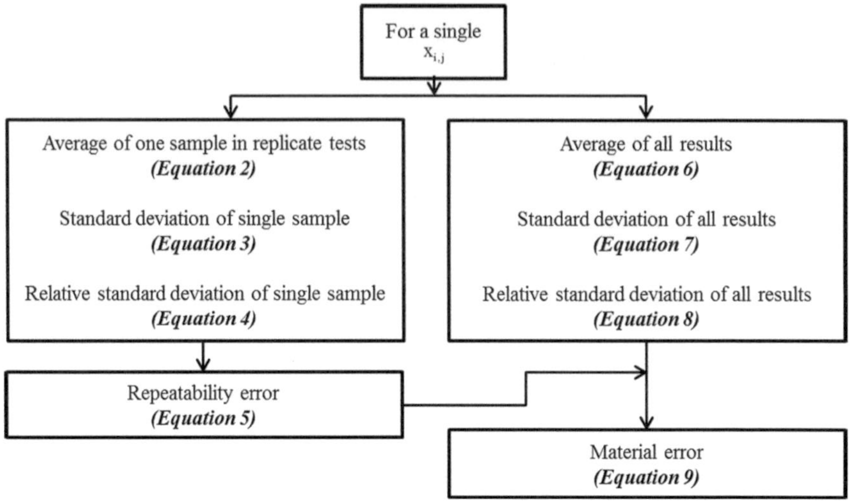

Fig. 29.2 Calculation process for repeatability and material errors (Feng et al. 2015)

Table 29.2 Details of relative standard deviation error for repeatability of thermal conductivity values and building material errors

Materials	$rs_{repeatability}$	$rs_{material}$	Samples (p)	Tests (q)
CSSB 31.6	10.123	0.555	6	3
CSSB 10.5	10.419	1.786	6	3
CSSB 5	11.159	–	6	3
CSSB 12	8.787	5.549	6	3
CSSB 16	9.756	6.069	6	3
CSSB 1.7	7.445	4.188	6	3
CSSB 1.8	6.845	–	6	3
CSSB 1.9	7.367	2.665	5	3

by materials heterogeneity is understood by repeatability errors. In that case, no $rs_{material}$ value is provided. Similar kind of observations were made by Feng et al. (2015) in their studies.

For the CSSBs of varying mix proportions, the materials and repeatability errors reveal the high random variations in resulted errors. The low cement content block CSSB 5 has a high-value $rs_{repeatability}$ of 11.159 and CSSB 1.8 block has $rs_{repeatability}$ of 6.845. Further, for these blocks the error caused by materials heterogeneity is understood by repeatability errors and no $rs_{material}$ value is obtained. The $rs_{material}$ has 0.555 and 6.069 for CSSB 31.6 and CSSB 16, respectively.

Figures 29.3 and 29.4 reveal the repeatability error and material errors plotted versus thermal conductivity of CSSBs. The $rs_{repeatability}$ value varies from 3.838 to 11.159. Similarly, $rs_{material}$ value varies from 0.555 to 8.297. These variations in error

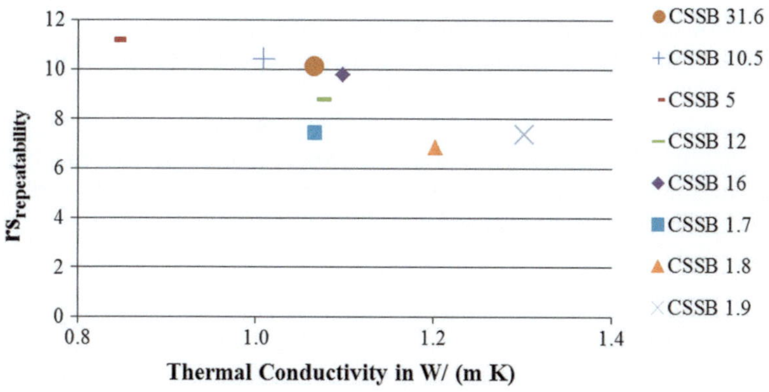

Fig. 29.3 Repeatability error versus thermal conductivity values of materials tested

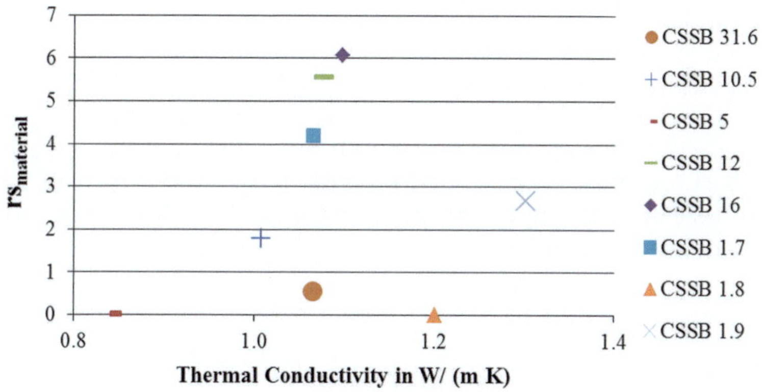

Fig. 29.4 Material error versus thermal conductivity values of materials tested

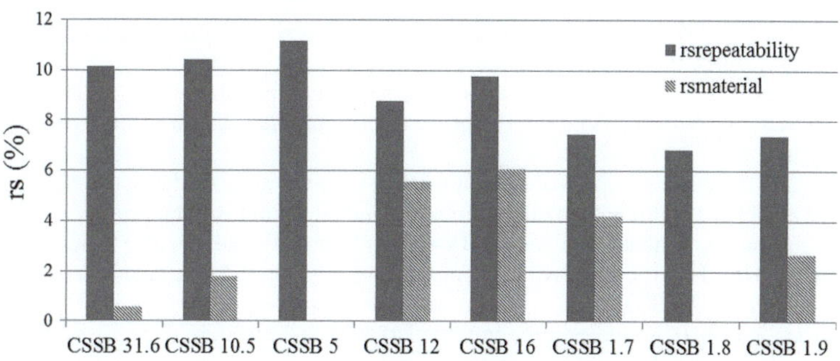

Fig. 29.5 Repeatability and material errors for thermal conductivity tests on different building materials

may be due to the material compositions, method of thermal conductivity testing, and the surface finish condition during repeatability tests. Figure 29.5 shows the repeatability and material errors for thermal conductivity tests of CSSBs. The $rs_{repeatability}$ and $rs_{material}$ values of CSSBs are plotted together for comparison. The $rs_{repeatability}$ values are 2.014, 4.951, and 3.838 for AAC, TMB, and FAL-G, respectively (Balaji 2016). The value of $rs_{repeatability}$ for CSSB materials varies from 6.845 to 11.159, which is higher than the other materials.

29.4 Conclusion

In this study, the CSSB material selected was low embodied energy material. The thermal conductivity of the CSSBs was experimentally determined at room temperature as per ASTM C 870–11. CSSBs at varying cement content, clay content, and dry densities were studied for its thermal conductivity error measurements. Detailed error analysis was carried to ascertain the repeatability and material errors ($rs_{material}$ and $rs_{repeatability}$) of the CSSBs for thermal conductivity measurements.

The repeatability errors were found to be low for all the materials tested except for CSSB 31.6, CSSB 10.5, and CSSB 5 materials; it can be attributed to the number of tests carried on same sample. The material errors can be attributed to the materials heterogeneity and the inherent characteristics of the materials.

References

Adam EA, Jones PJ (1995) Thermophysical properties of stabilised soil building blocks. Build Environ 30(2):245–253

ASTM C1113-99 (1999) Standard test method for thermal conductivity of refractories by hot wire (platinum resistance thermometer technique) p 6

ASTM C870-11 (2011) Standard practice for conditioning of thermal insulating materials, p 4

Bahar R, Benazzoug M, Kenai S (2004) Performance of compacted cement-stabilised soil. Cement Concr Compos 26(7):811–820

Balaji NC (2016) Studies into thermal transmittance of conventional and alternative building materials and associated building thermal performance. Indian Institute of Science

Balaji NC, Mani M, Venkatarama Reddy BV (2017) Thermal conductivity studies on cement-stabilised soil blocks. Proc Inst Civil Eng Constr Mater 170(1):40–54. Available at: http://dx.doi.org/10.1680/jcoma.15.00032

Bhattacharjee B (1989) A study on the influence of void and moisture contents of some building materials on their thermal transport properties and implications on thermal performance of building envelope. Indian Institute of Technology, Delhi

BIS 3792 (1978) Guide for heat insulation of non-industrial buildings. Bureau of Indian Standard, India

BIS 8112 (1989) Specification for 43 grade ordinary portland cement. Bureau of Indian Standards, India

Clarke JA, Yaneske PP, Pinney AA (1990) The harmonisation of thermal properties of building materials, report CR59/90 of the building research establishment. Watford, UK

Feng C et al (2015) Hygric properties of porous building materials: analysis of measurement repeatability and reproducibility. Build Environ 85:160–172
Hammond GP, Jones CI (2008) Embodied energy and carbon in construction materials. Proc Instit Civil Eng Energy 161(2):87–98
Hashemi A, Cruickshank H, Cheshmehzangi A (2015) Environmental impacts and embodied energy of construction methods and materials in low-income tropical housing. Sustainability 7(6):7866–7883
Hill (2015) Random versus systematic error. Available at: http://www.physics.umd.edu/courses/Phys276/Hill/Information/Notes/ErrorAnalysis.html. Accessed 17 Aug 2015
Horpibulsuk S et al (2011) Analysis of strength development in cement-stabilized silty clay from microstructural considerations. Constr Build Mater 24:2011–2021
Jakob Max (1949) Heat transfer, volume-1. Wiley, New York
Khedari J, Watsanasathaporn P, Hirunlabh J (2005) Development of fibre-based soil–cement block with low thermal conductivity. Cement Concr Compos 27(1):111–116
Koenigsberger OH (1975) Manual of tropical housing and building. Orient Longman Private Limited
Mosquera P et al (2013) Determination of the thermal conductivity in adobe with several models. J Heat Transfer 136(3):31303
Nagih ME, Ali AAA (1995) Strength and thermal properties of plain and reinforced soil-cement. J Islamic Acad Sci 8(3):107–118
Pryseley A et al. (2010) Estimating precision, repeatability, and reproducibility from Gaussian and non-Gaussian data: a mixed models approach. J Appl Stat 37(10):1729–1747. Available at: https://doi.org/10.1080/02664760903150706
Sabapathy A et al. (2011) Strategies for cleaner walling material in India, India
Sugawara A, Yoshizawa Y (1961) An investigation on the thermal conductivity of porous materials and its application to porous rook. Aust J Phys 14(4):469–480
Van Straaten JF (1967) Thermal performance of buildings. Elsevier Pub, Co
Venkatarama Reddy BV, Gupta A (2005) Characteristics of soil-cement blocks using highly sandy soils. Mater Struct 38(6):651–658
Venkatarama Reddy BV (2009) Sustainable materials for low carbon buildings. Int J Low-Carbon Technol 4(3):175–181

Chapter 30
Hygrothermal Behaviour of Cob Material

Tuan Anh Phung, Malo Le Guern, Mohamed Boutouil and Hasna Louahlia

30.1 Introduction

In recent years, there has been a renewed interest in earthen building materials for restoration/repair of historic and cultural heritage buildings and for use as a low-energy material in sustainable architecture (Aymerich et al. 2012). The renewal of the use of soil as a building material is related to the significant reduction of the environmental impact due to the use of local raw materials and simple manufacturing processes and energy-efficient (Minke 2000, 2009). In addition, earth-based materials allow better balance and thermal control and internal acoustic compared with conventional building materials. This is due to their performances in terms of moisture absorption/desorption, heat storage capacity (hydrothermal regulation) and sound transmission properties (Minke 2000, 2009; Binici et al. 2009). Currently, building owners and architects also choose earth as a building material because of the advantages in terms of aesthetics and sanitary quality of air (Röhlen et al. 2013).

Earth construction was influenced by the geographical situation and the local cultures. So many construction methods exist: adobe, compressed earth brick, cob and rammed earth. In Normandy, the most common technique is cob which is a mixture of raw earth and natural fibres (usually wheat straw). The cob building technique

T. A. Phung (✉) · M. Le Guern · M. Boutouil
ESITC Caen, Epron, France
e-mail: tuan-anh.phung@esitc-caen.fr

M. Le Guern
e-mail: malo.leguern@esitc-caen.fr

M. Boutouil
e-mail: mohamed.boutouil@esitc-caen.fr

H. Louahlia
LUSAC—Caen Normandy University, Saint-Lô, France
e-mail: hasna.louahlia@unicaen.fr

B. V. V. Reddy et al. (eds.), *Earthen Dwellings and Structures*,
Springer Transactions in Civil and Environmental Engineering,
https://doi.org/10.1007/978-981-13-5883-8_30

consists in the construction of a massive wall, often bearing, implemented by piling earth–fibre balls in plastic state, possibly by using formwork, compacted with the fork and the stick then cut to the wall. In order to ease mixing, largest elements are removed. Fibres allow to maintain the cohesion and to limit the shrinkage on drying (Galán-Marín et al. 2010). Usually, the mixed water content is between 10 and 20% in order to obtain a compact paste (which does not crumble). Typical soil used in cob contains about 30% gravel (2/20 mm), 35% sand (0.063/2 mm) and 35% silt (0.002/0.063 mm) and clay (<0.002 mm = 2 μm) with a variation of these parameters of more or less 10% (Akinkurolere et al. 2006).

In this study, straw fibre was used. Different cob formulations were made to evaluate the influence of different parameters (type of soil and fibre content). This study was conducted in CobBauge project which is an Interreg programme financed by European Union.

30.2 Materials

30.2.1 Soils

For this study, three different soils from the Lieusaint quarry (Normandy) were chosen: a sand (soil 1), a red clay (soil 2) and a silt (soil 3) (Fig. 30.1). Geotechnical properties of soils are presented in Table 30.1.

According to fines content, soil 1 has the largest percentage of coarse grains. Soil 1 will be used to form the material skeleton. Given their high fines fraction (<80 μm), soils 2 and 3 (between 70 and 90%) will act as binders between grains of soil 1. So two mixtures of soil 1, 2 and 3 were made: soil 1 with soil 2 (mix A) and soil 1 with soil 3 (mix B). The mass proportion retained is 2/3 of soil 1 and 1/3 of soil 2 or 3. This allows to have two mixes with a similar particle size distribution with a small difference between 0.3 mm and 80 μm (Fig. 30.2). The two curves are typical

(a) (b) (c)

Fig. 30.1 Soils used for the study: sandy soil 1 (**a**), red clayed soil 2 (**b**) and silty soil 3 (**c**)

Table 30.1 Soil properties

	Methylene blue value (g/100 g)	Liquid limit (%)	Plastic limit (%)	Plasticity index (%)	Proctor optimum moisture content (%)	Proctor optimum dry density (kg/m^3)	Fine fraction (<80 μm) (%)
Soil 1	0.30	48.9	28.5	20.4	9.5	2034	13.0
Soil 2	0.91	53.5	24.5	29.0	15.5	1770	72.7
Soil 3	0.55	34.1	20.4	13.7	14.0	1827	89.5

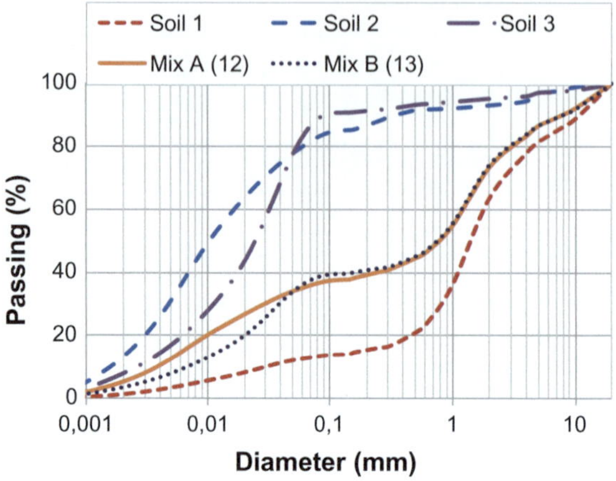

Fig. 30.2 Particle size distribution of the three soils and the two mixes

of traditional cob (Akinkurolere et al. 2006). Indeed, these two mixes have 26% of gravel, 35% of sand and 39% of silt and clay.

30.2.2 *Natural Fibres*

The use of natural fibres has different advantages when used as reinforcing materials. First, these fibres are widely available and cheap. Moreover, their use in construction will be new opportunities for agricultural materials. On the other hand, their valorisation allows a reduction of the environmental impact compared to mineral or polymeric fibres due to their renewable and biodegradable aspects, CO_2 emissions neutrality and low-energy products.

Furthermore, the addition of randomly distributed fibres into a material provides isotropic strength (Kumar et al. 2006). Fibres prevent cracking propagation during traction after initial deformation (Elizabeth and Adams 2005). Increasing fibre content allows to reduce cracks amount caused by shrinkage and to raise hydraulic conductivity of compacted soil (Millert and Rifai 2004; Tang et al. 2010).

As traditional cob, wheat straw was used in this study. Wheat straw comes from Laulne (Normandy). Wheat straw properties are presented in Table 30.2. Variability of wheat straw length is explained by its agricultural development but, also, its industrial processing and the harvest method.

An important characteristic of natural fibres is the water absorption. Indeed, this characteristic will influence, on one hand, the mix in the fresh state (absorption of available water) and, in the other hand, the long-term behaviour (change in fibre volume and fibre/soil interface modification). The water absorption coefficient was determined by fibres' immersion in water during several periods (1, 5, 15 min, 1, 4,

Table 30.2 Characteristics of straw fibres obtained in this study

Diameter (mm)	Density (g/cm^3)	Length (cm)	Initial water content (%)	Tensile strength (MPa)	Young modulus (MPa)
1–4	1.182 ± 0.073	15–50	8.4	28.83 ± 7.61	1734 ± 327

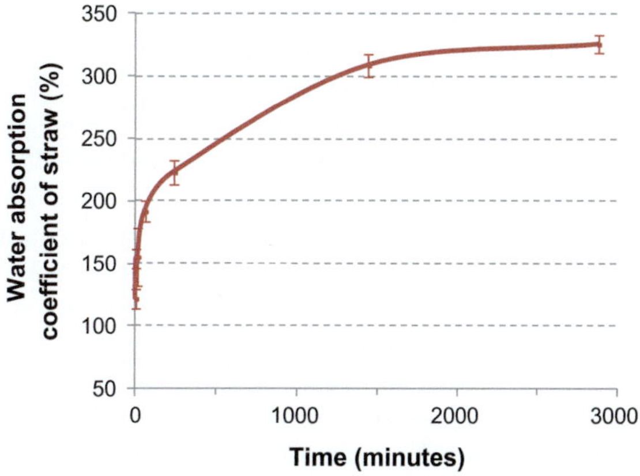

Fig. 30.3 Fibre water absorption coefficient through immersion time

24 and 48 h). After this, fibres were spun with a centrifuge at a speed of 500 rounds per minute for 15 s. Results show that straw water absorption coefficient is high (see Fig. 30.3). Indeed, absorption mean value at 24 h is 309% (each value is the average of six samples).

30.2.3 *Formulations and Samples' Preparation*

Two mixes A and B were used in this study. In order to study the influence of fibre, several mass fibre content were used: 1, 2 and 3%. The straw length used was 5 cm. Thus, fibres were cut to the required length, added randomly and mixed until getting a homogeneous composite. According to usual method, the water content used (18% for mix A and 16% for mix B) is 6% higher than mixtures Proctor optimum contents (between 11.7 and 12.1% for mix A and between 9.6 and 10.2% for mix B). All samples were stored at 20 ± 2 °C and 50 ± 5% relative humidity. Samples were removed from the mould after 2 days to have sufficient initial strength to stand alone. Initial mass (after manufacture) and mass after unmolding were measured in order to know the water content over time.

30.3 Methods

30.3.1 Thermal Conductivity

Concerning thermal conductivity, transitional method is often used for heterogeneous materials with high moisture content. For this method, a heat source is installed in soil and the soil temperature variations are recorded during measurement. As this method requires a short measurement time (a few minutes), change of the whole soil matrix/fibres/water (influence by the heat source–shrinkage for example) over time caused by measurement can be negligible. The hot-wire method is one of the most commonly used methods for measuring thermal conductivity of unsaturated soil (Tang 2005).

In this study, KD2 Pro device (Decagon Devices, INC) was used with TR-1 sensor. The standard used was ASTM D5334-08. Measurements were made on a cylindrical sample with a diameter of 152 mm and a high of 152 mm. For each sample, the experimental protocol consists of drilling 9 holes of the same sensor size, coating the sensor surface with a thermal paste and, then, install sensor in the hole before starting the test. Between each measurement, a minimum of 3 days is needed to dry (one day) and homogenise (two days) samples.

30.3.2 Hygrometric Properties

Dynamic vapour sorption is a gravimetric technique for measuring vapour interactions with solids. In this study, a ProUmid SPSx-1μ sorption/desorption analyser was used. According to NF EN ISO 12571, it is necessary to use at least four different atmospheres between 30 and 90% relative humidity. Here, for a greater accuracy, eighteen measurement points were made between 5 and 90% relative humidity for sorption and desorption.

For the fibres, samples of 2.5 cm length were used. For cob samples, they were approximately 40 × 40 × 30 mm in size obtained from cylindrical samples Ø150 × H60 mm. Samples were dried at 50 °C, then put in an airtight container containing silica gel in order to cool samples without letting the samples absorb moisture.

30.4 Results and Discussion

30.4.1 Thermal Conductivity

Results for thermal conductivity are presented in Figs. 30.4 and 30.5. These results show that thermal conductivity is linked to water content as found in previous studies (Tang 2005). Indeed, thermal conductivity decreases during drying. This can

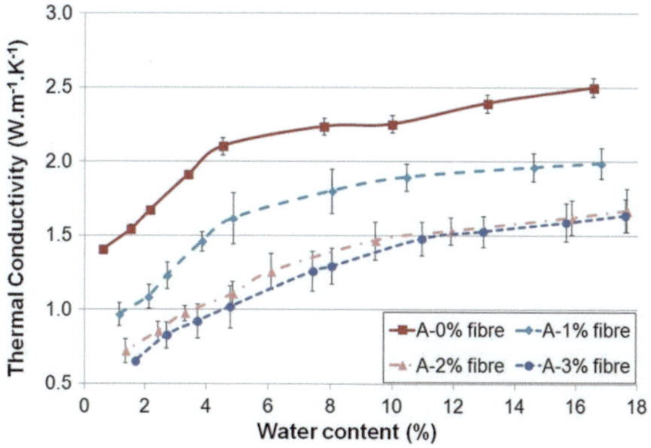

Fig. 30.4 Thermal conductivity changes of mix A with various fibre contents according to the water content during drying

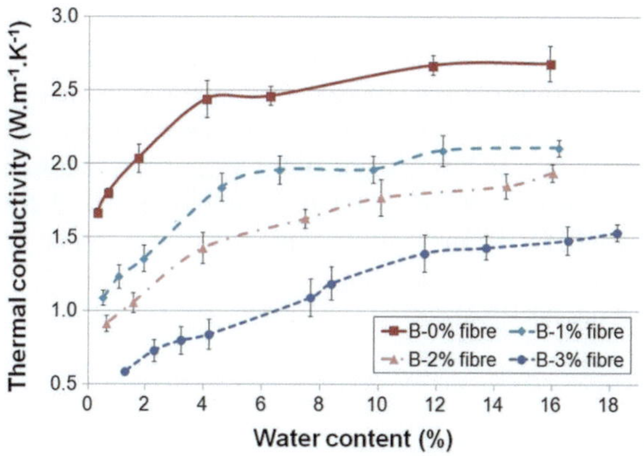

Fig. 30.5 Thermal conductivity changes of mix B with various fibre contents according to the water content during drying

be explained by the fact that water evaporates and is replaced by air which has a lower thermal conductivity than water 0.025 W m^{-1} K^{-1} instead of 0.6 W m^{-1} K^{-1}. However, it has to be noted that, above a water content of around 6%, the thermal conductivity decrease is lower (see Fig. 30.4 and 30.5). This can be due to shrinkage which increases slightly the sample dry density, therefore slightly increases the thermal conductivity. This leads to a competition between two phenomena (drying and shrinking) which have opposable effect on thermal conductivity. This explanation can be better seen in the case of mixture B. With 2% of fibre, there is less shrinkage, so the thermal conductivity decreases regularly from the beginning.

Moreover, results show that a higher fibre content leads to a lower thermal conductivity as seen by (Ledhem et al. 2000). Indeed, straw fibres have a thermal conductivity between 0.055 and 0.065 W m^{-1} K^{-1} compared to soil thermal conductivity. Figure 30.6 shows the evolution of thermal conductivity at the equilibrium (in storage

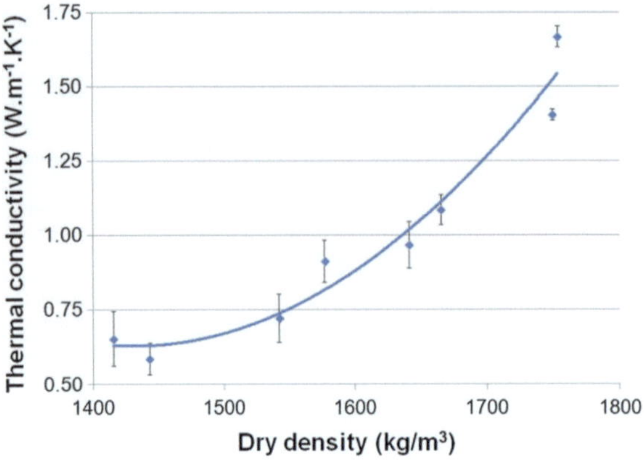

Fig. 30.6 Thermal conductivity versus dry density of mixes with various fibre contents

conditions at 20 ± 2 °C and 50 ± 5% relative humidity) according to dry density of the two mixes. An addition of fibre introduces an additional porosity and, thus, leads to a lower dry density of the soil/fibre mix. Results show that this porosity decreases thermal conductivity as seen by (Laurent 1986) Values obtained for mix A (0.72 ± 0.08 W m^{-1} K^{-1} with a density of 1560 kg/m^3) and for mix B (0.91 ± 0.07 W m^{-1} K^{-1} with a density of 1580 kg/m^3) with 2% fibre content are similar to characteristics observed in a nineteenth-century cob house (0.76 ± 0.12 W m^{-1} K^{-1} with a density of 1660 kg/m^3) (Laurent 1986).

30.4.2 Hygrometric Properties

Results for hygrometric properties are presented in Figs. 30.7 and 30.8. Firstly, these results show that values of standard deviation are lower for RH < 50% than for RH > 50%. The results for the fibre-free mixtures A and B show that the water content of the mixture A varies between 0.2 and 3.7% and that of the mixture B varies between 0.2 and 2.8%. Mixture B thus has a lower hygroscopic capacity than mixture A. This difference is probably due to the clay activity which is more important for mixture A. Indeed, since the clay activity is proportional to the specific surface area, a larger surface area means greater adsorption of water molecules.

These results also show that increasing fibre content increases the material hygroscopic capacity. Values obtained for RH = 90% show that, compared to mixes without fibre, the water content increase is equal to 3.5, 7.5 and 9.9% for an addition of 1, 2 and 3% fibre for mix A and 3.2, 7.1 and 11.1% for mix B. These results show that the water content increase seems to almost linearly depend on the fibre content. This phenomenon can be explained by the greater sorption/desorption capacity of wheat straw fibres compared to mixes A and B (Fig. 30.9). These results show

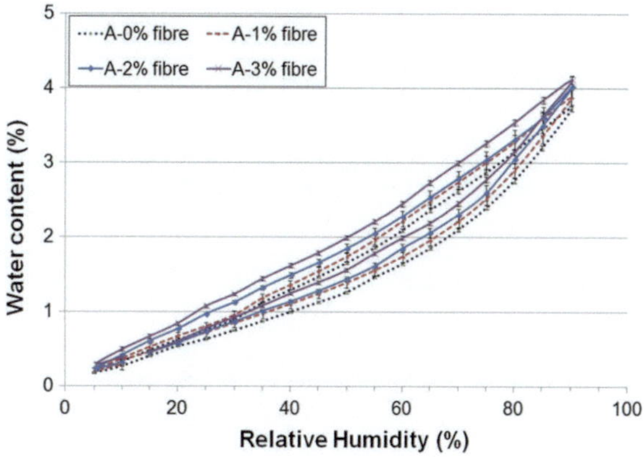

Fig. 30.7 Sorption/desorption curves of mix A with various fibre contents

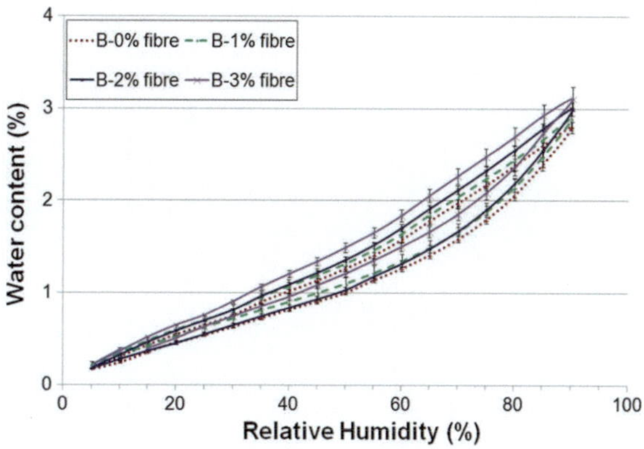

Fig. 30.8 Sorption/desorption curves of mix B with various fibre contents

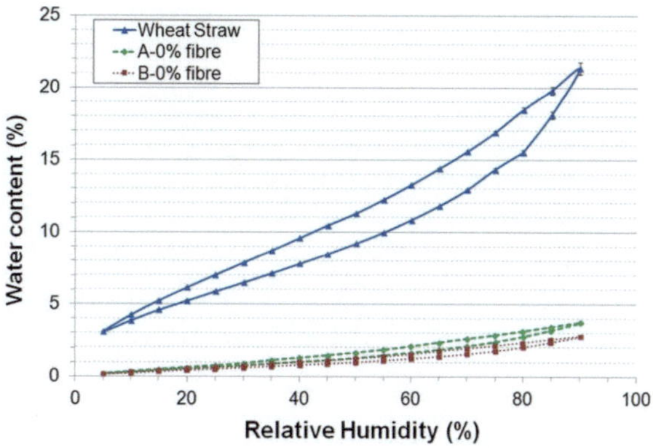

Fig. 30.9 Sorption/desorption curves of wheat straw, mix A and mix B

Table 30.3 Characteristics of straw fibres obtained in this study

Fibre-added content (%)		Mix A			Mix B		
		RH (%)		Variation	RH (%)		Variation
		90	50		90	50	
0	W (%)	3.7	1.7	2.0	2.8	1.3	1.5
	λ (W m^{-1} K^{-1})	1.969	1.568	0.401	2.220	1.933	0.287
1	W (%)	3.9	1.7	2.2	2.9	1.3	1.6
	λ (W m^{-1} K^{-1})	1.461	1.041	0.420	1.529	1.269	0.260
2	W (%)	4.0	1.9	2.1	3.0	1.4	1.6
	λ (W m^{-1} K^{-1})	1.038	0.785	0.253	1.277	1.025	0.252
3	W (%)	4.1	2.0	2.1	3.1	1.5	1.6
	λ (W m^{-1} K^{-1})	0.959	0.707	0.252	0.789	0.616	0.173

that hygrometric properties of cob can be almost determined knowing soil and fibre sorption/desorption curves.

30.4.3 Thermal Conductivity Versus Hygrometric Properties

Cob sorption/desorption curves allow to establish the influence of ambient hygrometry on material characteristics. Indeed, thermal conductivity has been followed during drying of mixes A and B. The influence of relative humidity (RH) on the thermal conductivity was thus determined and is presented in Table 30.3 for a decrease in RH from 90 to 50%. Results show that the RH decrease leads to a decrease of thermal conductivity around 20–30% for mix A and around 10–22% for mix B. Behaviour differences between mixes seem to come from clay activity. Indeed, clay activity is linked to specific surface area and, therefore, a lower clay activity leads to lower water content variations. Since thermal conductivity varies with water content, a lower water content variation results in a smaller change in thermal conductivity. It should also be noted that thermal conductivity evolution between 90 and 50% RH is greater for fibre materials than non-fibre materials. This is due to the fact that water content variation is similar for each formulation of each mixture and that thermal conductivity decreases as the amount of fibre increases.

30.5 Conclusions and Perspectives

In this study, thermal behaviour and hygrometric properties of cob formulations were presented. Results obtained show that thermal conductivity decreases when fibre content increases. This is due to material lower density and to fibre conductivity. With 3% fibre content, thermal conductivity is 0.65 ± 0.09 W m^{-1} K^{-1} for mixture A and 0.59 ± 0.05 W m^{-1} K^{-1} for mixture B. Concerning hygrometric properties, results show that knowing each component hygrometric properties will allow to predict cob properties. Knowing hygrometric properties allows to determine cob behaviour variation during the building life.

Thanks Authors thank the Interreg VA France (Channel) England programme financed by the European Union.

References

Aymerich F, Fenu L, Meloni P (2012) Effect of reinforcing wool fibres on fracture and energy absorption properties of an earthen material. Constr Build Mater 27:66–72. https://doi.org/10.1016/j.conbuildmat.2011.08.008

Akinkurolere OO, Jiang C, Oyediran AT, Dele-Salawu OI, Elensinnla AK (2006) Engineering properties of Cob as a building material. J Appl Sci 6:1882–1885. https://doi.org/10.3923/jas.2006.1882.1885

Binici H, Aksogan O, Bakbak D, Kaplan H, Isik B (2009) Sound insulation of fibre reinforced mud brick walls. Constr Build Mater 23:1035–1041. https://doi.org/10.1016/j.conbuildmat.2008.05.008

Galán-Marín C, Rivera-Gómez C, Petric J (2010) Clay-based composite stabilized with natural polymer and fibre. Constr Build Mater 24:1462–1468. https://doi.org/10.1016/j.conbuildmat.2010.01.008

Elizabeth L, Adams C (2005) Alternative construction: contemporary natural building methods. Wiley, New York

Kumar A, Walia BS, Mohan J (2006) Compressive strength of fiber reinforced highly compressible clay. Constr Build Mater 20:1063–1068. https://doi.org/10.1016/j.conbuildmat.2005.02.027

Ledhem A, Dheilly R, Benmalek M, Quéneudec M (2000) Properties of wood-based composites formulated with aggregate industry waste. Constr Build Mater 14:341–350. https://doi.org/10.1016/S0950-0618(00)00037-4

Laurent, J.-P. (1986). Contribution à la caractérisation thermique des milieux poreux granulaires optimisation d'outils de mesure "in situ" des paramètres thermiques, application à l'étude des propriétés thermiques du matériau terre, Ph.D. Thesis, Grenoble INPG

Miller CJ, Rifai S (2004) Fiber reinforcement for waste containment soil liners. J Environ Eng 130:891–895. https://doi.org/10.1061/(ASCE)0733-9372(2004)130:8(891)

Minke G (2000) Earth construction handbook: the building material earth in modern architecture. WIT Press, Southampton, p 2000

Minke G (2009) Building with earth: design and technology of a sustainable architecture. Birkhäuser, Basel

Röhlen U, Ziegert C, Mochel A (2013) Construire en terre crue: construction, rénovation, finitions. Éditions Le Moniteur, Paris

Sutton A, Black D, Walker P (2011) Straw bale: an introduction to low-impact building materials. IHS BRE Press, Watford

Tang C-S, Shi B, Zhao L-Z (2010). Interfacial shear strength of fiber reinforced soil. Geotextiles and Geomembranes 28(1):54–62. https://doi.org/10.1016/j.geotexmem.2009.10.001

Tang AM. (2005). Effet de la température sur le comportement des barrières de confinement. Ph.D. Thesis, Ecole Nationale des Ponts et Chaussées, Paris

Chapter 31
Light Earth Performances For Thermal Insulation: Application To Earth–Hemp

T. Vinceslas, T. Colinart, E. Hamard, A. Hellouin de Ménibus, T. Lecompte and H. Lenormand

31.1 Introduction

The building sector is nowadays one of the first sources of pollution. Taking Paris as an example, building heating is the second principal source of pollution with 20% of NOx emissions in the region and 25% of primary PM10 (Particulate Matter) emissions (International Energy Agency 2016). Taking these facts into account, nowadays eco-buildings efforts are encouraged by the public authorities. It leads to housing with lower in-service-energy consumption but in the other hand this gain is often made thanks to higher embodied energy of building materials.

Paying attention to the whole building life cycle, the choice of efficient materials with low embodied energy is a must (Thormark 2006). In addition, environmental impacts can be reduced further by the use of local materials, by the development of local distribution channel and by the evolvement of building professionals who use traditional or local ways to build (Morel et al. 2001; Floissac et al. 2009).

In this context, there is an interest since the nineties for bio-based materials such as straw and hemp for example. Lime–hemp mixes for thermal insulation were focused on numerous scientific studies (Collet 2004; Cerezo 2005; Pham et al.

T. Vinceslas (✉) · T. Colinart · T. Lecompte
Univ. Bretagne Sud, FRE CNRS 3744, IRDL, F-56100 Lorient, France
e-mail: theo.vinceslas@univ-ubs.fr

E. Hamard
IFSTTAR, MAST, GPEM, 44344 Bouguenais, France

A. H. de Ménibus
Eco-Pertica, Hôtel Buissonnet, 61340 Perche-En-Nocé, France

A. H. de Ménibus
Association Nationale Des Chanvriers en Circuits Courts, Bouquet, France

H. Lenormand
UniLaSalle, 3 rue du tronquet, 76134 Mont-Saint-Aignan, France

B. V. V. Reddy et al. (eds.), *Earthen Dwellings and Structures*,
Springer Transactions in Civil and Environmental Engineering,
https://doi.org/10.1007/978-981-13-5883-8_31

2012; Colinart et al. 2015; Tran Le et al. 2010). Preliminary estimations show that replacing lime-based binders by unfired clay could reduce by 20 the embodied energy of the insulation constructive system (Röhlen and Ziegert 2013; Keefe 2005), and lower the carbon impact by a factor 5 (Busbridge 2009).

Earth has a much longer story than commonly studied bio-based materials as it has been used for construction from 10,000 years ago in the Middle East. Earth was always used as a heavy material with density ranging from 1200 to 1800 kg/m^3 depending on the techniques. It is only recently that craftsmen and scientists tried to lower the earth density to improve its thermal insulation. The first experiments were carried out in Germany, using straw (Volhard 2016), but other resources can be used, such as coir rape and hemp.

Earth constructions can require almost no transport if earth from the construction site is used. Earth as a construction material needs no, or few, transformations (Keefe 2005). Moreover, clayey bonds are reversible (Röhlen and Ziegert 2013), require low deconstruction energy at the end-of-life and can be reused for a new construction.

Raw earth and hemp are subject of numerous variabilities: for earth, the particle size distribution, clay types and chemical components and for hemp: variety, particle size distribution, dust ratio, fibre ratio and crop practices. The aim of the ECO-TERRA project is to evaluate the impact of the components variability on the performance of light-earth and more specifically of hemp-earth.

This paper report the first results obtained regarding thermal and mechanical characterizations of some earth–hemp mixes, including earth variabilities such as particle size distribution and clay content.

31.2 Materials and Methods

31.2.1 Formulations

The studied material is only composed of raw earth and hemp shiv. Samples of raw earth were collected following one goal: to allow the observation of an important variability. Thus, different locations of collection were chosen: quarries, where it is possible to sample raw earths or washing sludge; private individuals' lands, where earth have been used during their home's construction; with craftsmen, where their know-how has led to the choice of earths with "extreme" characteristics. Finally, 26 raw earths have been collected in 10 departments in the west of France. Among these earths, 10 come from quarries (8 raw earths, 1 product in bag and 1 washing sludge) and the other 16 were given by individuals or craftsmen. A consistency matrix, i.e. earth at different water contents, was built in order to simply and quickly observe the variability of all collected earths. From these 26 raw earths, 5 have been chosen for sample fabrications: 3 earths visually classified with a high clay activity, 1 earth with average clay activity and 1 earth with low clay activity.

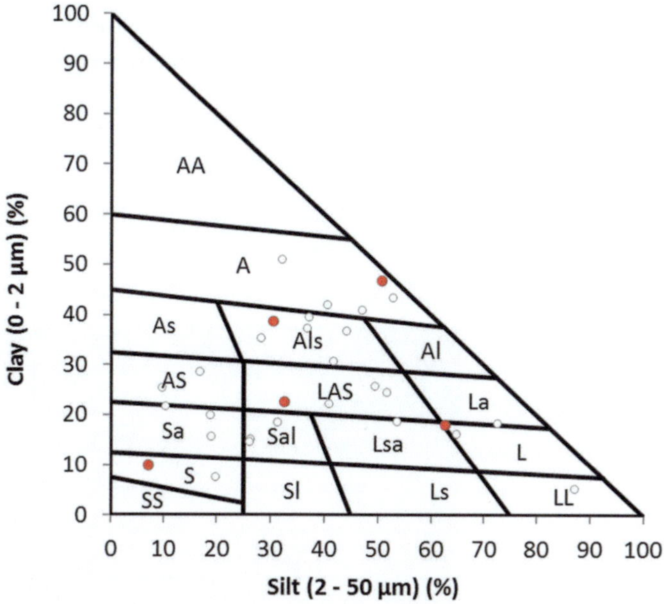

Fig. 31.1 Studied earths in the soil texture triangle

Materials for specimen production were dried at 50 °C and reduced to a powder using a mortar. A laboratory mortar mixer was used to prepare 4 mixes at different water contents (15, 30, 45, 60%). This protocol was used to measure clay activity and particle size distribution.

Particles size distribution was quantified by dry sieving for the coarse fraction (above 80 µm), according to French standard NF P 94-056, and and by sedimentation analysis for the fine fraction (below 80 µm), according to French standard NF P 94-057. Methylene blue value was measured according to the French standard NF P94-068, and specific weight was measured thanks to a water pycnometer according to NF P94-054.

Particle size distribution results are presented on a soil texture triangle in Fig. 31.1. Methylene blue value and specific weight are presented in Table 31.1. It can be observed that the studied earths range in the whole texture triangle and show very variable clayey behavior.

Figure 31.2 presents the hemp shiv employed in this study, coming from a short circuit hemp producer localized near Nocé, in the Perche territory, Normandy. The variety of this hemp shiv is Fedora 17. The seeding density was 50 kg/ha, and the harvesting year is 2013. Figure 31.3 presents the particle length and width distribution of used hemp shiv, measured by automated image analysis. Three different samples of shiv were analysed to produce a robust statistical characterization based on 700–850 particles. The repeatability level is satisfactory: for instance, average hemp length standard deviation between those three pictures was lower than 5%.

Tab 31.1 Used earth characteristics (specific density and methylene blue value)

Modification of the nomenclature	Specific density (kg/m^3)	Methylene blue value (g/100 g)
T1>HC1	2839	2.23
T2>HC2	2848	5.02
T3>HC3	2601	2.30
T4>AC	2860	0.87
T5>VLC	2756	0.68

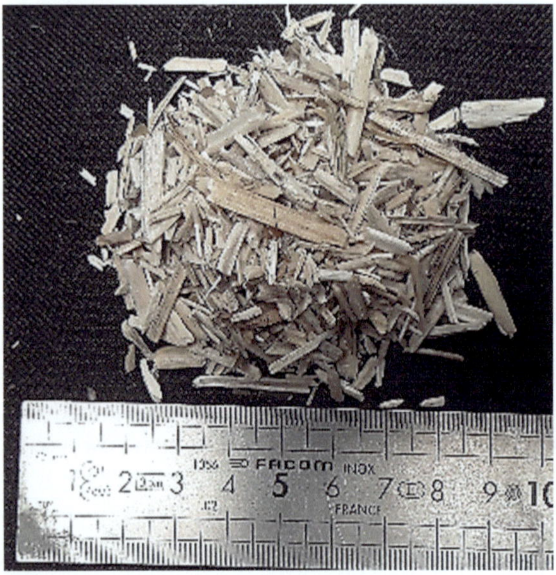

Fig. 31.2 Sample of hemp shiv

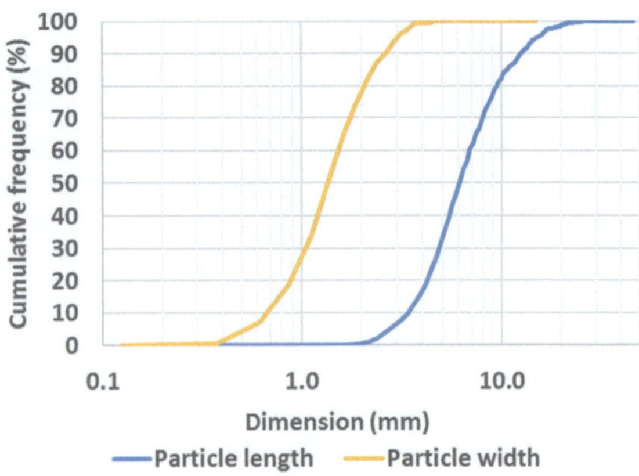

Fig. 31.3 Hemp shiv particles sizes distributions

Figure 31.4 presents the earth–hemp fabrication process used during the present study. The plan is divided into three parts:

- The first box contains the initial data that are compulsory to start the process;
- Grey boxes present the fabrication steps, from collecting raw earth to mixing slip with hemp shiv;
- Clear grey boxes describe data that has to be collected from each fabrication step to characterize the batch.

31.2.2 Specimens Preparation Method

Initial mixtures were prepared with a craftsman, to ensure its representativeness with regard to the material used in building works. In this study, 10 formulations with 5 different earths were tested. Specimen preparation method includes two different slip formulation methods: a first formulation with an equivalent water/earth ratio and a second with an equivalent rheology. This rheology is quantified by the yield stress, measured with a simple in situ apparatus.

Clay slip fabrication process is reported in Fig. 31.4. For every slips made, the process included sieving with 6 and 2.25 mm sieves.

Cubic specimens (10 × 10 × 10 cm) were made for thermal conductivity assessment, and cylindrical specimens (16 × 32 cm) were made for compressive strength assessment. Compaction of the material was controlled by defining a fresh-state-density. Then, moulds were filled up in five layers with equal mass of fresh material. In order to compare mix designs, every specimen presents exactly the same shiv mass. Specimens were unmoulded straight after their fabrication and left to dry in an ambiance controlled room at 23 °C, 50% RH.

Figure 31.5 presents mass proportions of each constituent of our 10 formulations. Variabilities due to the fabrication process are the following one:

- Sedimentation impacts the slip sample collection for density measurement, which has an influence on the water content estimation;
- The water content estimation calculus uses an average earth-specific density of 2650 kg/m^3. Then specific density variabilities impact this water content estimation.

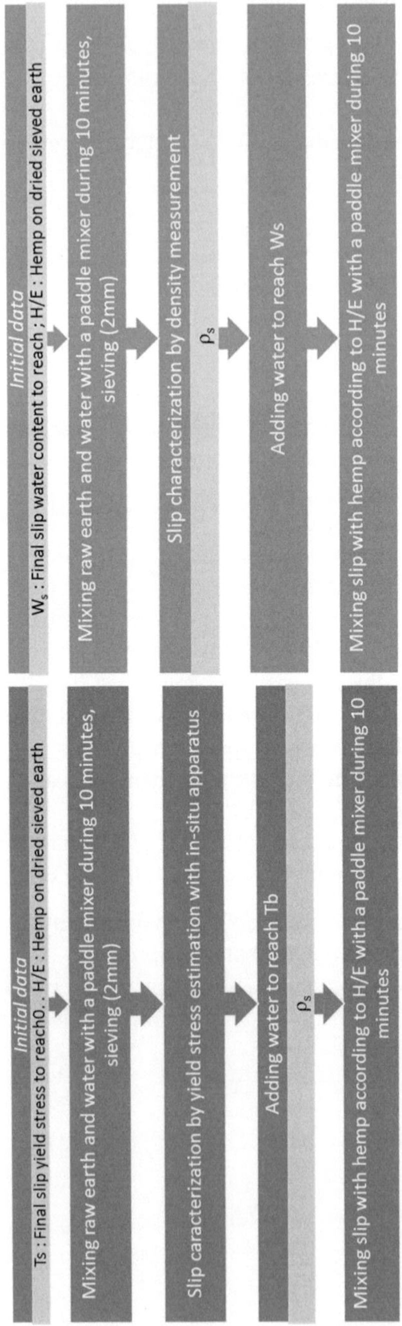

Fig. 31.4 Earth–hemp mix fabrication processes

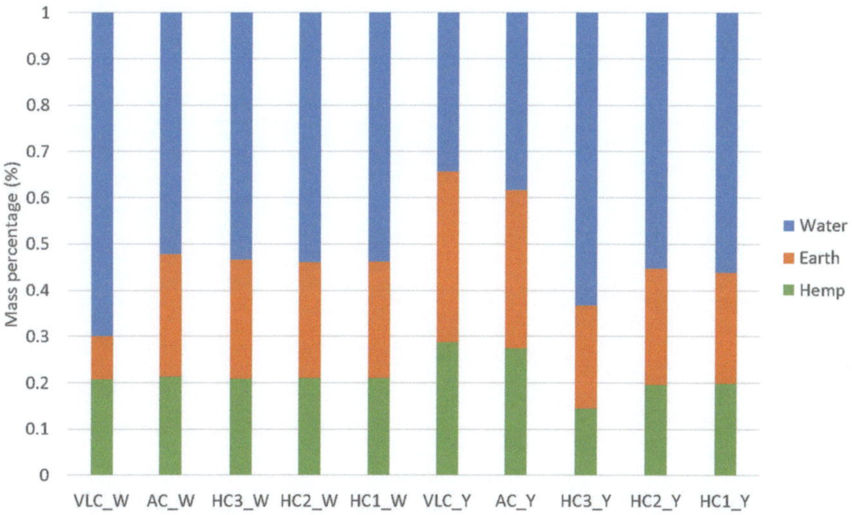

Fig. 31.5 Mass proportions of water, earth and hemp

31.2.3 Characterization Facilities

Thermal conductivity

The thermal conductivity of material is measured using home-made guarded hot plate device (Carré and Le Gall 1990) in accordance with the NF EN 12664 standard. Samples with dimensions of 10 × 10 × 4.8–4.9 cm^3 were dried at 70 °C in ventilated oven until constant weight before testing. Measurement are made at a mean temperature of 23 °C (cold and hot temperature are set respectively to 18 and 28 °C). By assuming one-dimensional heat transfer and negligible contact resistances, heat flow measurement allows assessing the thermal conductivity as:

$$\lambda = \frac{\Phi/S}{\left(T_{\text{sup}} - T_{\text{inf}}\right)/e}$$

The measurement uncertainties (±0.5 mm in sample thicknesses, ±1 mm in lateral dimensions, ±0.5 °C in temperature measurements and ±0.2 mW in dissipated or inlet power) lead to a thermal conductivity uncertainty of about 7% (Carré and Le Gall 1990).

Simple compression

Tests were carried out with an Instron 8803 press, with a 100 kN stress sensor. Strain is measured with the plate displacement. Cylindric specimens, once dry, were put during 48 h in a 40 °C chamber, according to the French Hemp building rules (Construire en chanvre 2010).

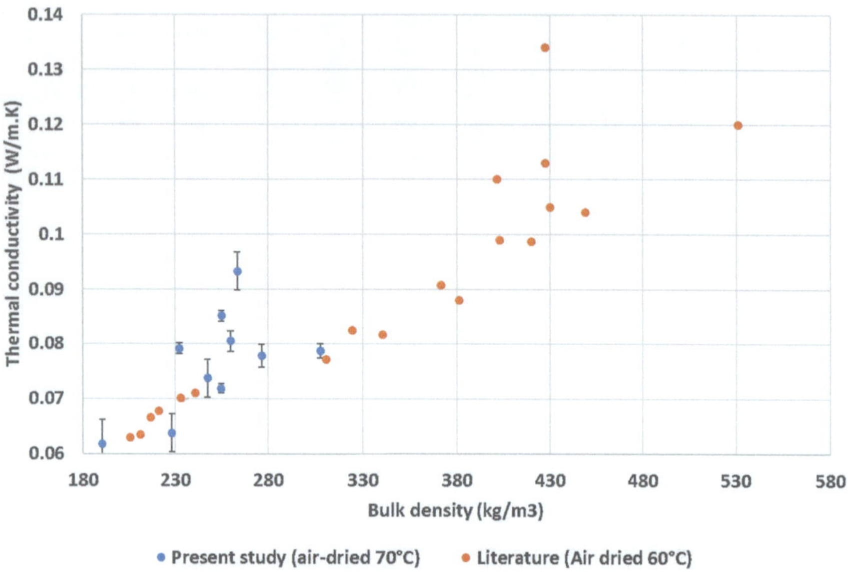

Fig. 31.6 Thermal conductivity values of present study formulas and literature (23 °C at dry state)

31.3 Results and Discussion

31.3.1 Thermal Conductivity

Results of thermal conductivity are presented in Fig. 31.6. A good level of repeatability was observed for some mixes (HC2 W, VLC W). But three mixes (HC1 W, HC3 W and AC W) show an important standard deviation: for a similar bulk density, thermal conductivity can range from 0.067 to 0.085 W m^{-1} K^{-1}.

Our results show:

- Influence of the earth nature is not clear. Surprisingly, at comparable bulk densities, mixtures with coarser aggregates have a lower thermal conductivity, while sand as a thermal conductivity higher than clay/silt (λsand = 2.000 > λclay/silt = 1.500 W m^{-1} K^{-1}) (Courgey and Oliva 2001);
- The higher the hemp shiv on dry sieved earth and sand ratio, the lower is the hardened material density, and thus the lower is the thermal conductivity.

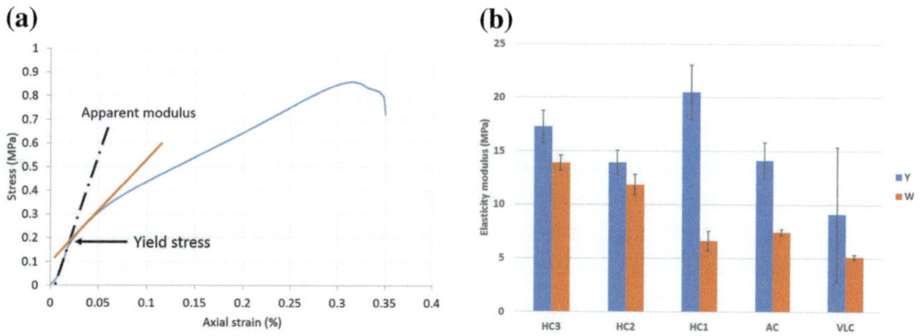

Fig. 31.7 **a** Stress strain curve of a HC2_Y specimen. **b** Elasticity modulus of the 10 studied mixes

Earth–hemp results are compared with lime–hemp results from previous studies, performed exclusively with permanent measure methods:

- Cerezo (2005): Guarded hot boxes
- Collet (2004): Hot Plate
- Amziane and Arnaud (2013): Guarded hot boxes. Finally, thermal conductivities of lime-hemp and a majority of our earth-hemp mixes are in the same order of magnitude. For similar measurement conditions, results variabilities observed on mixes lies in the 7% deviation due to the measurement apparatus.

31.3.2 Mechanical Behaviour

Figure 31.7 shows the curves of a simple compression tests on mix T2T. As presented by Tronet on lime and hemp concrete (Tronet et al. 2016), two phases are observed:

1. A very short pseudo elastic area;
2. An elasto-plastic deformation zone;
3. A densification zone, with rupture;

Data were analysed thanks to a graphic lecture developed in the paper of Tronet (Tronet et al. 2016) to obtain the maximum stress before irreversible deformations (σy) and the apparent elastic Young modulus (Ec).

Our results show that yield stress, varying between 0.15 MPa and 0.2 MPa, is similar to lime hemp results from the literature. Indeed, at comparable dry densities between 360 kg/m3 and 500 kg/m3, a yield stress between 0.18 MPa and 0.50 MPa was found on 16x32 cm cylindrical lime hemp specimens (Cerezo 2005; Arnaud and Gourlay 2012). Also, the observable yield stress variation between our different mixes is not directly linked to the two manufacture parameters, equivalent slip yield stress or equivalent slip water content. Figure 8 shows that elasticity modulus varies from 5 MPa to 20 MPa and is also similar to lime hemp results. In the same studies, an elastic modulus between 9 MPa and 44 MPa was found. Here we observe that "_Y" mixes always show a higher elasticity modulus than "_W" mixes, and the elasticity modulus is directly influenced by the clay activity of used earths.

Actually, the light earth as lime and hemp composites are not supposed to be load-bearing material. Then the stress threshold advised by construction rules for lime and hemp composites is arbitrary values and a yield stress of 0.15 MPa can be considered as largely enough for a self-bearing material. Further experiments should be made in tensile and shearing behaviour to complete the mechanical assessment of this kind of material. However, considering the large variability of raw materials, the compressive performances of the composites are very similar.

31.4 Conclusion

These preliminary results highlight the similarity between earth–hemp and lime–hemp insulation materials, regarding thermal conductivity and compression resistance. Even if not enough mechanical resistance tests were performed to draw conclusions, the preliminary results obtained show that earth–hemp might comply with the requirements of the French hemp building rules criteria for lime–hemp construction.

As lime should impact hydric properties (water resistance, vapour permeability or moisture transfer) of such mixes, additional hydric but also thermal and mechanical characterizations are currently being performed. In addition, new mixtures are being prepared, modifying the sample preparation (casted, sprayed), the component nature or the mixture mix design. These results would be analysed in the light of the feedback of real building thermal insulation with earth–hemp from the last five years, and of the building works that would be performed in parallel of the Eco-Terra R&D project.

Acknowledgments Eco-Terra Project is possible thanks to the financial support of the Normandy Region, the French state, the Fondation de France, the ADEME, the Bretagne Region and the Fondation d'Entreprise Legallais. The authors wish to thank Annick L'Alloret (IFSTTAR—Nantes), Mathilde Robin and Mathias Lemeurs (Eco-Pertica), Valentin Rochault and Hervé Bellegou (IRDL—Lorient) for their precious support.

References

Amziane S, Arnaud L (2013) Bio-aggregate-based building materials. ISTE Ltd and Wiley, Great Britain and the United States

Arnaud L, Gourlay E (2012) Experimental study of parameters influencing mechanical properties of hemp concretes. Constr Build Mater 28(1):50–56. https://doi.org/10.1016/j.conbuildmat.2011.07.052

Busbridge R (2009) Hemp-clay: an initial investigation into the thermal, structural and environmental credentials of monolithic clay and hemp walls

Carré P, Le Gall R (1990) Définition et détermination des conductivités thermiques dans la structures multicouches C.V.R.—balsa'. Revue générale de thermique, vol 340

Cerezo V (2005) Propriétés mécaniques, thermiques et acoustiques d'un matériau à base de particules végétales: approche expérimentale et modélisation théorique

Colinart T, Lelievre D, Glouannec P (2015) Experimental and numerical analysis of the transient hygrothermalbehavior of multilayered hemp concrete wall

Collet F (2004) Caractérisation hydrique et thermique de matériaux de génie civil à faibles impacts environnementaux

Construire en chanvre (2010) Protocoles d'essais pour la mesure des performances seuils des bétons de chanvre

Courgey S, Oliva JP (2001) L'isolation thermique écologique—Conception, matériaux, mise en oeuvre—Neuf et réhabilitation. Terre vivante

Floissac L, Marcom A, Colas AS, Bui QB, Morel JC (2009) How to assess the sustainability of building construction processes. In: Fifth urban research symposium, pp 1–17

International Energy Agency (2016) Energy and air pollution—world energy outlook—special report 9 rue de la Fédération, 75739. IEA Publications, Paris Cedex 15

Keefe L (2005) Eath building—methods and materials, repair and conservation. Taylor & Francis Group, Abingdon (UK)

Morel J-C, Mesbah A, Oggero M, Walker P (2001) Building houses with local materials: means to drastically reduce the environmental impact of construction. Build Environ 36(10):1119–1126. https://doi.org/10.1016/s0360-1323(00)00054-8

Pham TH, Férec J, Picandet V, Tronet P, Costa J, Pilvin P (2012) Etude expérimentale et numérique de la conductivité thermique d'un composite chaux–chanvre

Röhlen U, Ziegert C (2013) Construire en terre crue—construction—rénovation—finition. In: Claire Lefèvre (ed). Paris, Le Moniteur

Thormark C (2006) The effect of material choice on the total energy need and recycling potential of a building. Build Environ 41(8):1019–1026. https://doi.org/10.1016/j.buildenv.2005.04.026

Tran Le AD, Maalouf C, Mai TH, Wurtz E, Collet F (2010) Transient hygrothermal behaviour of a hemp concrete building envelope. Energy Build 42(10):1797–1806. https://doi.org/10.1016/j.enbuild.2010.05.016

Tronet P, Lecompte T, Picandet V, Baley C (2016) Study of Lime Hemp Concrete (LHC)—mix design, casting process and mechanical behaviour. Cement Concr Compos 67:60–72. https://doi.org/10.1016/j.cemconcomp.2015.12.004

Volhard F (2016) Construire en terre allégée. Actes sud

Part IV
Energy and Environmental Performance: Thermal Comfort and Indoor Air-Quality

Chapter 32
The Relevance of Earthen Plasters for Eco Innovative, Cost-Efficient and Healthy Construction—Results from the EU-Funded Research Project [H]house

Andrea Klinge, Eike Roswag-Klinge, Matthias Richter, Patrick Fontana, Johannes Hoppe and Jerome Payet

32.1 Introduction

In mostly industrialised countries and first of all in cold climates, where people spend the majority of their time in buildings, occupant's health, well-being and productivity are relying heavily on the indoor environmental quality (IEQ) of buildings. In Europe, more stringent energy efficiency standards led to increased airtightness levels of buildings and reduced air exchange rates, which in turn entail the application of mechanical ventilation in residential buildings to address unforeseen shortcomings with regard to increased relative humidity levels and higher concentration of airborne pollutants, which occur mainly during the winter period. Recent studies (Quinn and Shaman 2017) and experimental data (Klinge 2013), however, induce concerns amongst experts as increasingly drier internal relative humidity (RH) levels can be observed. The approach to introduce active systems in dwellings is discussed more and more controversially as a variety of negative side effects such as com-

A. Klinge (✉) · E. Roswag-Klinge
ZRS Architekten, Berlin, Germany
e-mail: klinge@zrs-berlin.de

M. Richter · J. Hoppe
Bundesanstalt für Materialforschung und –prüfung (BAM), Berlin, Germany
e-mail: matthias.richter@bam.de

P. Fontana
RISE Research Institutes of Sweden, Göteborg, Sweden
e-mail: patrick.fontana@ri.se

J. Payet
Cycleco, Ambérieu-En-Bugey, France
e-mail: jerome.payet@cycleco.eu

B. V. V. Reddy et al. (eds.), *Earthen Dwellings and Structures*,
Springer Transactions in Civil and Environmental Engineering,
https://doi.org/10.1007/978-981-13-5883-8_32

promised occupant's health and well-being, system inefficiencies, etc. are becoming increasingly important in light of a holistic approach. However, common practice is to install mechanical ventilation systems despite associated constraints, such as additional space requirements, increased construction and maintenance cost as well as compromised occupant comfort and control. The main criteria for ventilation can be summarised as follows:

- Control of indoor air humidity
- Flash off of harmful substances
- Provision of fresh air.

It is assumed that through the application of appropriate materials the above-listed requirements can be satisfied so that such an approach can reduce the need for active systems and promote natural ventilation.

The EU-funded project [H]house aimed to develop alternative, sustainable and affordable solutions for widespread application in residential buildings (new construction and renovation) that follow a low-tech rather than a high-tech approach. The project investigated materials with regard to their hygrothermal and air-purifying properties and developed innovative systems for internal partition wall application. Natural building materials were compared to conventional materials in order to identify their potential to improve the indoor air, while limiting the use of technology. A special focus was placed on earthen materials with regard to moisture buffering and air-purifying properties. Sorption velocity and also maximum uptake were investigated in greater detail.

Comprehensive experiments at material and component level have been undertaken to provide a database for the numerical prediction of appropriate material combinations that are able to react to different scenarios, since available surface area, occupation density, air volumes but also the construction of the building envelope might differ significantly between projects.

Accompanying experiments with regard to the emissions of the materials as well as to their potential adsorption of airborne pollutants have been conducted to identify materials that generate the most positive effects on the living environment.

In addition, experimental data coming from a monitoring of dwellings in Berlin fitted out with either earth plasters or conventional materials have been evaluated in relation to indoor air temperatures and relative humidity levels.

To assess the overall cost of the developed solutions and not only capital cost, a comprehensive life-cycle costing (LCC) analysis has been carried out.

Table 32.1 Extract of investigated materials in [H]house

Function	Material	Thickness [mm]
Finishing materials	Marble powder paint, casein primer, dispersion paint, deep sealer and joint filler	0.5–2
Render	EPFF (Earth plaster fine, final coat), EPRF (Earth plaster with straw, final coat)	3–10
Reinforcement	Flax fibre reinforcement, glass fibre reinforcement	0.5
Adhesive	Earth adhesive	2–3
Wall lining boards	Earth dry board, wood fibreboard with integrated earth plaster, wood fibreboard, gypsum fibreboard and gypsum plasterboard	12.5–31
Insulation	Wood fibre insulation boards and mats, mineral wool (eco), mineral wool	40–80
Internal partition walls	Non-load bearing, dry lining walls based on wall lining boards and insulation materials listed above	75–126

32.2 Materials and Test Methods

32.2.1 Material Selection

The material selection was based on an in-depth market analysis focused on natural building materials for internal partition walls, characterised through hygroscopic properties. In addition, natural building materials with no or limited data collection were shortlisted to close a scientific gap. However, special emphasis was placed on earth materials. Acoustic properties, although not presented in this contribution but equally important for the design of innovative partition walls, have been investigated and were taken into consideration for the proposed solutions. For benchmarking purposes, also conventional construction materials have been included. In total a selection of approx. 100 materials, grouped into their function of assembly, has been investigated. Table 32.1 provides an extract of materials that are relevant to this study.

32.2.2 Water Vapour Sorption Tests

To identify the capacity of earth plasters to adsorb moisture from the air via the specimen's surface within set time intervals a voluntary test procedure, determined in DIN 18947 (2013) was applied. The test requires to precondition three material samples (50 cm $\times$ 20 cm $\times$ 1.5 cm) in a climate chamber at a temperature of (23 $\pm$ 1) °C and (50 $\pm$ 5) % relative humidity (RH) until constant weight is achieved. The RH level is then increased to 80% and the weight of the samples is measured at specific time intervals (0.5, 1, 3, 6 and 12 h). Based on the results, the water vapour adsorption class of plasters is determined. For entire wall build-ups, the test has been modified in such way that five adsorption/desorption cycles were conducted and sorption was enabled from both sides of the specimen in order to take into account the increased thickness of the elements.

32.2.3 Emission Tests

The materials listed in Table 32.1 were screened for potential emissions prior to the standard tests to estimate the compounds to be expected. Final emission tests ((S)VOCs, radon) for single materials (nine types) and 13 combinations of them were carried out in specially designed test chambers over a testing period of 28 days. The tests were conducted following the requirements of EN 16516 (2015) and evaluated against the German AgBB scheme (2015), in the absence of harmonised evaluation procedures. Formaldehyde and VOC-analyses were carried out according to ISO 16000-3 (2011) and ISO 16000-6 (2011) and radon measurements in accordance with a procedure developed by Richter et al. (2013).

32.2.4 Adsorption of Indoor Pollutants

For adsorption tests according to ISO 16000-24 (2009), the chamber supply air was spiked with 1- pentanol, hexanal, butyl acetate, *n*-decane and α-pinene representing important indoor air contaminants. The test gas mixture was generated with a gas mixing device published by Richter et al. (2013). Concentrations ranged between 200 and 500 $\mu g/m^3$ and were specified of being higher than usually measured in indoor air to ensure a distinct determination of the reduction of the test chamber air concentration caused by the material. The air-purifying performance of the material was determined by monitoring the difference of the inlet and outlet concentration of the test chamber. Tests were carried out for more than nine materials (six single earth plasters with and without additions, three multi-layer specimens composed of different materials).

32.2.5 Monitoring

Monitoring data has been obtained from two different flats located in Berlin from August 2012 to September 2012 (summer period) and from November 2012 to January 2013 (winter period). Flat 01 was fitted out with natural materials and naturally ventilated whereas Flat 02 was constructed from conventional building materials and mechanically ventilated. Measurements were carried out with a miniature sensor and data logging system (iButton®) i-buttons, measuring temperatures (internally and externally as well as relative humidity internally and externally) (Klinge 2013).

32.2.6 LCC Analysis

The LCC analysis was carried out for a case study building, consisting of six storeys, where two different scenarios for the internal partition walls were considered, taking into consideration all associated cost. Scenario 1 is based on an interior fit-out with internal partition walls made out of natural building materials including wood fibres and earthen plaster (Fig. 32.2). Exhaust vents are provided for internal bathrooms and the kitchen. Due to the outstanding moisture sorption capacity of this construction, no further ventilation is required. Scenario 2 is based on an interior fit-out with internal partition walls made out of conventional building materials, such as gypsum plasterboard and mineral wool (Fig. 32.2). Exhaust vents are provided for internal bathrooms and the kitchen; however, due to the limited capacity to adsorb moisture, additional ventilation units incorporated into the external façade are anticipated, which does not require additional pipework or ducting though. Both solutions have been planned by a mechanical engineer with the briefing for cost-efficient solutions and assessed with regard to global and cumulated cost, taking into consideration different growth rates for energy prizes and discount rates for market adaptation for innovative materials.

32.3 Results

32.3.1 Water Vapour Sorption Tests

Figure 32.1 shows an extract of experimental results from different specimens that demonstrate common build-ups for wall finishes of internal partition walls. Casein painted earth plasters on top of wall lining boards have been compared with conventional, gypsum-based boards that were painted with the same colour. Results (Fig. 32.1) demonstrate that earth plasters in combination with wood fibre boards are characterised through an outstanding water vapour adsorption capacity, which is up to three times higher in comparison with gypsum plasterboards, as evidenced also in (Minke 2012; Eckerman and Ziegert 2006). Gypsum fibreboards range between earth plasters and gypsum plasterboards. Interestingly, it can also be observed that the casein paint increased the adsorption capacity of the gypsum fibreboard.

Additional tests have been performed at component level, investigating the potential of entire wall build-ups not only for the immediate moisture uptake but also their potential to provide a comfortable and healthy environment due to seasonal changes. An overview of the most relevant results is presented in Fig. 32.2. Materials were

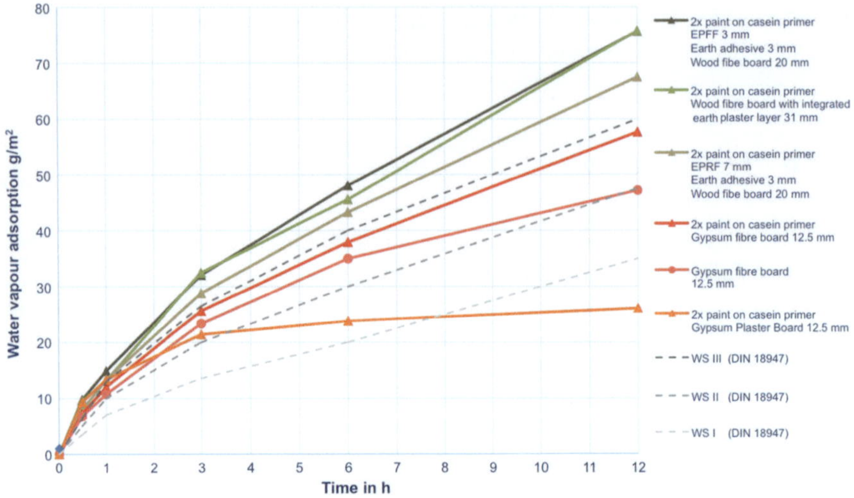

Fig. 32.1 Results of water vapour adsorption test (DIN 18947) of finishing layers of internal partition walls

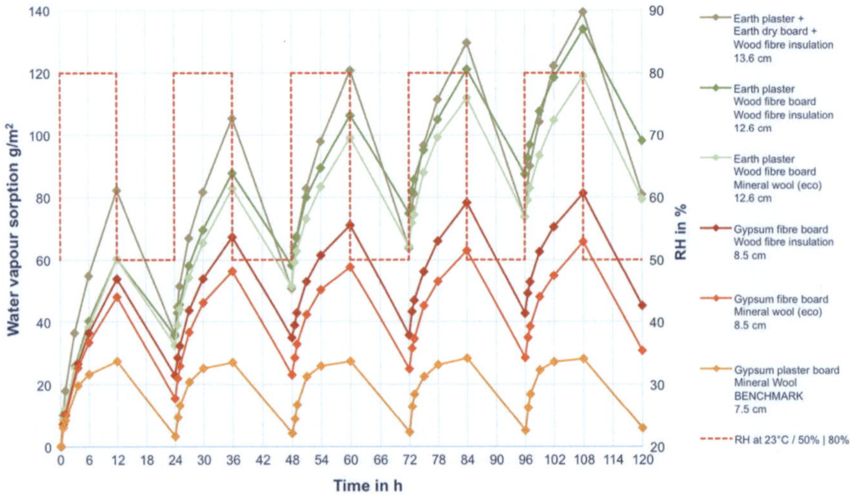

Fig. 32.2 Results of water vapour sorption test (5 adsorption/desorption cycles) of wall build-ups

tested in the most common thickness used for standard partition wall applications and although they differ, a direct comparison of specimens seems useful to identify the most capable materials and their combination.

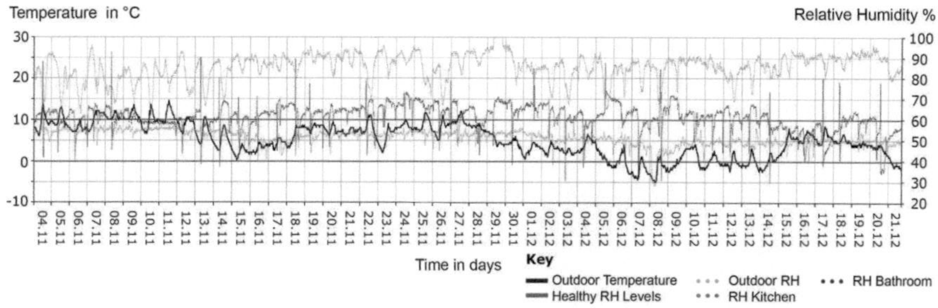

Fig. 32.3 Flat 1 results of monitoring RH (winter, natural building materials + natural ventilation)

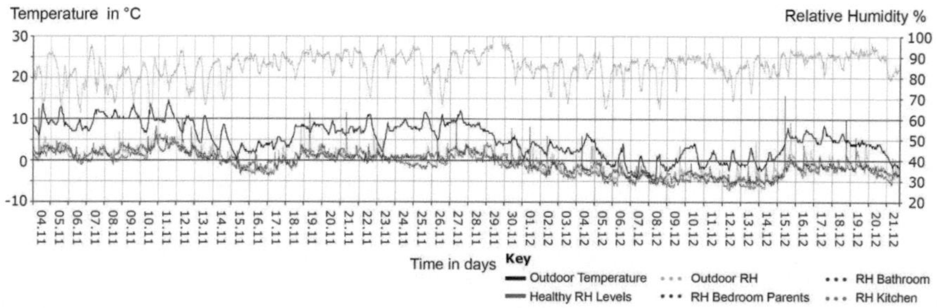

Fig. 32.4 Flat 2 results of monitoring RH (winter, conventional building materials+ mechanical ventilation)

32.3.2 Adsorption of Airborne Pollutants

The results from adsorption tests demonstrated slightly better results for earthen plasters in comparison with wall lining boards specifically designed for the adsorption of airborne pollutants. The best performance was observed for the earth plasters similar to those performing best for the water adsorption behaviour. In general, the polar compounds showed the strongest affinity for all the materials, the nonpolar compounds have hardly attached.

32.3.3 Monitoring

Figures 32.3 and 32.4 show measurements of the relative humidity (RH) of two apartments in Berlin carried out in Berlin in winter 2012–2013. In Flat 1 (natural building materials + naturally ventilation), measurements were taken in the kitchen and bathroom, while in Flat 2 (conventional building materials + mechanically ventilation), measurements were additionally taken in the master bedroom. The measurements show that in Apartment 1 the level of RH remains more stable than in Apartment 2, lying mostly in a healthy and comfortable range of 50–60%. Certain periods

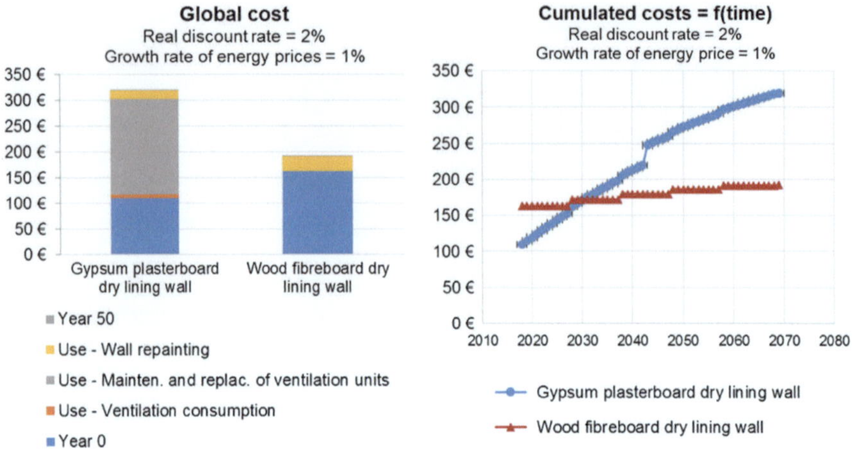

Fig. 32.5 Global costs and cumulated costs $= f(\text{time})$ of the gypsum plasterboard scenario and the wood fibreboard scenario

in the bathroom exceeded, however, 60%, which related to occupant behaviour and insufficient ventilation after showering. After the users were informed and adapted their ventilation behaviour, levels of RH were generally below 60%. In Apartment 2 instead, RH levels are influenced much more by the outdoor condition, which can be seen in the measurements where indoor levels curves outdoor curves. This is caused through the mechanical ventilation system, which takes constantly air in from the outside.

32.3.4 LCC Analysis

Figure 32.5 presents the comparison of the global cost and the cumulated costs $= f(\text{time})$ of both scenarios. It can be observed that, taking into account a real discount rate of 2% and a growth rate of energy price of 1%, the global cost of the gypsum plasterboard construction scenario is higher than the one of the wood fibreboard construction scenarios. More specifically, the investment cost for the wood fibreboard construction is higher than the one for the gypsum plasterboard construction, but this additional cost is compensated by the expenses of the additional ventilation in the conventional scenario. It is worth to note that it is the maintenance of the ventilation units that contribute the most to the global cost, not their energy consumption. The payback time of the investment the wood fibreboard constructions is 13 years.

A sensitivity analysis to both the real discount rate and the growth rate of energy price was carried out. When the discount rate increases, the global cost of the gypsum plasterboard construction scenario decreases. Nevertheless, even taken into account a real discount rate of 4%, the global cost of the gypsum plasterboard construction scenario remains above the one of the innovative partition wall scenarios.

When the growth rate of energy price decreases, the global cost of the gypsum plasterboard construction scenario decreases. Nevertheless, even taken into account a growth rate of energy price of 0%, the global cost of the gypsum plasterboard construction scenario remains above the one of the innovative partition wall scenarios. Taking into account a real discount rate of 4% and a growth rate of energy prices of 0%, the global cost of the gypsum plasterboard construction scenario remains higher than the one of the wood fibreboard construction scenarios and the payback time after investment in the latter is 22 years.

32.4 Discussion

32.4.1 Water Vapour Adsorption Tests

The comparison of the water sorption performance of natural materials in comparison with conventional ones demonstrated the outstanding sorption capacity of the earth plasters in combination with wood fibreboards in comparison with gypsum plaster or gypsum fibreboards. The tests revealed that the performance of the gypsum fibreboard could be increased through a chalk-based paint that was added as a finishing layer. The performance of the earth-based specimen can be attributed in the first instance to the clay minerals, while the qualities of the wood fibreboard lie in the performance of the timber but mainly in the high degree of porosity and therefore the large surface area.

Material investigations at component level demonstrated similar results and showed the superior performance of natural building materials in comparison with conventional wall build-ups. Figure 32.2 demonstrates the impact of earth plasters in combination with wood fibreboards and wood fibre insulation boards in comparison with conventional wall build-ups with gypsum plasterboards and mineral wool. The exact benefit on indoor air quality with regard to seasonal changes has to be determined; however, it can be assumed that buildings fitted out with such walls will benefit from evaporative cooling processes during hot summer months.

32.4.2 Emission Tests

It is important to note that the AgBB criteria were developed for individual building materials, for the analysis of wall systems a different set of criteria would be more appropriate. The results can therefore only have an orienting character. Most of the materials and material combinations that were tested had earth plaster facing surfaces of different thickness. Nevertheless, it can be established that all other tested natural building materials were uncritical with respect to their emission properties, which means that they are suitable for indoor use.

32.4.3 Adsorption of Airborne Pollutants

The adsorption tests revealed that earth plasters have a good adsorption capacity, which was better than that of specifically designed wall lining boards.

32.4.4 Monitoring Results

The results of the monitoring of real spaces (Figs. 32.3 and 32.4) show that the relative humidity of the rooms in the apartment fitted with earth plaster was consistently in the region of 50–60%, which can be attributed to the buffering capacity of the earth, as discussed also in () and (Klinge 2013). The low level of relative humidity of ~30% in Apartment 2 fitted out with conventional building materials can be attributed in part to the mechanical ventilation system that draws in dry air from outdoors all day, and in part to the materials used, that are unable to adsorb significant quantities of moisture arising within the apartment as a result of cooking or showering. It can be stated that in Flat 1 stable RH levels were measured, which ranged between 40 and 60%, which not only increase occupant comfort and health but also do not require mechanical ventilation. In Flat 2, instead unhealthy and uncomfortable values below 35% were measured, in fact, due to the provision of a high-tech approach.

32.4.5 LCC Analysis

An innovative internal partition wall based on earth plasters and other natural building materials (Scenario 1) has been compared to a conventional dry lining wall system based on a gypsum plasterboard and mineral wool (Scenario 2) on their global cost, taking into account 50 years use. The assessment accounted for the fact that additional mechanical ventilation is needed in case the conventional partition wall is used, whereas the innovative partition wall can follow a low-tech approach, based on mainly natural ventilation due to the relatively good humidity buffering capacity.

Taking into account different real discount and growth rates of energy prices, the global cost of the conventional partition wall scenario is always higher to the one of the innovative partition wall scenario. Although the investment cost of the innovative wall is higher than the one of the gypsum plasterboard wall plus ventilation equipment, this additional cost is compensated by the yearly maintenance and replacement expenses of the ventilation units during use (185 actualised € over 50 years of use) taken into account in the conventional scenario. The payback time after investment differs in respect of the assumptions for real discount and growth rates of energy prices. However, the study demonstrates that LCC is a useful to predict more realistic costs for a construction as all associate costs are considered.

32.5 Conclusion and Outlook

The results presented in this study indicate that low emitting, natural building materials with enhanced hygroscopic properties such as earth plasters, wood fibreboards and other natural building materials in combination with natural ventilation offer robust alternatives to mechanical ventilation. Through application of materials able to adsorb airborne pollutants, indoor air quality can be enhanced further. The extract of the LCC analysis demonstrated that such solutions are even more cost-effective than standard construction, if associated costs are considered. This result is especially important and relevant, since constructions based on natural building materials are often not implemented due to cost arguments and a supposedly cheaper option, requiring a high approach.

Although numerical simulations have been undertaken to translate current findings into hygrothermal models for the evaluation of indoor environment quality of residential buildings, additional research is needed as these models still lack an appropriate profile of earthen materials. First of all the adsorption velocity is often not correctly implemented so that results are slightly distorted. Also, the impact of air-purifying materials has to be investigated further.

Acknowledgements This research study was made possible with the support of the European Union's 7th Framework Programme for research, technological development and demonstration under grant agreement no. 608893 ([H]house, www.h-house-project.eu).

References

AgBB (2015) Health-related evaluation procedure for volatile organic compounds emissions (VVOC, VOC and SVOC) from building products

DIN 18947: Earth plasters—terms and definitions, requirements, test methods, August 2013

Eckerman W, Ziegert C (2006) Impact of earthen construction materials on indoor air humidity (October 2006)

EN 16516 (2015) Construction products—assessment of release of dangerous substances—determination of emissions into indoor air

ISO 16000-part 3 (2011) Indoor air—determination of formaldehyde and other carbonyl compounds in indoor air and test chamber air—active sampling method

ISO 16000-part 6 (2011), Indoor air—determination of volatile organic compounds in indoor and test chamber air by active sampling on Tenax TA® sorbent, thermal desorption and gas chromatography using MS or MS-FID

ISO 16000-part 24 (2009), Indoor air—performance test for evaluating the reduction of volatile organic compound (except formaldehyde) concentrations by sorptive building materials

Klinge A (2013) Natural material with high hygroscopic properties in naturally ventilated buildings, Master thesis, London Metropolitan University
Minke G (2012) Handbook earth construction 8th edition. Staufen bei Freiburg: ökobuch press
Quinn A, Shaman J (2017) Indoor temperature and humidity in New York City apartments during winter. Science Direct, New York
Richter et al (2013) System to generate stable long-term VOC gas mixtures of concentrations in the ppb range for test and calibration purposes. Gefahrstoffe—Reinhaltung der Luft 73:103–106
Richter M et al (2013) Determination of radon exhalation from construction materials using VOC emission test chambers. Indoor Air 23:397–405

Chapter 33
Indoor Air Quality Regulation Through the Usage of Eco-Efficient Plasters

Maria Idália Gomes, João Gomes and Paulina Faria

33.1 Introduction

The current use of the Earth resources has led to a development level on Western society that is now considered to be not sustainable. Noticeable environmental disruptions clearly indicate that, unless urgent measures are to be taken, human beings will have considerable difficulties for adapting to its global habitat. These environmental disruptions, mainly caused by anthropogenic activity, comprise climate changes, non-balanced ecosystems, shortage of mineral resources and diminution of soil fertility. It is expected that environmental pressure will tend to be increased in the forthcoming years, as problems related to pollution and shortage of resources are currently being aggravated by population increase. Also, the majority of manufacturing activities will be increasingly located in urban areas.

In 1994, the International Council for Building (CIB) defined sustainable construction as "the creation and responsible management of a healthy built environment, based on the efficient use of resources and ecological principles" (Kibert 2005). In what regards buildings, Agenda 21 for Sustainable Building (UN 1992) has identified the improvement of environmental parameters and the re-engineering of the

M. I. Gomes (✉)
Civil Engineering Department, Lisbon Engineering Superior Institute (ISEL), Lisbon Polytechnic Institute (IPL), Rua Conselheiro Emídio Navarro 1, 1959-007 Lisbon, Portugal
e-mail: idaliagomes@dec.isel.pt

J. Gomes
Chemical Engineering Department, Lisbon Engineering Superior Institute (ISEL), Lisbon Polytechnic Institute (IPL), Rua Conselheiro Emídio Navarro 1, 1959-007 Lisbon, Portugal
e-mail: jgomes@deq.isel.pt

P. Faria
CERIS and Civil Engineering Department, NOVA University of Lisbon (FCT NOVA), 2829-516 Caparica, Portugal
e-mail: paulina.faria@fct.unl.pt

B. V. Reddy et al. (eds.), *Earthen Dwellings and Structures*, Springer Transactions in Civil and Environmental Engineering,
https://doi.org/10.1007/978-981-13-5883-8_33

building process, including sustainable development, as one of the major challenges for the building sector. Sustainable (or eco-efficient) building appears as the response of building industry to the aim of achieving global sustainability. It is also expected that built environment will be as healthy and eco-friendly as possible, thus contributing to an improved indoor air quality. As a result of many issues related to energy, environment, ecology, as well as economics, building with earth materials can be considered as an alternative. Globally, building techniques using earth are currently being revived, and the Portuguese situation is not an exception. In several regions, building with earth is largely used as it uses indigenous materials, incorporates low energy and, thus, is economic. This also results in other advantages in what regards sustainability: earth as a building material is natural, non-toxic, ecological, recyclable and with low embodied energy cost, being non-combustible and contributing for increased thermal and acoustic performances (Gomes et al. 2018). However, in less developed countries, earth building is, nowadays, associated with low-quality construction when no other materials are available. Nevertheless, this situation is bound to be altered if the previously mentioned advantages will be considered. Experimental tests, currently underway on this subject, could act as good examples to change this attitude.

Also, the level of indoor air pollutants is, frequently, much higher than exterior ambient air. Therefore, it is important to understand the type and concentrations of indoor air pollutants and also to develop materials that are able to absorb these pollutants, thus reducing its concentration in the indoor environment, simultaneously being able to contribute to balancing temperature and relative humidity. This paper intends to divulge the preliminary work being developed within the INDEED project, on the effect of eco-efficient plasters on indoor air quality. Within the scope of this project, earth-based mortars will be developed and tested to assess its effect on the air quality of indoor environments.

33.2 Ecologic Sustainability—Indoor Air Quality

It should be noticed that buildings also contribute to degradation of the environment as they are responsible for 50% of fossil fuel consumption and 50% of greenhouse gas emissions (Smith 2005). According to the UN Programme for the Environment (SBCI 2009), buildings, at world scale, are responsible for:

1. 25–40% of energy consumption and 30–40% of carbon dioxide (CO_2) emissions;
2. in what concerns natural resources, they are responsible for: materials and extracted minerals—30%; water—20%; surface use—10%.

Gustavsson and Joelsson (2010) in the 3rd Report on Assessment of Climate Change, claim that, in Europe, the residential sector is accountable for a major part of the primary use of energy, thus generating CO_2 emissions and a negative environmental impact. In order to have an environmental sustainable building, it is important to assess the environmental impact throughout its whole life cycle. Building techniques should be optimised according to this aim, considering, in particular, operational aspects, maintenance and end of life. Certain priorities have to be taken into consideration during the preliminary project stages, including low consumption of non-renewable materials, low production of wastes and pollutants, the use of eco-efficient materials, protection of water resources, achieving a healthy and friendly indoor environment, assessment of the efficiency of adopted solutions, cost reduction during the life cycle and optimised techniques involving construction and maintenance. The interaction between the building and its surroundings is also a very important issue. The life cycle of a building is, in fact, a balance between costs and resources: ecological, social, human and energetic. The life cycle starts with the selection of materials to be used in its construction, the construction and operational phases, until its demolishing and comprises even the waste management, as well as all operational, utilisation, maintenance and rehabilitation consumptions. Along these phases, all resulting environmental impacts must be assessed in order to allow for newer solutions to be provided. The operational phase plays a very significant part of the life cycle assessment of a building in what concerns its energy consumption. For a housing or office building, the primary energy consumption is about 150–400 and 250–550 kWh/m^2 year, respectively, whereas 80–90% correspond to the operational phase, and only 10–20% are related with other life cycle phases (Ramesh et al. 2010). Several studies indicated that, for buildings located in moderated to cold climate regions, the major energy consumption occurs in the operational phase (Winther and Hestnes 1999; Scheuer et al. 2003; Gustavsson and Joelsson 2010).

During the last decades, both active and passive strategies have been explored in the project of buildings with low energy consumption (Chwieduk 2003; Guy and Farmer 2001). The passive approach concerns a very concise and careful approach in what regards project development strategy, utilising bioclimatic concepts, such as geometry and solar orientation, which play important roles in terms of capture, storage and further distribution of solar and wind energy, instead of focusing on building maintenance (Sadineni et al. 2011; Loonen et al. 2013). To perform a project in a bioclimatic way consists in analysing the building bearing in mind the specific climatic features in the location, as well as the environmental features and the use of natural resources available in the region, in order to attain the maximum energy efficiency and also a friendly indoor environment.

Nowadays, benefiting from mentality change, there is the necessity to develop healthy and friendly environments, as much as possible incorporating positive inputs for indoor air quality. Bearing these considerations in mind, a rammed earth wall was built in 1993 in the Feldkirch hospital in Austria. This wall acts as a climatic regulator and, simultaneously, creates some contrast with the aesthetics of the reception hall of the hospital. As previously mentioned, a healthy indoor environment must be a priority for sustainable construction. The indoor air quality depends mainly of (EPA CPSC 1995; Bonn 2006): pollutant emissions inside buildings derived from building materials and furniture, carpets, insulation degradation, combustion process, chemicals used in hygiene and cleaning activities, air-conditioning systems and bioeffluents; infiltration of external air pollutants, such as radon, ozone, carbon

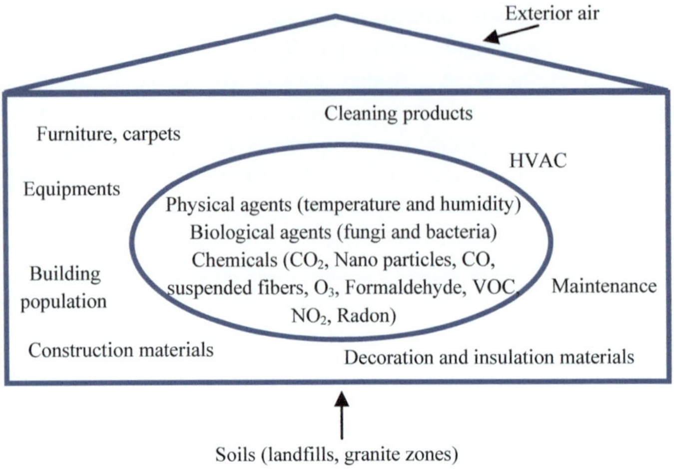

Fig. 33.1 Representation of the parameters affecting the indoor air quality in buildings (adapted from ADENE et al. 2009)

monoxide and pesticides; accumulation of pollutants inside building due to low or no ventilation.

According to ADENE et al. (2009), the major contributors for low indoor air quality are the climatisation like heating, ventilation, and air conditioning (HVAC), and also the building population. The concentration of local pollutants depends from (EPA CPSC 1995; Bonn 2006): the emission rate; air renovations; specific characteristics of the rooms (geometrical dimensions, types of furniture and coatings), as well as its occupancy. Figure 33.1 shows the main parameters affecting indoor air quality.

Natural ventilation was used because of the bioclimatic architecture until air-conditioning systems appeared. The use of natural ventilation is one of the basic principles of sustainable architecture, or good architecture, after all wind is a natural, free and renewable resource. The proper use of this source brings several advantages to the buildings, maintaining the internal quality of the air through the constant exchange—deliver fresh air into buildings, creating healthy and comfortable environments, also reducing the energy expenses, mainly the decrease of the use of air conditioning that is one of the main consumers of energy.

As a consequence of increased energy and its cost, natural ventilation has become an increasingly attractive method for reducing energy use and cost rather than the more prevailing approach of using mechanical ventilation. This change in behaviour has led to an increase in air quality, maintaining a healthy, comfortable and productive indoor climate. According to the National Renewable Energy Laboratory in the USA is possible to save 10–30% of total energy consumption in favourable climates and buildings types, using natural ventilation as an alternative to air-conditioning plants (Walker 2016).

Ventilation is required to remove odours from human physiological activity, tobacco use, domestic activities such as food preparation, the need for aeration to combustion appliances. In order for the environment to be adequate for people to remain, the air must be permanently renewed in order to avoid the formation and accumulation of substances that create an unhealthy environment for the occupants, even when the outside temperature forces the windows to be closed.

The easiest and least expensive way of ventilation and cooling of dwellings is undoubtedly by wind. Natural cross-ventilation is one of the most widely used techniques in buildings and is used in different opening spans in an environment, whether in opposite or adjacent elements. It is necessary to identify the predominant wind of the region (frequency, direction and speed), since the natural ventilation can cause discomfort and unwanted cooling if not analysed properly. The important thing is to allow fresh air to enter, either by opening aperture near the floor, windows, doors, pushing hot air to another part with opening like patio, ceiling, skylight, cast element, wind towers or ventilation tiles in the covers.

Effective ventilation allows good thermal comfort and adequate fresh air levels in the compartments. The solution is the correct location of the building, the strategic location of the spans and its adequate design, with little or no mechanical ventilation. Openings in the building or dwelling must be provided that work permanently and without obstruction. These openings may regulate the renewal of the air, but in no case should they inhibit ventilation.

Nowadays, around 50% of the world population lives in major towns, and people spend 85–90% of their time inside buildings (considering house, work and leisure activities). Therefore, they are greatly affected by these environments (EU—Joint Research Centre 2003). Thus, it is possible to derive a cause–effect relationship between housing conditions and the health of inhabitants, which calls for the increase of sustainable building.

Stieb et al. (2003) pointed out the need for monitoring pollutant concentration levels in all micro-environments. In fact, the concentration of certain pollutants inside buildings is, sometimes, higher than outside. Indoor air has several pollutants such as NO_2, SO_2, CO_2, CO, particulate and micro-organisms, as well as bacteria living in foamy materials or transmitted by moisture inside buildings. Volatile organic compounds (VOC) can also be found. These compounds are highly volatile compounds, mainly aromatics, easily oxidisable in air and heat reactive, which are present in paints, solvents, foamy materials and phenolic products. In indoor environments, products containing VOCs can last about a year to be completely eliminated, as its degradation is usually 100 times slower inside buildings. This happens with synthetic paints containing VOCs. Most common VOCs include formaldehyde, xylene, benzene and toluene. Formaldehyde is a toxic compound that is usually found in indoor environments released from materials such as phenolic adhesives used in the production of agglomerated wood panels (OSB—oriented strand board, MDF—medium-density fibreboard), paints and wood coatings and carpets made of synthetic fibres. Also, a high CO_2 level usually appears in highly populated indoor environments such as classrooms, hospital waiting rooms, sports pavilions. CO_2 in ambient air is usually 400 ppm, while in indoor environments could reach 600–800 ppm, which is mainly due to human breathing. Therefore, the reduction of CO_2 levels in highly populated environment is quite important as CO_2 levels could rise up to 1000 ppm, or more, resulting in symptoms such as headaches, sleepy states and loss of attention. Gomes et al. (2007) pointed out that the knowledge on the real concentrations of specific pollutants, such as VOCs, inside buildings, as well as the knowledge of its health effects on human beings is essential to derive specific protection measures for building occupants. Therefore, there is a considerable need to perform indoor air quality measurements and assess the level of toxic compounds release by coating materials, in order to achieve a better indoor air quality.

The effects of pollutants on human health can be classified as (ADENE et al. 2009): nuisance effects—smells (after 5–60 min of exposure), irritation reactions of eyes, throat, mouth and nose; acute effects—immediate; prolonged effects—allergic or infectious reactions, lung cancer.

Table 33.1 sums up the main sources and effects of the most important pollutants affecting indoor air quality. According to the American Society of Heating, Refrigerating, and Air-Conditioning Engineers (ASHRAE), indoor air quality is acceptable if: toxic concentration levels for pollutants are not exceeded; more than 80% of people exposed to a certain indoor air quality feel comfortable.

The materials and technologies used in a building are typically selected in accordance with the project of the building, its availability and commonly used practices in the construction site. They should fulfil the needs of society development and the user, nevertheless minimising its environmental impact. In order to control the negative environmental impact, great attention should be devoted to minimisation of greenhouse gases emissions. Therefore, the production processes for obtaining of building materials should also minimise these emissions, which also requires accurate research on the energy consumption and greenhouse gas emissions related to the production processes themselves. To minimise all energy spent in building construction, which is the net sum of all energy needed to assemble a building, extraction of raw materials, manufacture, transportation, construction process, operation and maintenance, demolition and recycling after this shelf life (Sartori and Hestnes 2007), it is, thus, important to construct buildings having a high recycling potential. Thormark (2006) mentions that a very important amount of energy can be saved by reusing construction materials. In fact, it is not enough to conclude that a certain material is recyclable, but all processes related to its valorisation should be accounted for. In order to reduce the total consumption of energy in buildings as a hole, particular attention should be given to the selection of materials and the operations related to its end of life.

Earth construction can be an effective alternative to certain problems in what concerns sustainability and indoor air quality. As previously mentioned, the use of earth as construction material brings several advantages such as: earth is a natural and eco-efficient material; earth is non-toxic and can improve air quality; earth is ecological, producing very low wastes, as it is reusable; earth is non-polluting, as there implies only low CO_2 levels during processing; earth is a low cost material, in what concerns extraction because it is frequently obtained as a waste in construction sites, thus reducing costs and energy for transportation, and for processing. Furthermore,

Table 33.1 Major sources and effects on human health from pollutants affecting indoor air quality (APA 2009; DGEG, APA and ADENE 2009)

Pollutant: CO (Carbon monoxide)	
Sources	Health effects
Combustion processes (heating, stoves, fireplaces), exhaust gases from vehicles. Tobacco smoke	Carboxyhemoglobinemia (hinders oxygen absorption) Headaches, nausea, dizziness Effects on nervous and cardiovascular systems
Pollutant: CO_2 (Carbon dioxide)	
Occupants (sweat, respiration, stomach and intestines). Tobacco smoke	Effects on nervous and cardiovascular systems Headaches, Dores de cabeça, eye and throat irritation Dizziness, asthma
Pollutant: HCHO (Formaldehyde)	
Disinfectants, pesticides Wood preservers, construction materials, insulation foams, furniture, textiles, adhesives, paints, glues, solvents, resins. Tobacco smoke	Eye, throat and skin irritation Respiratory problems Dizziness Headaches
Pollutant: VOC (volatile organic compounds)	
Paint, solvents, adhesives, resins and varnishes, construction materials, agglomerated cork, furniture, cleaning products, disinfectants, deodorants, fragrances, insecticides, pesticides, fungicides, tobacco smoke, proximity to gas filling stations	Smells Allergy symptoms Headaches, nausea, dizziness Leukaemia Lung and skin cancer Throat and nose dryness, eye irritation
Pollutant: O_3 (Ozone)	
Photocopy machines LASER printers Cleaning activities Photochemical reactions Water disinfectant	Respiratory problems, allergic reactions, asthma Eye irritation, headaches Lung oedema for prolonged pain repeated exposure Mouth and throat dryness Cough
Pollutant: PM_{10} (Particle)	
Combustion processes, tobacco smoke Occupants AVAC systems Paper	Respiratory problems, cough, sneezes Eye irritation, asthma, allergies Dryness of skin and nose Professional diseases (metals)

(continued)

Table 33.1 (continued)

Pollutant: Bactérias, fungos e *legionella*	
Sistema AVAC systems, construction materials, textiles (carpets), pollens, wet construction areas, occupants (bacteria), hair, and insect droppings, still waters (Legionella and fungi)	Allergies—rhinitis, sinus, asthma Infections—tuberculosis, pneumonia, criptococose Irritation—eyes, nose, throat and skin (fungi) Headache, fever Legionary diseases and Pontiac fever—*Legionella*
Pollutant: Radon	
Construction materials, soil from granitic regions	Increases risk of lung cancer
Pollutant: C_6H_6 (Benzene)	
Wood-derived products Tobacco smoke	Cancer
Pollutant: NO_2 (Nitrogen dioxide)	
Combustion process	Respiratory problems, chronic bronchitis Irritation of eyes, throat, cough and dizziness
Pollutant: Naphthalene	
Tobacco smoke Naphthalene	Eye irritation Respiratory irritation

during the operational phase of the building their utilisation may bring advantages such as contribution towards comfort improvement by the regulation of thermal and hygrometric equilibrium.

In Portugal, there is a strong tradition regarding construction with earth. The main utilised techniques are rammed earth (earth compacted between formworks) and adobe (earth blocks sun dried) (Gomes et al. 2014). The use of these techniques suffered a decay after 1950–60, when more modern techniques using newer construction materials, such as cement, started to be use all over the country. The construction technique using compressed earth blocks started to appear in Portugal during the 1950s, although this technique did not spread because it was by that period that earth construction in Portugal started to decline. However, in the last decades, earth construction started, again, to be used in new buildings and the interest to retrofit and repair old earthen-based buildings also started due to their environmental benefits, contribution for improving indoor air quality, as well as good thermal and acoustic characteristics, as referred before

Ventakarama-Reddy and Kumar (2010) also pointed out other advantages related to rammed earth construction: low energy intensity and low carbon emissions, the recycling feature of the materials, the aspect of being locally available thus reducing transportation costs, flexibility for use in the geometry of buildings, allowing a great variety of finishing's and textures, and also the fact that wall thickness can be easily adjusted, when stabilised rammed earth is used.

33.3 Use of Earth Mortars

Mélia et al. (2014) assessed the energy incorporated when using earth mortars which compared very favourably with other alternatives. One of the most important characteristics of porous building materials is the way they transport and react to the presence of moisture, i.e. their hydric behaviour. This is particularly significant for earth-based materials in which the binder is clay (Gomes et al. 2016). Some authors pointed out that mortars based on clayish earth used in wall plasters can contribute to the improvement of air quality—clay can act as a passive absorption material, thus decreasing ozone concentrations and, therefore, reducing the probability of occurrence of ozone reactions with other materials existing inside buildings (Lima and Faria 2016). However, certain aspects related to the possibility of development of biological agents are also to be accounted for (Santos et al. 2017).

Earth, as a construction material, acts as a protection against major humidity variations, thus contributing to balance relative moisture inside buildings (Minke 2006; Kirsima and Maddison 2009; Liuzzi et al. 2013; Bui et al. 2014; Lima et al. 2016). This capacity is due to the exchange mechanism of water vapour with air: it releases humidity when the air is more dry, absorbing it when it is more humid. This hygroscopic equilibrium depends on factors such as the type of clay, its thickness (Fionn et al. 2017) or the eventual stabilisation obtained by the finishing, that will influence the adsorption velocity and the capacity of water vapour release. It is important to note that these mortars, when organic solvents are not incorporated, do not release toxic organic compounds to the indoor air, as their components are clay and sand. One of the aims of project *INDEED* is to analyse if—and to what extent—earth mortars, when in contact with toxic substances and high carbon dioxide concentrations, are able to reduce smells and the concentration of certain pollutants such as particulate (PM_{10} and $PM_{2.5}$), carbon monoxide, ozone, and volatile organic compounds, such as formaldehyde and BTEX.

It is also important to assess the behaviour of earth mortars in the presence of particulate aerosols as the latter are potential condensation nuclei for micro-organisms such as virus, pollen, bacteria and fungi, mainly as these micro-organisms can be developed and favoured by the presence of high moisture conditions indoor (Lima 2013). As earth plasters can act on the balance between moisture and indoor air, the drive for development of micro-organisms could be greatly reduced.

Therefore, it is paramount to analyse the behaviour of earth mortars when exposed to different pollutants existing in indoor air, as a way to reduce human exposure to aggressive indoor environments, as well as to perform the monitoring of comfort indoor conditions in Portugal. Within INDEED project, five test chambers will be built and pollutants will be injected, namely aerosol particles (PM_{10} and $PM_{2,5}$), carbon dioxide (CO_2), carbon monoxide (CO), ozone (O_3) and volatile organic compounds (VOCs). Inside each test chamber, different earth-based plasters and a cement plaster will be placed, simulating walls and ceiling coatings. The main objective is to collect the air inside the chambers and to determine the changes in pollutants concentrations, as well as temperature and relative humidity. It is expected to be able to quantify the de-pollutant capacity of the earth-based plasters in comparison with the cement-based one.

33.4 Final Considerations

There is currently enough scientific evidence relating complaints and environmental discomfort from building occupants in what regards construction materials used inside buildings. Both the hygienic aspects as well as the toxicological ones have been studied in order to guarantee the existence of pleasant, healthy and comfortable environments. As there is evidence that earth-based mortars have an active and positive effect on the regulation of thermal and hygrometric equilibrium, it is also important to study its behaviour when put into contact with toxic, pollutants and smelly compounds. *INDEED* project aims to quantitatively answer these queries so that a better indoor air quality can be achieved and, as expected, an important contribute to the use of earth plasters may be presented.

Acknowledgements The work is within the project 23349–02/SAICT/2016 – INDEED: Indoor air quality regulation through the usage of eco-efficient mortars.

References

ADENE, DGEG, APA (2009) PQ intervention of the buildings covered by RCESE—Indoor air quality (in Portuguese). Portugal

APA (2009) Indoor air quality: a technical guide (in Portuguese). Amadora, Portugal

Bonn G (2006) Development of WHO guidelines for indoor air quality. Report on a working group meeting. Germany

Bui T, Bui Q, Limam A, Maximilien S (2014) Failure of rammed earth walls: from observations to quantifications. Constr Build Mater 51:295–302. https://doi.org/10.1016/j.conbuildmat.2013.10.053

Chwieduk D (2003) Towards sustainable-energy buildings. Appl Energy 76(1–3):211–217. https://doi.org/10.1016/S0306-2619(03)00059-X

DGEG, APA, ADENE (2009) Technical Note NT-SCE-02. Methodology for periodic audits of indoor air quality in existing services buildings under the RSECE (in Portuguese)

EPA CPSC (1995) the inside story: a guide to indoor air quality. United States of America

Fionn M, Fabri A, Ferreira J, Simões T, Faria P, Morel JC (2017) Procedure to determine the impact of the surface film resistance on the hygric properties of composite clay/fibre plasters. Mater Struct 50(4):193–206. https://doi.org/10.1617/s11527-017-1061-3

Gomes J, Bordado J, Sarmento G, Dias J (2007) Measurements of indoor air pollutant levels in a university office building. J Green Build 2(4):123–129. https://doi.org/10.3992/jgb.2.4.123

Gomes MI, Gonçalves TD, Faria P (2016) Hydric behavior of earth materials and the effects of their stabilization with cement or lime: study on repair mortars for historical rammed earth structures. J Mater Civil Eng 28(7). https://doi.org/10.1061/(asce)18MT.1943-5533.0001536

Gomes MI, Faria P, Gonçalves TD (2018) Earth-based mortars for repair and protection of rammed earth walls. Stabilization with mineral binders and fibers. J Clean Prod 172:2401–2414. https://doi.org/10.1016/j.jclepro.2017.11.170

Gomes MI, Gonçalves TD, and Faria P (2014) Unstabilised rammed earth: characterization of the material collected from old constructions in south Portugal and comparison to normative requirements. Int J Architectural Heritage, Taylor & Francis, 8(2):185–212. https://doi.org/10.1080/15583058.2012.683133

Gustavsson L, Joelsson A (2010) Life cycle primary energy analysis of residential buildings. Energy Build 42:210–220. https://doi.org/10.1016/j.enbuild.2009.08.017

Guy S, Farmer G (2001) Reinterpreting sustainable architecture: the place of technology. J Architectural Edu 54(3):140–148. https://doi.org/10.1162/10464880152632451

Kibert CJ (2005) Sustainable construction: green building design and delivery. Wiley, New Jersey, United States of America

Kirsima K, Maddison M (2009) The humidity buffer capacity of clay—sand plaster filled with phytomass from treatment wetlands. Build Environ 44:1864–1868. https://doi.org/10.1016/j.buildenv.2008.12.008

Lima J (2013) The contribution of clay mortars to the quality of the interior environment of buildings: the case of clays from the Eastern Algarve (in Portuguese). 2° Congresso Internacional da Habitação no Espaço, Lisboa

Lima J, Faria P (2016) Eco-efficient earthen plasters. The influence of the addition of natural fibers. In Fangueiro R (ed) 2nd international conference on natural fibres. Azores, Portugal, 27–29 April: advances in science and technology towards industrial applications, Vol 12. Springer, RILEM Book Series, pp 315–327

Lima J, Faria P, Silva AS (2016) Earthen plasters based on illitic soils from barrocal region of Algarve: contributions for building performance and sustainability. Key Eng Mater 678:64–77. https://doi.org/10.4028/www.scientific.net/KEM.678.64

Liuzzi S, Hall MR, Stefanizzi P, Casey SP (2013) Hygrothermal behaviour and relative humidity buffering of unfired and hydrated lime-stabilised clay composites in a Mediterranean climate. Build Environ 61:82–92. https://doi.org/10.1016/j.buildenv.2012.12.006

Loonen RM, Trcka M, Cóstola D, Hensen JM (2013) Climate adaptive building shells: State-of-the-art and future challenges. Renew Sustain Energy Rev 25:483–493. https://doi.org/10.1016/j.rser.2013.04.016

Melià P, Ruggieri G, Sabbadini S, Dotelli G (2014) Environmental impacts of natural and conventional building materials: a case study on earth plasters. J Clean Prod 80:179–186. https://doi.org/10.1016/j.jclepro.2014.05.073

Minke G (2006) Building with earth—Design and technology of a sustainable architecture. Birkhäuser—Publishers for Architecture. https://doi.org/10.1007/3-7643-7873-5

Ramesh T, Prakash R, Shukla K (2010) Life cycle energy analysis of buildings: an overview. Energy Build 42:1592–1600. https://doi.org/10.1016/j.enbuild.2010.05.007

Reddy Venkatarama B, Kumar Prasanna P (2010) Embodied energy in cement stabilised rammed earth walls. Energy Build 42(3):380–385. https://doi.org/10.1016/j.enbuild.2009.10.005

Sadineni SB, Madala S, Boehm RF (2011) Passive building energy savings: a review of building envelope components. Renew Sustain Energy Rev 15(8):3617–3631. https://doi.org/10.1016/j.rser.2011.07.014

Santos T, Nunes L, Faria P (2017) Production of eco-efficient earth-based plasters: influence of composition on physical performance and bio-susceptibility. J. Cleaner Prod 167:55–67. https://doi.org/10.1016/j.jclepro.2017.08.131

Sartori I, Hestnes AG (2007) Energy use in the life cycle of conventional and low-energy buildings: a review article. Energy Build 39:249–257. https://doi.org/10.1016/j.enbuild.2006.07.001

SBCI U (2009) Buildings and climate change: summary for decision makers

Scheuer C, Keoleian G, Reppe P (2003) Life cycle energy and environmental performance of a new university building: modelling challenges and design implications. Energy Build 35(10):1049–1064. https://doi.org/10.1016/S0378-7788(03),00066-5

Smith PF (2005) Architecture in a climate of change: a guide to sustainable design, 2nd edn. Architectural Press an imprint of Elsevier

Stieb DM, Judek S, Burnett RT (2003) Meta-analysis of time-series studies of air pollution and mortality: update in relation to the use of generalized additive models meta-analysis of time-series studies of air pollution and mortality: update in relation to the use of generalized additive. Air & Waste Manag Assoc 53(September):258–261

Thormark C (2006) The effect of material choice on the total energy need and recycling potential of a building. Build Environ 41:1019–1026. https://doi.org/10.1016/j.buildenv.2005.04.026

UN (1992) Agenda 21 - Rio Declaration. United Nations conference on environment & development, Rio de Janeiro, Brazil

Walker A (2016) Natural ventilation. National Institute of Building Sciences, National Renewable Energy Laboratory—Whole Building Design Guide®. Retrieved from https://www.wbdg.org/resources/natural-ventilation

Winther B, Hestnes A (1999) Solar versus green: the analysis of a norwegian row house. Sol Energy 66(6):387–393. https://doi.org/10.1016/S0038-092X(99),00037-7

Chapter 34
Full-Scale Simulation of Indoor Humidity and Moisture Buffering Properties of Clay

Valeria Cascione, Daniel Maskell, Andy Shea and Pete Walker

34.1 Introduction

Indoor environment humidity levels have an important influence on occupant health and well-being. Low- or high-humidity levels influence thermal comfort, the perception of indoor air quality, and increase risks of exposure to bacteria, viruses and mould spores (Arundel et al. 1986). Therefore, relative humidity (RH) should ideally be maintained between 40 and 60% (Rode et al. 2005) to optimise indoor quality.

Energy consuming mechanical devices, such as air-conditioning systems, are commonly used to maintain optimal RH levels. However, low-energy design strategies are needed to provide more comfortable and healthier indoor climates whilst minimising overall energy consumption. One potential solution is the wider use of hygroscopic materials on indoor surfaces, which have the ability to moderate indoor humidity fluctuations through exposure to the room air, potentially reducing operational energy use. In particular, earthen materials, such as finishing plasters and clay masonry, reduce the peaks of internal relative humidity due to their excellent moisture buffering capacity (McGregor 2014).

In recent years, clay plasters have been studied and characterised (Liuzzi et al. 2013; Faria et al. 2016; Thomson et al. 2015), showing an increased interest in its dynamic water absorption property. Experimental tests, based on the step-response

V. Cascione (✉) · D. Maskell · A. Shea · P. Walker
Department of Architecture and Civil Engineering, University of Bath, Bath, UK
e-mail: V.Cascione@bath.ac.uk

D. Maskell
e-mail: D.Maskell@bath.ac.uk

A. Shea
e-mail: A.Shea@bath.ac.uk

P. Walker
e-mail: P.Walker@bath.ac.uk

B. V. V. Reddy et al. (eds.), *Earthen Dwellings and Structures*,
Springer Transactions in Civil and Environmental Engineering,
https://doi.org/10.1007/978-981-13-5883-8_34

method such as the NORDTEST protocols (Rode et al. 2005), ISO 24353 (ISO 2008) and JIS A 1470-1 (JIS 2002), were developed to characterise the moisture buffering effects of materials. These tests methods are applied in a controlled environment, in which temperature is constant and humidity varies cyclically. Material mass variations are continuously recorded to quantify the amount of water absorbed and desorbed during the humidity cycles. The step-response tests do not represent the moisture buffering behaviour of finishing materials in real buildings, due to the influence of ventilation, occupant's behaviour and other factors on the moisture exchange between materials and the environment. It is important to observe the effects of these phenomena on the moisture balance in real buildings. For this purpose, numerical simulations can be useful to evaluate the impact of finishing materials in real buildings.

This paper presents the results of an investigation to quantify moisture buffering capacity of clay and lime plasters. Moisture buffering capacity was experimentally determined, using the NORDTEST step-response tests (Rode et al. 2005). Numerical simulations, based on measured materials characteristics, were used to investigate the response of earthen plasters, and compared to lime plasters, to influence the indoor environmental conditions in a test room. The aim of this study is to highlight the discrepancy between moisture buffering predictions of the numerical model and experimental results.

34.2 Materials and Methods

Samples of commercially available clay and natural hydraulic lime (NHL 3.5, moderately hydraulic lime) were prepared and tested for the study. Finishing surfaces are usually composed of up to 3 mm topcoat and a 10–15 mm basecoat. Maskell et al. (2018) demonstrated with clay plasters in the density range of 1700–1870 kg/m^3, and only the first 6–10 mm of the earthen materials is involved in moisture buffering, which means the sub-layer has a larger contribution to buffer indoor water vapour. Therefore, only undercoat plasters were selected due to their greater volume of involvement to the dynamic water absorption process.

34.2.1 Specimen Preparation

The clay and lime plasters were prepared by mechanical mixing in the laboratory. To the air-dry clay plaster (mix of natural clay <5 mm and sand 0–2 mm), a further 20% mass of water was added. The lime plasters comprised 1.2:5 (lime: fine aggregate sand) by volume with 30% water added. The mixing water amount was set according to the workability of the plasters. Specimens were cast in 150 × 150 × 20 mm moulds made with phenolic-faced plywood (Fig. 34.1).Thereafter, the specimens were stored for 28 days before testing in an environmental chamber at 20 °C and 60% RH.

Fig. 34.1 Moisture buffering specimen: **a** clay, **b** lime

Table 34.1 Dry bulk density (ρ_{dry}), coefficient of variation (CoV) and water vapour diffusion resistance factor (μ)

Material	ρ_{dry} (kg/m^3)	CoV (%)	μ (–)
Clay	1258	2.32	24.45
Lime	1590	1.69	19.62

34.2.2 Density and Vapour Diffusion Resistance Factor

The dry bulk density was measured for the plaster specimens after drying them at 105 °C for three days. The vapour diffusion resistance factors were determined in accordance with BS EN ISO 12572 (ISO 2008), using the dry cup method. Specimens were pre-conditioned in climatic chamber at 23 °C and 50% RH for 24 h. Specimens were afterwards sealed on the top of a plastic container, which contained calcium chloride ($CaCl_2$) with 1.5 cm thickness of air layer between salt and internal sample surface. Aluminium tape was used to ensure a vapour tight seal. The assembly was then placed in a climatic chamber at 23 °C and 50% RH and weighted by balance (±0.01 g readability) until constant mass was achieved. Table 34.1 presents the measurement results.

34.2.3 Vapour Absorption/Desorption Curve

The sorption isotherm was measured at 23 °C and a RH range of 0 to 90% with a dynamic vapour sorption (DVS) apparatus. The experimental accuracy of mass change was ±0.1 mg; temperature was maintained to within ±1% °C and RH to ±1%. Specimens with a mass of approximately 0.3 mg were placed on the DVS scale, and the surrounding humidity was increased gradually in steps, measuring the change in mass of the plasters samples.

34.2.4 Moisture Buffering

The moisture buffering test was performed in a climatic chamber, following the NORDTEST protocol (Rode et al. 2007). Three specimens for each plaster type were pre-conditioned to 50% and 23 °C for 24 h, until the mass varied by less than 5%. Specimens were exposed to three cycles of 75% RH for 8 h and 33% for 16 h. Specimens were placed on a mass balance inside the climatic chamber and covered by a screen to reduce the airspeed to less than to 0.1 m/s. The mass of each specimen was measured every minute. The moisture buffering value (MBV) is expressed in g/(m^2 %RH).

34.3 Hygrothermal Simulation

34.3.1 Simulation Theoretical Model

For thermo-hygrometric simulation of the building, WUFI®Plus V3.0.3 (Wärme-Und Feuchtetransport Instationär) was chosen. In this model, the transport of heat and humidity in the building structures is described with the following equations (Künzel 1995):

$$\frac{\mathrm{d}H}{\mathrm{d}T}\cdot\frac{\mathrm{d}T}{\mathrm{d}t}=\nabla\cdot(\lambda\cdot\nabla T)+h_v\cdot\nabla\big(\delta_p\cdot\nabla(\varphi p_{\mathrm{sat}})\big) \tag{34.1}$$

$$\frac{\mathrm{d}w}{\mathrm{d}\varphi}\cdot\frac{\mathrm{d}\varphi}{\mathrm{d}t}=\nabla\cdot\big(D_\varphi\nabla\varphi+\delta_p\cdot\nabla(\varphi p_{\mathrm{sat}})\big) \tag{34.2}$$

where φ is the relative humidity (–), t is the time, T is the temperature (°C), w is the moisture content (kg/m^3), p_{sat} is the saturation vapour pressure (Pa), λ is the thermal conductivity (W/mK), H is the enthalpy (J/m^3), D_φ is the liquid conduction coefficient (kg/ms), δ_p is the vapour permeability (kg/ms Pa), and h_v is the latent heat of phase change (J/kg).

The indoor absolute moisture ratio (c_i [kg/m^3]) is calculated from the following water vapour mass balance equation:

$$V\frac{\mathrm{d}c_i}{\mathrm{d}t}=\sum_j A\cdot\dot{g}_{wj}+nV(c_a-c_i)+\dot{w}_{\mathrm{Imp}}+\dot{w}_{\mathrm{Vent}}+\dot{w}_{\mathrm{HVAC}} \tag{34.3}$$

where c_a and c_i are, respectively, the absolute moisture ratio of the exterior and interior air, $\dot{g}_{wj}$ is the moisture flux from the interior surface into the room, $\dot{w}_{\mathrm{Imp}}$ is the moisture production, and $\dot{w}_{\mathrm{Vent}}$ and $\dot{w}_{\mathrm{HVAC}}$ are, respectively, the moisture gains or losses due to ventilation and HVAC systems.

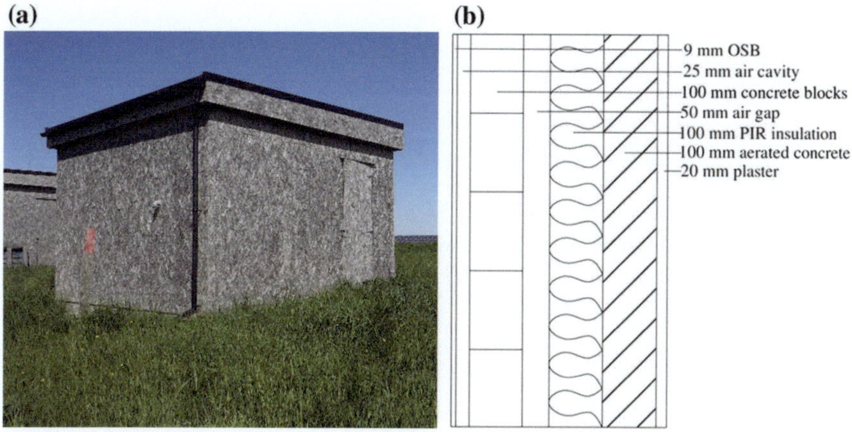

Fig. 34.2 **a** Test building, **b** vertical section of the concrete block masonry wall

34.3.2 Study Case

A single test room, located at the University of Bath's Building Research Park, Wroughton, UK, was modelled (Fig. 34.2a). Its internal dimensions are 4 m × 3 m and 2.5 m high. The envelope of the structure is a lightweight cavity wall of 404 mm total thickness and comprising of, from outside to inside, exterior concrete block, PIR insulation and aerated concrete blocks (Fig. 34.2b). Floor and ceiling are timber sandwich panel structure of PIR insulation and particle board (total thickness of 350 mm). Walls have a U-value of 0.15 W/m^{2}K, whilst floor and ceiling have a U-value of 0.10 W/m^{2}K. The internal aerated concrete block surface was coated alternatively with clay and lime plaster of 20 mm thickness, whilst the floor, ceiling and door were covered with an impermeable layer ($s_d = 1500$), in order to ensure that the room moisture balance is not affected by the particle board.

The test room was simulated over a two-year time frame. The first-year simulation was run without any moisture generation source, as the assembly took one year to reach again equilibrium with the surrounding environment. During the second year, the test room had a simulated occupancy of two people during the night, 7 days a week, between 11.00 p.m. and 7.00 a.m. For each occupant, a water vapour generation rate in the room was defined as 60 g/h. The natural ventilation rate, in terms of air changes per hour (ACH), was set to a constant 0.5 h^{-1}. The infiltration rate applied to the model was based on the infiltration rate as measured in a previous experimental study of this test cell (Latif et al. 2016), which measured a value equating to 0.0162 h^{-1}. Temperature was kept constant at 23 °C. The outdoor climatic data were taken from Lyneham, UK, a weather station at around 10 miles from the location of the test room.

The computational analyses were performed for three simulated wall internal finishes, namely clay plaster or lime plaster or a vapour barrier ($s_d = 1500$). Plaster

Table 34.2 Porosity (ε), thermal conductivity (λ) and specific heat (c)

Material	ε (%)	λ (W/mK)	c (J/kg K)
Clay	27.4	1.0	849
Lime	33	1.0	929

hygric properties are reported in Table 34.1, whilst porosity and thermal properties (Table 34.2) were obtained from a review of published literature, for clay Faria et al. (2016) and Thomson et al. (2015) and, for lime, Delgado et al. (2013) and Černý et al. (2006).

34.4 Results

34.4.1 Moisture Buffering and Vapour Sorption Curve

The sorption isotherm for clay and lime is shown in Fig. 34.3a. Two cycles were performed to ensure the stability of the sorption curves, but only the second curve was analysed. The increase in water content from 0% RH to 90% RH for both materials can be approximated as linear. Clay reached higher values of moisture content, approximately 1.5% mass at 90% RH, compared to the lime (0.7%). However, in each cycle lime gained around 0.08% mass more than the previous cycle, as the lime binder present may not have completed its hydration. Moisture buffering test results also highlighted that lime did not reach a balance after three cycles. Tests were repeated, and an increase of around 3% weight every cycle was observed. Nonetheless, clay has significantly higher buffering performance (Fig. 34.3b): after 8 h of humidification, it reached a MBV of 1.67 g/m^2 %RH, whereas lime reached only 0.95 g/m^2 %RH.

34.4.2 Moisture Buffering Simulation

Simulation results were carried out on all internal walls to compare possible effects of solar radiation, rain and wind direction on the moisture balance of the enclosure. It was noticed that the percentage variation of the moisture content between walls was less than 3%. For this reason, only the east wall has been reported here. Figure 34.4 shows the test room absorbed and desorbed moisture in response to humidity generation quite effectively with the lime or clay are applied on internal surfaces. Moisture fluctuation was reduced to 30% RH compared to the impermeable walls case.

The internal clay plaster, unlike lime plaster, adsorbed a greater proportion of the generated moisture. Figure 34.4a presents the RH variation in the simulated room and indicates that for clay the average indoor relative humidity peak-to-peak amplitude

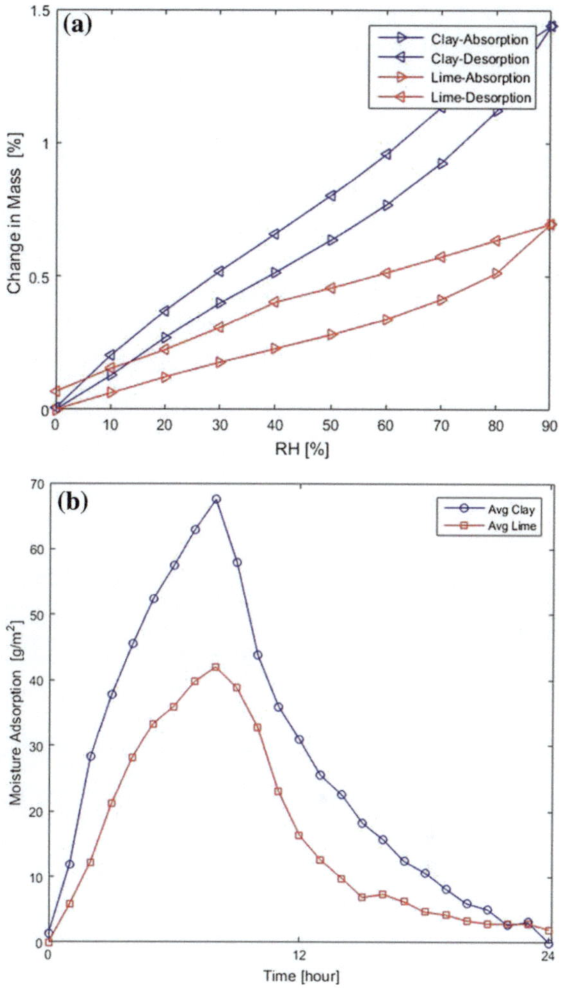

Fig. 34.3 Sorption isotherm curve and moisture buffering profile of clay and lime

is reduced to only 10.91% RH in the indoor environment, adsorbing and desorbing each cycle 10.60 g/m^2, whilst lime has a lower buffer potential of 9.24 g/m^2, which corresponds to humidity fluctuation of 13.18% RH. Figure 34.4b shows also that the average RH in the room varies at a faster rate with lime, than with clay, due to the inferior moisture buffering capacity of lime.

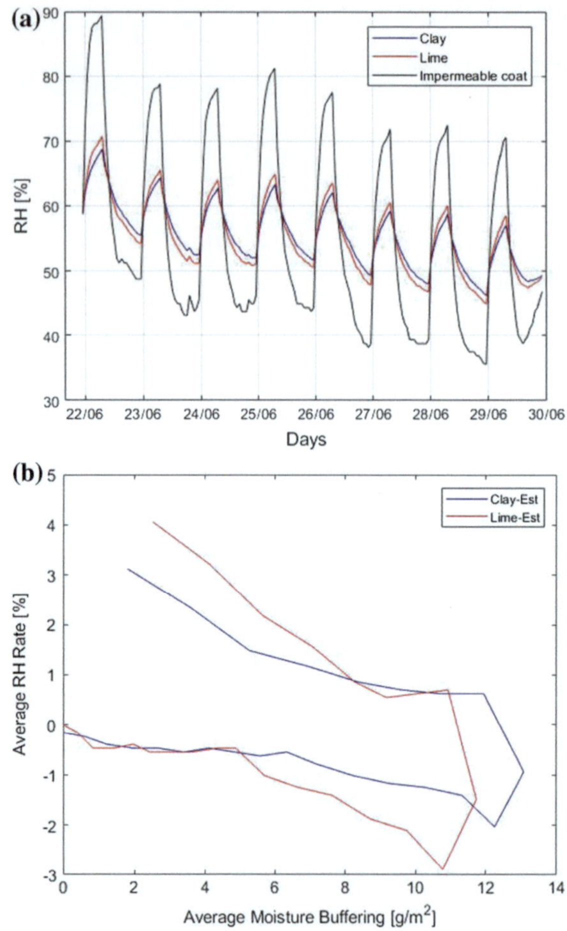

Fig. 34.4 **a** RH fluctuation in the simulated room, **b** RH variation rate over time

34.5 Analysis and Discussion

Experimental results and simulations present similar trends: both indicate the improved moisture buffering behaviour of clay material compared to lime plaster; however, the step-response test overestimated the moisture buffer potential of both materials, as shown in Fig. 34.5. Conceptually, the two analyses are not comparable, as the NORDTEST is performed in a climatic chamber, where relative humidity is controlled and water vapour control inside the chamber depends on the amount of water in the air necessary to keep RH constant. Conversely, in the simulated test room a fixed amount of water is released, independent of relative humidity variations. Consequently, the released moisture amount in the two cases may differ and a volume correction may be necessary, as the climatic chamber has a volume of 0.33 m^3 compared with the room of 30 m^3. However, it was observed that relative

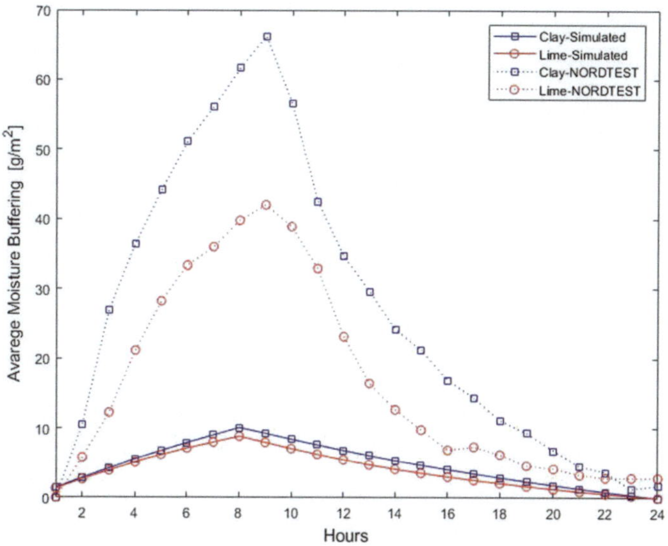

Fig. 34.5 Moisture buffering comparison: step-response test and simulation results

humidity fluctuations in the test cell with the impermeable membrane (Fig. 34.4) are similar to the humidity variations recorded in the climatic chamber (around 40% RH fluctuation). For this reason, results were considered comparable.

In Fig. 34.5, significant differences between lime and clay moisture buffering performance are highlighted. The simulation output presents both low average moisture buffering values (g/m^2) and small differences between the two plasters, whilst the results of experimental methods present a much greater average moisture buffering value (g/m^2) and a much higher relative performance of clay plaster, compared with lime plaster. The reasons for such a discrepancy are linked to the interference of other phenomena, e.g. ventilation, infiltration and humidity transport mechanisms in the test room enclosure.

Moisture buffering in a building is strongly dependent on ventilation rate. Depending on vapour pressure differential between the indoor and outdoor, ventilation and infiltrations carry moisture in and out, influencing the maximum absorbed moisture content of the finishing materials. Figure 34.6a illustrates how ventilation moisture exchange can increase and decrease the moisture buffering potential of the finishing materials: When outdoor vapour pressure is much lower than indoor vapour pressure, ventilation moisture uptake reduces the adsorption potential of materials by 70%.

In contrast with values reported in Maskell et al. (2018), simulations also show that the autoclaved aerated concrete blocks (AAC), underneath the plaster, are participating in the moisture buffering process. In Fig. 34.6b, the moisture exchange between the indoor and AAC is illustrated: the lime plastered simulation case presents daily moisture content fluctuations in the AAC (around 2 g/m^2), which is less marked in the case with clay plaster, due to the higher capacity of the earthen material to absorb

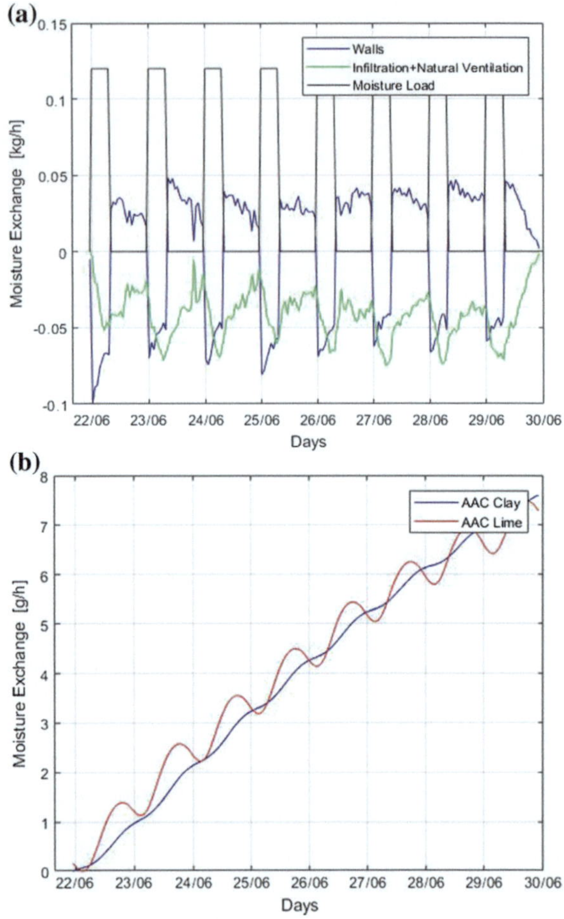

Fig. 34.6 **a** Influence of ventilation and walls in the moisture exchange with clay plaster, **b** contribution of the aerated concrete block to moisture buffering

water and lower water vapour permeability. The deeper penetration of water into the wall can be linked to the vapour pressure and temperature difference between the indoor and outdoor, as well as to water transport mechanism between different materials. Therefore, further verification using full-scale testing of buildings is necessary.

34.6 Conclusions

Moisture buffering capacity of finishing materials was determined both experimentally and through hygrothermal simulations. For the experimental measurement, the step-response method, following the NORDTEST procedure, was performed in a climatic chamber, whilst a full-scale test room was simulated in WUFI®Plus. Two plasters, clay and lime, were investigated, and it is observed that the clay plaster exchange more humidity than lime. However, the step-response test overestimated the moisture potential of both materials, as it does not consider the effect of ventilation on the moisture absorption and desorption of the building surfaces.

Simulation also highlighted that other wall components participated, together with the finishing coatings, to buffer the indoor moisture. According to the computational model, within the daily 8-h moisture production, moisture reached the AAC block layer, probably due to the vapour pressure and temperature differential between the indoor and outdoor environment. However, this behaviour is in contrast with other studies on the moisture buffering penetration depth, and more analysis on the moisture buffering potential of wall assemblies is recommended.

The significance of a direct comparison between full-scale analysis and laboratory-scale material results is to improve moisture buffering testing, optimising material use to increase indoor environment quality. A precise quantification of the material absorbed/desorbed moisture can lead to an improved understanding of the impact of earthen plaster on the hygrothermal comfort and to the consequent reduction of "active" conditioning system energy consumption.

References

Arundel AV, Sterling EM, Biggin JH, Sterling TD (1986) Indirect health effects of relative humidity in indoor environments. Environ Health Perspect 65:351

Černý R, Kunca A, Tydlitát V, Drchalová J, Rovnaníková P (2006) Effect of pozzolanic admixtures on mechanical, thermal and hygric properties of lime plasters. Constr Build Mater 20(10):849–857. https://doi.org/10.1016/j.conbuildmat.2005.07.002

Delgado JMPQ, Barreira E, Ramos NMM, de Freitas VP (2013) Inputs for hygrothermal simulation tools. Appl Sci Technol 7–20. https://doi.org/10.1007/978-3-642-35003-0_2

Faria P, dos Santos T, Aubert J-E (2016) Experimental characterization of an earth eco-efficient plastering mortar. J Mater Civ Eng 28(1):1–9. https://doi.org/10.1061/(ASCE)MT.1943-5533.0001363

ISO 24353 (2008) Hygrothermal performance of building materials and products—determination of moisture adsorption/desorption properties in response to humidity variation. Internal Standardization Organization (ISO)

JIS A 1470-1 (2002) Test method of adsorption/desorption efficiency for building materials to regulate an indoor humidity-part 1

Künzel HM (1995) Simultaneous heat and moisture transport in building components. One-and Two-Dimensional Calculation Using Simple Parameters. IRB-Verlag Stuttgart

Latif E, Lawrence M, Shea A, Walker P (2016) In situ assessment of the fabric and energy performance of five conventional and non-conventional wall systems using comparative coheating tests. Build Environ 109:68–81. https://doi.org/10.1016/j.buildenv.2016.09.017

Liuzzi S, Hall MR, Stefanizzi P, Casey SP (2013) Hygrothermal behaviour and relative humidity buffering of unfired and hydrated lime-stabilised clay composites in a Mediterranean climate. Build Environ 61:82–92. https://doi.org/10.1016/j.buildenv.2012.12.006

Maskell D, Thomson A, Walker P, Lemke M (2018) Determination of optimal plaster thickness for moisture buffering of indoor air. Build Environ 130:143–150. https://doi.org/10.1016/j.buildenv.2017.11.045

McGregor FAP (2014) Moisture buering capacity of unfired clay masonry. University of Bath

Rode C, Peuhkuri R, Mortensen LH, Hansens K, Time B, Gustavsen A, Ojanen T, Ahonen J (2005) Moisture buffering of building materials. Department of Civil Engineering Technical, University of Denmark

Rode C, Peuhkuri R, Time B, Svennberg K, Ojanen T (2007) Moisture buffer value of building materials. J ASTM Int 4(5):100369. https://doi.org/10.1520/JAI100369

Thomson A, Maskell D, Walker P, Lemke M, Shea A, Lawrence M (2015) Improving the hygrothermal properties of clay plasters. In: 15th international conference on non-conventional materials and technologies, 8

Part V
Architecture/Design

Chapter 35
Climate Responsive Earthen Architecture of Chigule

Amit C. Kinjawadekar

35.1 Introduction

Earthen Architecture has been used for generations for a variety of uses across the globe. One such settlement is a small village named 'Chigule' in Khanapur Taluka of Belgaum District in Northern Karnataka. It is located around 50.0 km south-west of district headquarters Belgaum surrounded by hills and forests all around. Its location is unique bordering three states with Karnataka–Maharashtra border around 2.0 km away and Karnataka–Goa border 7.5 km away. It lies in tropical monsoon climatic region in the Western Ghats presenting a unique case study of indigenous traditional construction techniques which are climate responsive. The village has maintained its original character, though present-day developments have started influencing it. Climatic, environmental, ecological and social concerns have influenced the planning of this settlement along with providing solutions to the lifestyle of the inhabitants.

35.2 Background

Folklore describes this settlement for more than 1000 years old. The native inhabitants believe in the story of one sage *Kulak* who is believed to have an ashram in this area around 1000 A.D. A village named Kankumbi which is 5 km from Chigule is named after the sage. It is said a tribal disciple named '*Malli*' devotedly served sage *Kulak*. River Malaprabha, one of the principal tributaries of river Krishna originates at village Kankumbi is named in memory of *Malli*. It is said that the sage granted *Malli's* death wish of always serving the people which marked the birth of the river Malaprabha at Kankumbi.

A. C. Kinjawadekar (✉)
Faculty of Architecture, MIT, MAHE Manipal, Manipal, India
e-mail: amit.ck@manipal.edu

B. V. V. Reddy et al. (eds.), *Earthen Dwellings and Structures*,
Springer Transactions in Civil and Environmental Engineering,
https://doi.org/10.1007/978-981-13-5883-8_35

Over a period of time, people started settling on the banks of river Malaprabha which gave rise to many small villages. Due to the availability of water, people from nearby areas like present-day Goa, Maharashtra and Karnataka started settling and gave rise to an amalgamated culture. Fear of political persecution in the hands of British and Portuguese rulers made people settle in remote areas of jungle.

35.3 Present Scenario

As per census data of 2011, village Chigule has a population of 656 people. This village is accessible by a single road which connects it to neighbouring village Kankumbi 5 km away. It is served by state-owned bus services with one trip per day. Most of the inhabitants are Hindus of Maratha community and speak Konkani language. They are small-scale farmers and dependent for a livelihood on nearby jungles. They produce rice, ragi and *vari* for self-consumption. Agriculture is highly dependent on monsoon due to lack of adequate irrigational facilities. The much talked Mahadayi river project is located nearby. The average rainfall is around 4000 mm and sometimes during the four-month-long monsoon, this village is cut off from rest of the world. All these factors have influenced the lifestyle of the inhabitants and have given rise to indigenous climate responsive earthen shelters.

35.4 Lifestyle

35.4.1 Religious

The chief deity of this settlement is Goddess *Mauli*, who is believed to have seven sisters spread across seven temples in neighbouring Goa and Maharashtra with main *Mauli* Temple at Kankumbi some 5 km away. *Holi* is the main festival of this village which is celebrated with great enthusiasm at Mauli Temple. This festival is celebrated uniquely. A 15–18-m high tree trunk brought from surrounding forests is the main attraction of the Holi festivities. It is decorated with mango leaves, and cultural programmes are performed around it for almost a week after it is erected near the Mauli Temple. The Mauli Temple located on the edge of a cliff on the western side of the village is the nerve centre of the village. It is the venue for all public meetings, religious festivities, social gatherings and cultural celebrations. The location of the temple approximately divides the village into old and new parts. Every year, a village fair during Holi is conducted and once in 12 years, a big Jatra is arranged at village Kankumbi when all surrounding seven *Maulis* meet at Kankumbi.

35.4.2 *Lifestyle*

The inhabitants of Chigule have a simple lifestyle. The primary occupation is farming and collecting forest products—firewood, construction materials, honey, fruits, roots and palms. They sell these collected products and charcoal to nearby markets like village Kankumbi (5 km), Belgaum (50 km) and Panaji (80 km) away. Many own small pieces of land, while some work as labourers in fields.

35.5 Houses at Chigule

35.5.1 *House Types*

Houses in Chigule can be broadly classified into four typologies based on the use of local materials for construction of various building components (Tables 35.1, 35.2 and Fig. 35.1).

Table 35.1 House structure

Type	Walls	Roof support	Roof covering
A	Superstructure mud walls over laterite plinth	Jungle wood members	Jungle grass thatch (*ghar gavat*)
B	Superstructure mud walls over laterite plinth	Jungle wood members	Country tiles
C	Sun-dried mud blocks	Jungle wood truss	Mangalore tiles
D	Laterite (*Jamba*)	Jungle wood truss	Mangalore tiles/R.C.C. Slab

Table 35.2 House finishing

Type	Flooring	Wall finishing	Toilet
A	Mud flooring finished with cow dung	Mud plaster finished with local clay varieties and cow dung. West-facing walls protected by *Karvi* stick frame covered with thatch	Locally available stone flooring
B	Mud flooring finished with cow dung	Mud plaster finished with local clay varieties and cow dung. West-facing walls protected by *Karvi* stick frame covered with thatch	Locally available stone flooring
C	Mud flooring finished with cow dung	Mud plaster finished with local clay varieties and cow dung. West-facing walls protected by *Karvi* stick frame covered with thatch	Tiles
D	Natural tiles	Cement plaster	Tiles

Fig. 35.1 Location of different types of houses in village Chigule. *Source* Chigule, Karnataka (28 Feb. 2018). Google Maps. Google. Retrieved from https://www.google.co.in/maps/place/Chigule,+Karnataka/@15.7439425,74.2064881

House types A, B and C are mostly on the southern side of the village which is the earlier part of the village. House type D is on the northern side of the village which is an extension of the original settlement (Fig. 35.2).

1. **Soil type**—Red murum mixed with soft laterite particles. As per standards, safe-bearing capacity of such soils can be estimated around 300 KN/m^2. Hard strata for foundation are available at depth around 60 cm deep.
2. **Foundation**—Laterite stone masonry construction in mud mortar used for the foundation of such structures. Soil moisture due to heavy rains during the rainy season does not affect the foundation since—

   A. Structures are limited only to ground floor single storey construction,
   B. Availability of hard strata at a reasonable depth of around 30–90 cm and
   C. Foundations are usually provided sufficient width of around 90 cm.

3. **Walls**—They are prepared in thick layers of 60 cm width and 60 cm height with mud mixed with thatch. Next layer around the periphery of the house is laid once the bottom layer is completely dry in 7 days. Construction activities are carried out during non-rainy days, i.e. October to May. Superstructure walls are generally painted with cow dung mixed with charcoal powder till sill level of around 90 cm to give dark shades and with *shedu* (clay) to give lighter shades of yellow above.

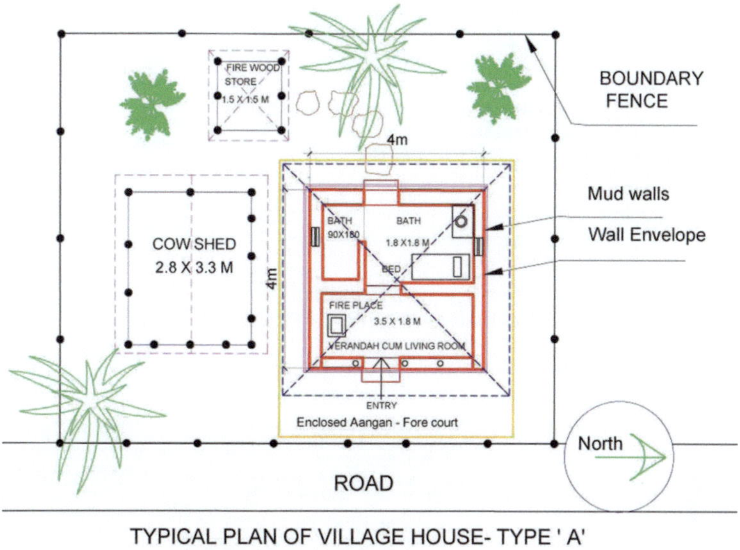

Fig. 35.2 Typical plan of Chigule village house. Type 'A'. *Source* Author

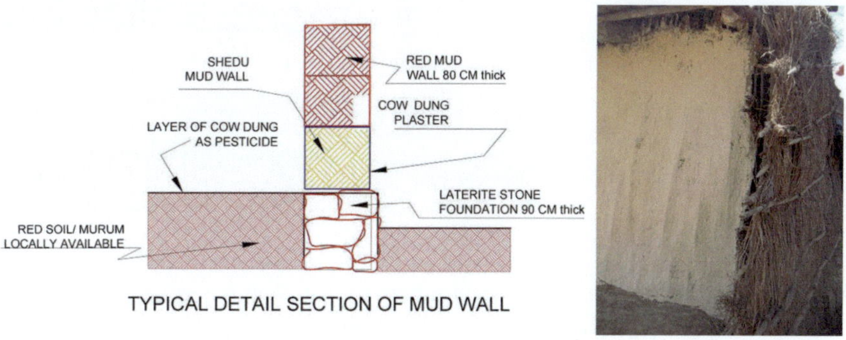

Fig. 35.3 Typical detail section and image of mud wall. *Source* Author

4. **Plinth and flooring**—Laterite is used for plinth construction in the mud. Local boulders and red murum soil are used in filling plinth and for finishing of flooring. Cow dung available from household cattle is used as the final coat which needs regular maintenance (Figs. 35.3, 35.4 and 35.5).
5. **Plastering**—

   a. Internal walls—Walls are covered with smooth mud plaster finished with local clay varieties and cow dung. Red, yellow and white clay varieties are available locally which are used for plaster finishing instead of paints.
   b. External walls—West-facing walls protected by *Karvi* stick frame covered with thatch.

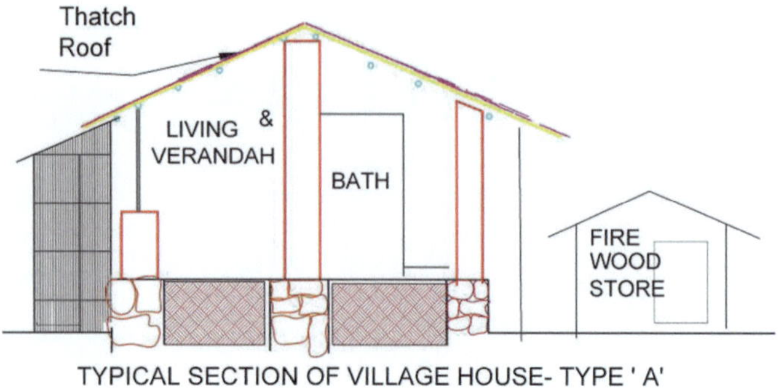

Fig. 35.4 Typical section of village house—Type 'A'. *Source* Author

Fig. 35.5 View from typical *Angan* (Forecourt). *Source* Author

6. **Attic**—Prepared with jungle wood rafters for the central portion of the houses which has got more height compared to front and rear portions. It is used for storing hay, fodder and firewood during the four-month-long rainy season.
7. **Roofing**—Entire roof structure is supported by jungle wood posts, rafters and central ridge covered with *karvi* sticks instead of wooden battens and covered with thatch or black curved country tiles. Locally available cashew nut oil is applied as preservative and termite protection for the embedded wooden part in the ground as well as wooden components in the superstructure (Figs. 35.6 and 35.7).
8. **Doors and windows**—Simple type of doors and windows are prepared from jungle wood with the help of local carpenter and blacksmith who are the only skilled people required for construction activities. All other works are carried out by people themselves which significantly reduces construction costs and results in highly economical houses (Fig. 35.8).

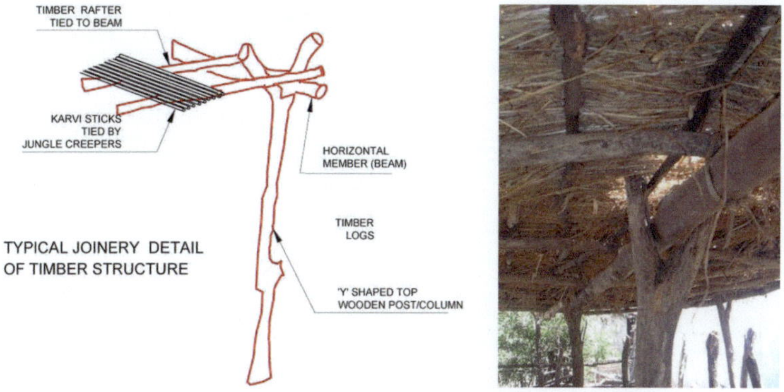

Fig. 35.6 Typical joinery detail and image of a timber structure. *Source* Author

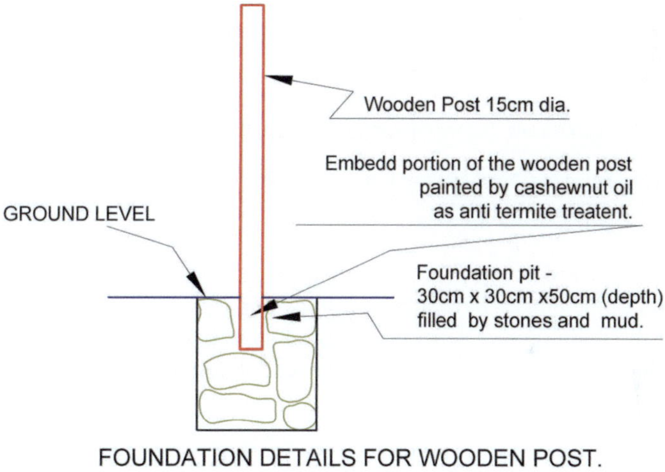

Fig. 35.7 Typical foundation detail for the wooden post. *Source* Author

35.5.2 Functional Planning of Houses

1. *Angaan*-front yard—(public zone)—Used for communicating with visitors, neighbours, drying of grains, used by children for playing, used for household functions in short and a multi-functional interactive space.
2. Living/verandah—(semi-public zone) used for family and community discussions, many houses maintain a fireplace which is used for almost nine months in a year for warmth.
3. Kitchen—(private zone)—Used for cooking and serving food. In most of the houses, bath attached with kitchen used for utensil and dishwashing.
4. Bed room (private zone)—Type A houses do not have a separate bedroom. Kitchen and living room double up as a bedroom during the night.

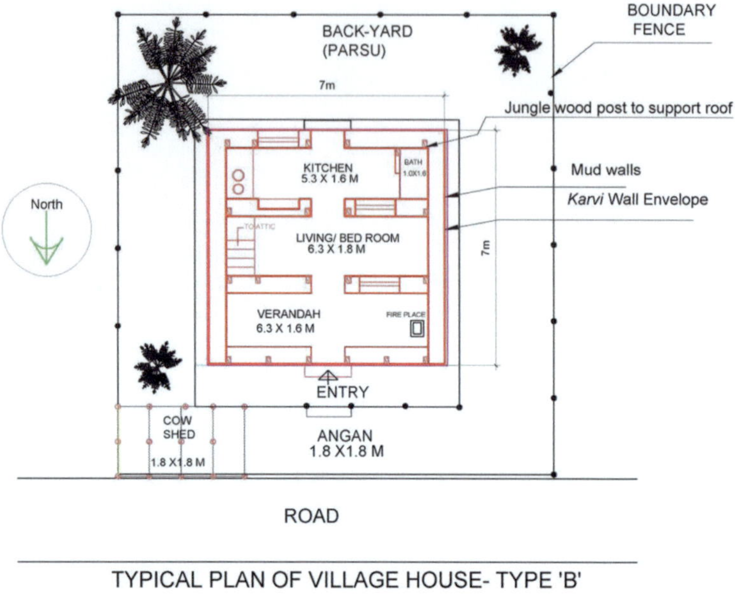

Fig. 35.8 Typical plan of Chigule village house. Type 'B'. *Source* Author

5. *Parsu*—Backyard (semi-private zone) used for drying clothes, stacking farm products, some houses have firewood store, and cattle shed either on the side or behind the main house. Separate firewood store is provided in this area if the house does not have an attic to store firewood during the rainy season.

35.5.3 Climatic Trends in Chigule

Indoor: Warmth inside the houses is maintained with the use of local firewood during rainy season and winter. During summers, excessive heat is controlled due to thick mud walls and country tile roof insulation.

Outdoor: The entire village surrounded by paddy fields and forest area with pollution-free pleasant atmosphere (Fig. 35.9).

35.5.4 Climatic Responses in Planning of Typical Houses

a. Enclosed ***Angaan*** (Front yard) acts a multi-functional for a variety of activities. It is one of the most active zones of the houses. It has a flexible screen made up

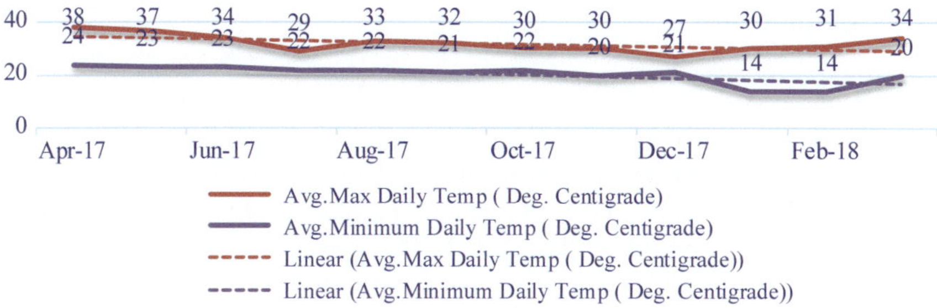

Fig. 35.9 Average maximum and minimum temperature yearly graph. *Source* graph based on temperature data from https://www.accuweather.com

of *karvi* sticks which form a screen used to cover top and sides depending on the season and privacy requirement.

b. **Fireplace in verandah** or living area provides warmth to residents and visitors after they enter the house. It is used for the entire year due to the four-month-long monsoon season with heavy rains along with chilly winters for another four long months. It is also used to dry wet clothes, humid beddings and blankets among other things.
c. **Wall envelope**—One of the most critical and unique elements of these houses. They are made up of *karvi* sticks in the form of screens and used all around the houses and for covering cattle shed. *Karvi* stick screens are invariably used on western sides to protect the mud walls from solar radiation and heat gain during summer. In long rainy season, the *karvi* screen protects the mud walls from heavy rains. The gap between *karvi* screen and mud walls acts as a thermal insulator during chilly winters.
d. **Fireplace in the kitchen** which uses firewood for cooking keeps the whole house warm. Smoke escapes from the air gaps present in the thatch or black curved country tiles roofing provided above such spaces. Also, dry firewood, dried during summer months is used all around the year which reduces indoor air pollution with a reduction in emission of smoke from the kitchen stove.
e. The **attic** if provided reduces the height of rooms and helps in insulation guarding against heat gain from the roof during summers and providing warmth during chilly rainy season and winters.
f. Mud walls act as a thermal insulator keeping out heat during summer and providing warmth during long rainy and winter season.
g. Bed room (private zone)—Type A houses do not have a separate bedroom. Kitchen and living room doubles up as a bedroom during the night.
h. *Parsu*—Backyard (semi-private zone) used for drying clothes, stacking farm products, some houses have firewood store and Cattle shed either on the side or behind the main house.

35.6 Conclusion

- These houses are great treasures of innovative ways of dealing with harsh climates in sustainable, economical, efficient and environment-friendly ways.
- They are sources in knowing how highly efficient indigenous mud construction techniques have developed in remote areas over a period of thoughtful innovations.
- The enveloping *karvi* screen due to its light weight provides lots of flexibility along with the efficient response to climatic concerns. It gives a unique character not only to the individual household but entire village as a whole.
- However, today, these houses are facing severe threats due to issues like—material shortage, the requirement of regular maintenance, lack of time and urban influence on a younger generation. They are now being used only for secondary uses like cattle sheds and firewood store.
- Serious efforts are required to efficiently conserve and preserve this unique mud construction style by making people of Chigule understand the real worth of this invaluable asset.
- In the absence of such measures, climate responsive, highly efficient, environment-friendly earthen construction techniques of Chigule will be lost forever.

References

Chigule-Google Maps (n.d.) Retrieved 13 May 2018, from https://www.google.co.in/maps/place/Chigule,+Karnataka/@15.7439425,74.2064881

Chigule May Weather 2018—AccuWeather Forecast for Karnataka, India (EN-GB). (n.d.). Retrieved 13 May 2018, from https://www.accuweather.com/en/in/chigule/2874478/may-weather/2874478

Chapter 36
Exploring Attributes of Vernacular Assam Type House Design Techniques in Contemporary Setting

Shiva Ji and Ravi Mokashi Punekar

36.1 Introduction

A typical ATH (Or *Ikra* house, named after reed which is called *Ikra* locally) is constructed on plot leaving free land for various activities. Generally has one floor structure (sometimes 2 floors). The structure constitutes of wooden members as main structural frame and filled in by reed reinforced (mud covered) thin walls in between the frames. Sometimes, house carries a symbol like lotus or chakra. The light materials used in construction make it earthquake resistant and remain largely minimal damaging for life in cases of earthquake. It houses one unit family and generally stands alone without sharing structural linkages with another. The plot bears beetle nut trees, banana, coconut, etc. The groundwater level is high with two types of waterbodies, one mainly for rearing fish and second as a recycler of household solid organic waste.

36.1.1 General Information

Housing type: Timber (wooden) building; **housing Sub-type**: Bamboo/reed walling system in mesh and post (Wattle and Daub); **construction**: Generally self-erected and fabricated using local materials and tools; **region**: Assam, India; **climate**: Warm

S. Ji (✉)
Indian Institute of Technology Hyderabad, Hyderabad, India
e-mail: shivaji@iith.ac.in

R. M. Punekar
Indian Institute of Technology Guwahati, Guwahati, India
e-mail: mokashi@iitg.ac.in

B. V. V. Reddy et al. (eds.), *Earthen Dwellings and Structures*,
Springer Transactions in Civil and Environmental Engineering,
https://doi.org/10.1007/978-981-13-5883-8_36

temperate, winter dry, hot and humid monsoon summer; **sub-climate**: Deciduous vegetation, sweet water Himalayan river-fed region; **seismographic region**: Zone 5 (high-risk zone) (Kaushik and Babu 2009).

36.2 Background

36.2.1 Context

Geographically and climatically, the states of north-east India has things like heavy rain pattern, soil, topography, landscape, climate and agriculture in common in more than general average sense. The typology used in this area is somewhat in use in this entire region and has been in exchange for centuries. The present architecture language is there in practice for last 200 years at least.

36.2.2 Seismic Context

Seismic activities have largely shaped the architecture of place and have evolved over time using local materials, tools and techniques. The region has seen three great earthquakes and several big ones over last century. It has almost negligible dependence on any foreign object or technique (Fig. 36.1).

Fig. 36.1 Corrugated galvanized iron sheet roofing of Integrated Craft Complex of Craft Based Resource Centre and Material Bank, Shillong

36.3 Observations

36.3.1 Rural Set-up

This construction type is prevalent in rural areas specifically. This technique was evolved as a result of frequent and huge impact earthquakes, which have always given challenges to the people to erect again from the rubbles of damage and reduce damages on life and goods.

36.3.2 Socio-economics

The society is agrarian and has remained agrarian for centuries, it has maintained its intrinsic, concentrated but open culture.

36.3.3 Demographics

The women have the major role in society and equal say and respect from old times and have maintained it even today.

36.3.4 Occupation

The majority of people are engaged in agriculture and related activities as Fisheries. The land is fertile and does not require fertilizers much, and produces huge paddy crop, mustard, coriander, jute, vegetables and some fruits such as pineapple.

36.3.5 Urban Setup

This technique was practiced in ancient city of *Pragjyotishpur* (present Guwahati) but due to advent of technology, changing demographics and fast lifestyle; people nowadays prefer contemporary methods of construction over traditional methods.

36.4 Methodology

36.4.1 Chosen Approach

Inspect traditional vernacular Assam-type buildings to study its design and details, discuss with inhabitants and enquire about same from engineers to understand engineering, structural and earthquake resistance aspects.

36.4.2 Literature Review

Studied research papers and books.

36.4.3 Interviews

Civil engineers: (1) Er. Haridoss, Building Department, North Eastern Council, Shillong, (2) Mr. Mendon Pariat, Craft Based Resource Center and Material Bank, Shillong, (3) Prof. Hemant B. Kaushik, Department of Civil Engineering, Indian Institute of Technology Guwahati and (4) Mr. Vishwajit, Ph.D. Scholar, Department of Civil Engineering, Indian Institute of Technology Guwahati.

36.4.4 Survey of Buildings for Primary Data Sourcing

(1) Integrated Craft Complex, Craft based Resource Centre and Raw Material Bank, Scheme of DC (Handicrafts), Lower Nongrim Hills, Shillong, (2) Comoros Linestone Mining Company Ltd., Upper Lachumiere, Shillong, (3) Mr. Blah's house, Shillong and (4) Mr. XY's House, Ujan bazaar, Guwahati, pictures of various houses in Guwahati and Shillong.

36.5 Architectural Design Aspects

36.5.1 Handling Topography of Place

The plinth of house is normally raised from ground to avoid moisture, air and sufficient space to pass of stray animals, run-off water, insects and reptiles on water-rich soil or marshy piece of land. The vertical posts are made of sal wood and fitted with iron nails.

36.5.2 Dimension and Sizes

The typical sizes for residential building construction length are between 6 and 12 m and width between 6 and 12 m. Majority are ground floor structures only; with very few with G+1. Typical roofing span is 3 m, and typical height is 3.5 m.

36.5.3 Structure

Structure and skeleton of the house constitute sal wood and beetle nut tree timber as beams. The sal wood takes compressive stresses effectively and beetle nut tree wood takes tensions efficiently. The framing of building takes shape at plinth, lintel, sill and ceiling levels, hence creating a four layer of horizontal rings on periphery. Similarly, vertical members of columns and partitions come on the corners (primary load-bearing column) and partition column (secondary load bearing) at an internal of 3′. These entire connections of columns, beams and tie members create a portal frame sort of structure which can withstand normal storms and earthquakes. Even if they succumb to wind load or seismic loads, they remain largely less harming in comparison to concrete structures. Vertical posts are of size 150–250 mm diameter spaced at 2,000–3,000 mm and intermediate ones of 125 * 125, 125 * 100, 125 * 75, 100 * 100 and 100 * 75 mm of sal wood or Nahar wood placed at a distance of 1,200 mm c/c and stopped at sill level (Fig. 36.2).

36.5.4 Planning and Costing

The small unit has drawing room, a bedroom kitchen, toilet bathroom and a lobby for common purpose. It grows as per site, requirement and economic level of family.

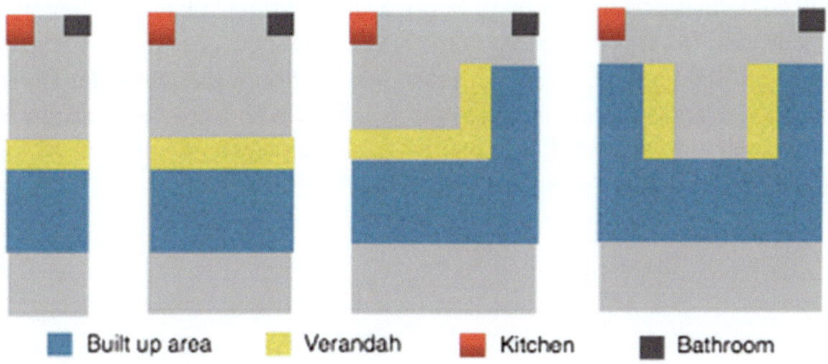

Fig. 36.2 Plan variations of typical Assam-type house

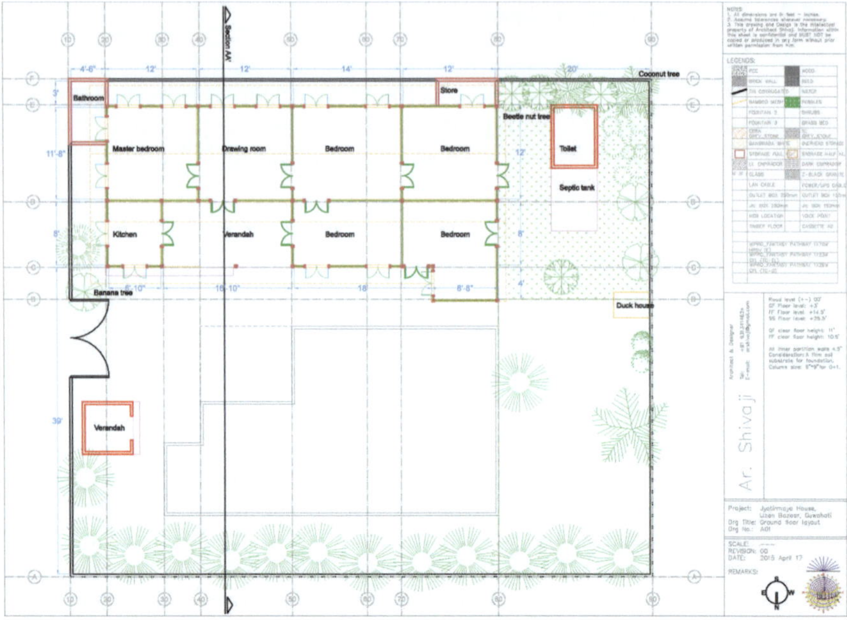

Fig. 36.3 Layout of Mr. XY's House, Ujan bazaar, Guwahati, shows openings, interconnected rooms and open land with trees and waterbody

A medium house can accommodate 4–6 bedrooms and long verandas in front of the house to accommodate ventilation and light intake. This type of construction is non-engineered but authorized as per development control rules of local authorities. The unit cost of construction (as in year 2009) was approximately at Rs. 5,000/m^2 and labour cost at Rs. 1,250/m^2, totalling to Rs. 6,250/m^2 while cost of a similar size brick/concrete structure comes approximately at Rs. 18,000/m^2.

The placement of rooms is linear in straight line or in L shape, running along with corridor. Windows open in the open space in both sides (see Fig. 36.3). Kitchen and verandah come in the front portion of house as in traditional planning along with the master bedroom. Other private areas come later in the sequence. Toilets are generally placed outside the main structure of house, preferably in the rear side. Density of beetle nut and banana trees is more in the rear side with wastewater accumulation pond in the rear side of land. Location of fowls' area where they can also consume kitchen waste (Figs. 36.4, 36.5, 36.6 and Table 36.1).

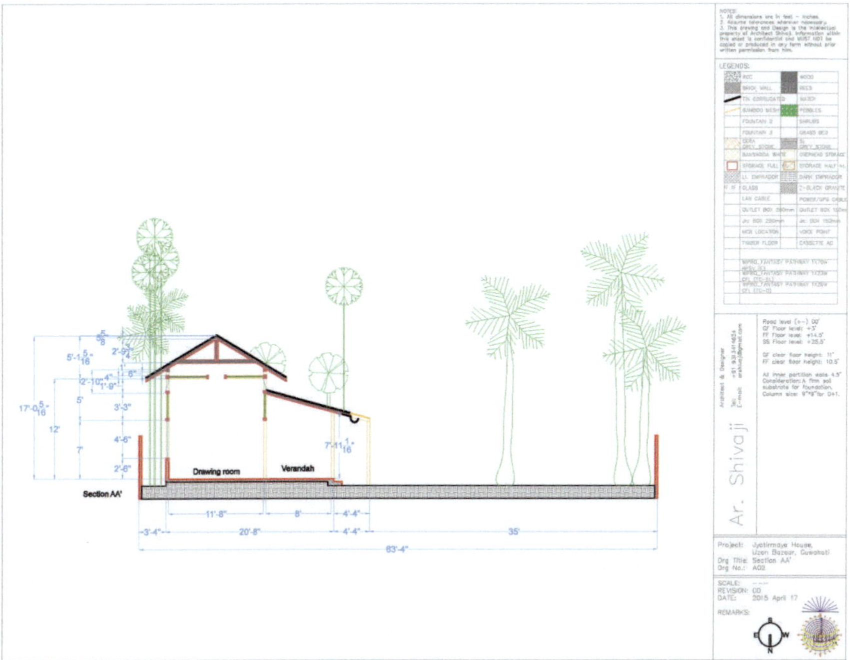

Fig. 36.4 Section of Mr. XY's House, Ujan bazaar, Guwahati, shows ventilation openings, insulator air chamber, corridor along rooms and open land with different types of trees

Fig. 36.5 *Ikra* wall portion and its internal structure

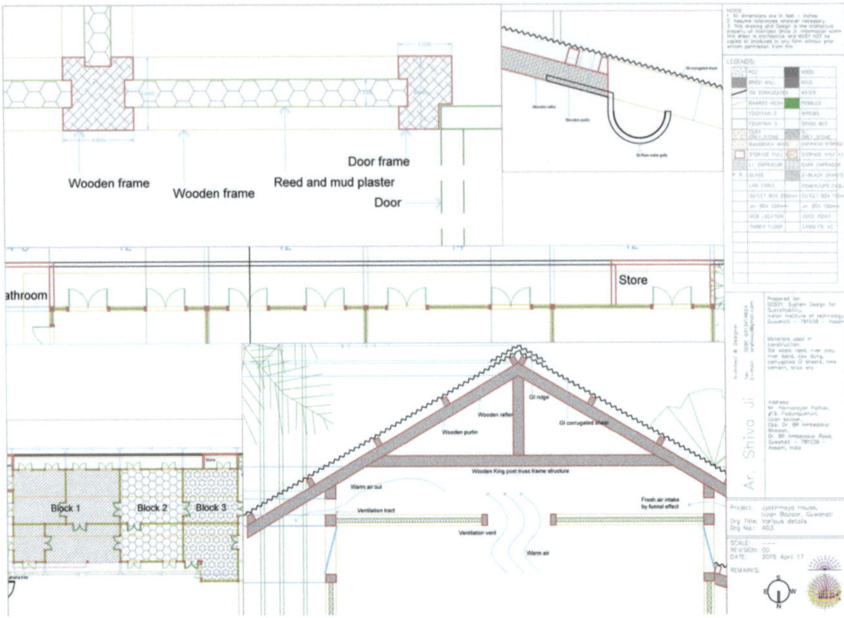

Fig. 36.6 Details of different components of ATH

Table 36.1 Cost of construction in year 1959

Description	Area (in ft^2)	Cost	Rate of construction per ft^2
Block 1 (refer page 21)	510	₹2,000	
Block 2	296	₹1,500	
Block 3	292	₹2,000	
Total	1,098	₹5,500	₹5.01

36.5.5 Foundations

Burnt brick foundations: Mostly used in houses of affluent people, either in shape of running wall along the periphery of building or as piers on corners of building at a shallow depth of not more than 900 mm deep mostly on flat lands. Brick is preferred for being more stable and resistive to water content in soil compared to bamboo or wood.

Wooden post foundations: Initially, there were only wooden or bamboo posts used in foundations in plains as well as hill slopes too up to a depth of 600–900 mm in ground. They were used generally in water-logged areas and to cope up with the slope of land so as to avoid cutting and levelling of land hence protecting land form.

36.5.6 Flooring

The flooring generally made up sal wood beams and floored with wooden members as planks on top. The height of plinth and gap from soil gives protection in case of water logging, moisture, dampness, etc. In water-logged areas, brick piers are erected with wooden or bamboo posts resting in a conical tin cap which prevents the wood/bamboo from weathering damages.

36.5.7 Roofing

Mostly, pitched roofing made of sal wood posts underneath and covered with tin corrugated sheets to drain off rainwater on top. Beams of beetle nut tree span the size of rooms, halls and verandas etc., supported by queen posts in wood from below. They generally form ridge in the middle and have opening for kitchen exhaust on the slopes. The inner side of roofing is hung with the help of steel nut bolts and U clamps etc. Roof truss is generally made of sal wood posts and rafters (150 mm diameter) c/c 600–750 mm. Bamboo purlins (100 mm) are placed at 300 mm c/c. Earlier, even roofs were made of *Ikra* weed by forming a thatch roof system.

36.5.8 Walls

The walls are made of vertically aligned reed bamboo (*Ikra*) in Wattle and Daub technique, plastered on both sides with mud-dung mixture. *Ikra* is a weed commonly growing in river plain and hilly regions of north east and is used in building construction, furniture, craft, etc. The stiffness and flexible nature of reed comes useful in walling system as it takes wind load and sticks around the mud forming a natural walling system. The infill of *Ikra* reed panelling varies in thickness from 25–100 mm, plastered in mud or cow dung, sand plastered and/or cement lime fine sand plastered and painted. In some cases, *Ikra* walling is done after sill level, and till that brick masonry wall was erected. Lime and flour of a particular rice type along with clay were used earlier for plastering. Presently common substitute of lime is cement.

36.5.9 Different Materials Used in Walls

***Stone masonry walls**: encasing of rubble stone in mud/lime mortar (wooden roofing), dressed stone masonry (in lime/now cement mortar). ***Adobe/earthen walls**: clay/mud walls using horizontal wooden members, adobe walls, rammed earth walling system. ***Unreinforced masonry walls**: brick masonry in mud/lime mortar using vertical wooden members, concrete block masonry. *__Confined masonry__: clay brick masonry using wooden posts and beams, with concrete posts/tie columns and beams. *__Reinforced masonry__: stone masonry in cement mortar. *__Load-bearing timber frame__: thatch and reed walling system, walling with horizontal beams/planks at intermediate levels, post and beam frame, stud-wall frame with plywood/wooden/bamboo base, wooden/bamboo panelling and walling.

36.5.10 Openings

Windows in walls to take in light and air for ventilation in humid months of year. Placed at regular level of sill and goes till lintel, made of sal wood frame. Window sizes vary from 600 to 1,500 mm in width and 900–1,500 mm in height depending upon location and requirement.

36.5.11 Services

The building services like toilets, etc., remain out of main structure in the end of plot and leave waste in the khal or connecting sewer line behind the site. Water supply from individual hand pipes or municipal water supply line.

36.6 Case Example of Mr. XY Pathak, Guwahati

36.6.1 Cost of Construction in Year 1959

It includes material, labour and other expenses incurred towards construction. The materials used were sal wood, reed, river clay, river sand, cow dung, corrugated GI sheets, lime cement, brick etc. Address: Mr. XY Pathak (personal identity concealed), Uzan bazaar, Guwahati—781039—Assam, India. Today's average cost of construction in brick, mortar and concrete approximately is Rs. 2,000/ft^2. Consider inflation while comparison.

36.7 Sustainability Aspects

The typical habitat of ATH works like an independent unit and water has a key role to play in it. The population density is evenly distributed and there is ample space around a house to grow trees, plants, breed fowls, rear fish in the backyard pukhri and rear domestic cattle like cows and goats. Since society is agrarian and the land is fertile, it does not require external aid and has remained self-sufficient. This has played a vital role in keeping this society intrinsic and concentrated for centuries. The cycle of water closes well here from rains, ponds, groundwater exhaustion, household affluent, agricultural fields, wells and pukhri/khal. The cycle of nutrient also closes well as the area produces crop, people consume, create waste which degrades and goes back to its molecular form or become feed for others.

36.7.1 Afterlife of House

The discarded components of ATH are columns, walls, partitions, windows, king posts, rafters and purlins; since they are all made up of natural materials without any treatment, they go back to nature again in their molecular form without leaving any residue and aftereffects. The discarded components of ATH are reused, ultimately closing the cycle of material lifecycle (Fig. 36.7).

36.8 Seismic Aspects

The seismic performance of ATH is excellent, i.e. F: VERY LOW VULNERABILITY, The ATH has remained extremely good in earthquakes occurred in NE India

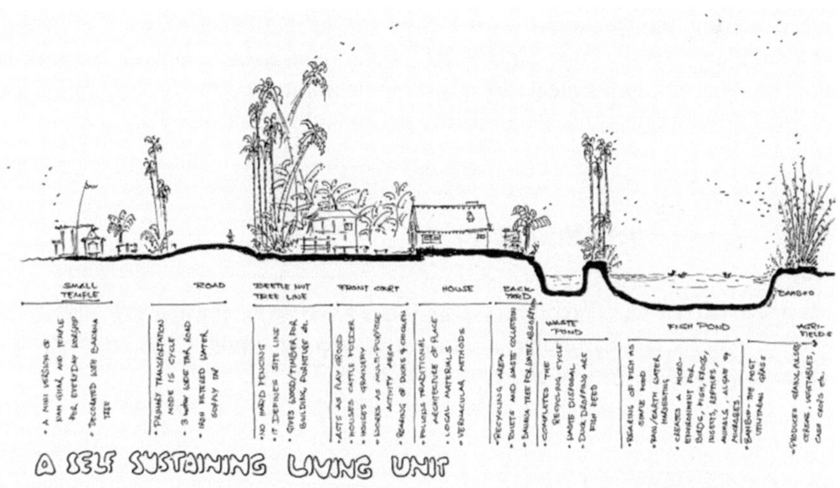

Fig. 36.7 A typical self-sustaining living unit in a village setting, and sketch shows various sections of the ATH and its role in the overall habitat design

(Kaushik et al. 2006). The latest was Sikkim earthquake occurred on 18 September 2011 (M6.9), severe damages were reported in concrete buildings but vernacular ATH bore only damages. Although there are rare G+2 structures in ATH; but even in those minimal damages were observed due to lightweight materials. The falling *Ikra* walls do not create big damage to life and property.

36.8.1 Seismic Strengthening in ATH

- The lateral load path of structure from horizontal directions serves to transfer inertial forces from superstructure to substructure.
- The exterior joints of roofing and walling are anchored out-of-plane with metal anchors and fasteners at each diaphragm level to dissipate seismic forces.

The arrangements of horizontal force transfers are proved to be of high seismic performance rendering ATH as low vulnerability structure.

36.9 Environmental Aspects (Using Lifecycle Analysis)

36.9.1 Minimizing Resources

Minimizing Resources in ATH is largely the focus to exert minimal resource consumption. Material contents/efficiency on consumption system/flexibility in material system/transportation (as it locally sourced)/efficient resource consumption/minimizing energy consumption, etc., were found to be largely on satisfactory side. The areas of improvements are achieving flexibility in energy consumption and material consumption system/minimizing scraps and waste (refer Fig. 36.8).

36.9.2 Low Impact Resources

Low impact resources and processes are found to be on the positive side in areas of reducing toxicity and harmfulness/optimizing biocompatibility. The points of biocompatibility and toxicity of energy sources need to be looked at.

36.9.3 Optimizing Life Span

Optimizing lifespan is the keyword for ATH as reliability/appropriate life span/renewability/refabrication/intense usage, etc. are found to be on positive side,

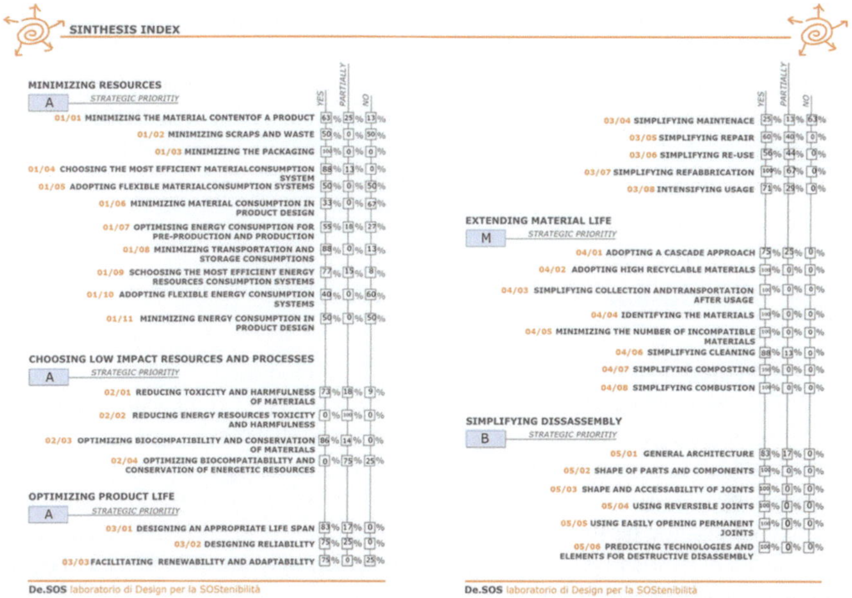

Fig. 36.8 Synthesis index Vezzoli (2007) of various points checked for lifecycle analysis of Assam-type house

making ATH a preferred style of construction and persistence over ages. This particular point has played a vital role in establishing ATH as a strong vernacular architectural system. Improvements can be done in points of simplifying maintenance.

36.9.4 *Extending Material Life*

Extending material life is found to be achieved up to a larger extent as recyclability/collection/transportation/composting of materials are handled efficiently. Simplifying cleaning could be an area of improvement.

36.9.5 *Simplifying Disassembly*

Simplifying disassembly is found to be achieved quite nicely as the ATH has very minimal components using a simple assembly technique. The dimensions and details are shown in Fig. 36.6. Whether it is joint shape/size/accessibility/ease of opening joints, etc., it is found to be minimal and simple which could be a learning.

(Detailed sheets (34) of LCA analysis are available with author as additional enclosure for this paper).

LCA Tool used: The LCA was carried out using tool for the 'design of low impact for the environment' by Lab of Design for Sustainability (De.SOS) by Carlo Proserpio and Prof. Carlo Vezzoli from Durando 10 Edificio 7 (POLIteca) 20158 Milano, Italy under an EU funded research project.

36.10 Conclusion

The wood-based materials such as sal, reed and bamboo used in lighter proportion are the key thing about ATH. Another key observation is dividing walls in multiple smaller sections by wooden frames resulting into overall smaller segments of house structure which tackles earthquake by resisting and resulting into minimal damage in case of any lateral or shear forces effectively. This has made ATH a low vulnerable structure on seismic performances. Sustainability aspects of ATH design and planning considers various factors on and off site which play crucial roles and largely dissipate impacts on surroundings.

36.11 Summary

The Assam-type houses are the best examples of vernacular technique of building construction developed over several centuries in the region of north-east India. The region falls under high-risk seismic zone and remains prone to small, medium and bigger devastating earthquakes over time; the performance of Assam-type house has maintained its reputation and has been followed by millions of families. The properties of this building type is using largely wood-based materials in its parts in thinner and lighter ratios which keeps it light weight and friendly in case of cyclones, earthquakes etc., remains airy and connected with elements resulting into a healthy, habitable and nature friendly living style. The advent of new construction materials such as brick and concrete has resulted into large following but remains highly prone to several damages for life and property. A combination of traditional and contemporary materials and construction technique is recommended to cater to safety, ease of construction, degradable materials, less energy consuming, suiting the lifestyle and aspiration of people. The study found certain components which can be taken forward to incorporate in contemporary designs.

References

Emanuel W et al (2011) Clarifying societies' need for understanding sustainable systems. J Appl Glob Res JIBMR 2(4):29–39

Kaushik H, Babu RKS (2009) Housing report Assam-type House, Report #154. World Housing Encyclopedia Earthquake Engineering Research Institute (EERI) and International Association for Earthquake Engineering (IAEE)

Kaushik HB, Dasgupta K et al (2006) Performance of structures during the Sikkim earthquake of 14 February 2006. Curr Sci 91(4):449–455

Vezzoli C (2007) System design for sustainability. Theory, methods and tools for a sustainable "satisfaction-system" design, 2nd edn. Maggioli Editore (for LCA), Milan

Part VI
Heritage: Conservation, Repair and Reuse

Chapter 37
Role of Earthen Materials in Rural Vernacular Architecture: The Case of Anavangot Ancestral Home

Sridevi Changali and Rosie Paul

37.1 Introduction—Defining the Vernacular

Over the many years that conservation of historic buildings has been in existence, a myriad of issues have been the central focus of many scholarly discussions. Unfortunately, indigenous architecture or the vernacular has not gained as much consideration as it rightfully deserves in academia. This building type, though at first sight looks inconspicuous, has a wealth of information hidden away in its structure. Rapoport (1991) recognises the neglect this line of research faces but at the same time states that the interest in the vernacular is definitely increasing especially from an archaeological and anthropological standpoint as it is an unintentional attempt at building construction and therefore a true reflection of the culture and way of life. Accurate numbers are unavailable, but it is estimated that a massive 90–98% of all building stock consist of these "lesser" dwellings (Vellinga et al. 2007), yet it continues to receive relatively little attention in academic circles. It is common to associate the prevalence of vernacular building materials and construction with developing nations or poorer communities where traditional practices have persisted longer than in the developed nations. Also, these kinds of buildings are often the first to be destroyed in the name of development. Using this aspect of vernacular to our advantage, this paper will study the use of mud in vernacular practices through a specific case study of a village dwelling in the South Indian state of Kerala during its dismantling.

S. Changali (✉)
Masons Ink, Bangalore, India
e-mail: sridevichangali87@gmail.com

R. Paul
CRA terre-ENSAG, Grenoble, France
e-mail: rosiepaul87@gmail.com

B. V. V. Reddy et al. (eds.), *Earthen Dwellings and Structures*,
Springer Transactions in Civil and Environmental Engineering,
https://doi.org/10.1007/978-981-13-5883-8_37

37.2 Mud and the Vernacular

The close relationship that the vernacular has to its surroundings reflects greatly in the material palette. What is considered as primitive architecture draws greatly from its context in terms of climate, availability of resources, livelihood and many other factors deeply rooted in both tangible and intangible aspects. At the same time, these anonymous builders are working under scarcity and have little or no margin for error as they are vulnerable to the vagaries of nature and other unpredictable forces. Yet this architecture has seen to perform well in comparison with modern technology despite the constraints (Fitch and Branch 1960) or possibly due to the sensitive approach to solve the issues. In light of this, around the world, mud has been seen commonly as part of this robust material palette. Is this due to the ready availability of the material irrespective of its performance or has mud as a material been performing consistently well over the many years.

37.3 Aims and Objectives

Now that we have set the background of the theme of the paper, this section will briefly look at an overview of what will unfold in the upcoming sections. Before delving into the physical case study, the context of the study area will be set. This will range from the history of the area to the architectural features of the vernacular in the region. The research process and methodology will then be discussed through technical drawings and images. This will be followed by a conditional assessment and pathology study which will be the fulcrum of inquiry for the research paper.

37.4 Context

37.4.1 Introduction to the Region

The state of Kerala is located in the south-western coast of peninsular India. Puthiyankam is a quaint agrarian village in the district of Palakkad and is where the dwelling is located. It lies at the foothills of the Western Ghats and the combination of the lush paddy fields and hills at a distance offers some great views and vistas (Fig. 37.1).

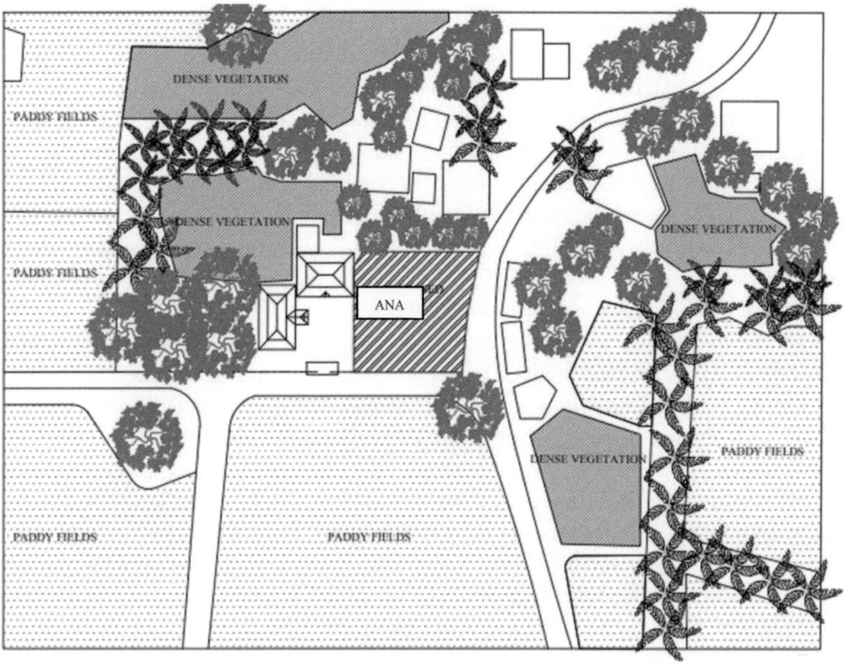

Fig. 37.1 Map showing location of dwelling and surroundings

37.4.2 History

It goes without saying that like any other region in India, Kerala too has had its fair share of a long and complicated history. The region was ruled over by the powerful local kings dating back from prehistory belonging to the Rama Varma family. Through trade and colonisation, Kerala has had a wide range of western influences as well. Although there is not any written history of this village specifically, Palakkad district has a distinct historic timeline starting from megalithic sites dating back to the Palaeolithic age.

37.4.3 Architectural Characteristics

One main aspect that was not touched upon in the historical context was the existence of tribal communities in Kerala. The communities have habituated the forests of the Western Ghats in settled clans, from time immemorial. Therefore, this makes the houses constructed by these communities a great place to start looking for clues to the evolution of the style of vernacular houses in the region. Factors like family structure, social norms and belief systems, terrain features, availability of local building materials and occupational factors all influence the overall design of the dwelling. What is interesting to note here is that no modern materials are used in the construction. Only locally sourced materials like mud, bamboo, wood and straw are used. Special techniques are used to split bamboo and weave it into a panel form to be used as screens. In terms of the layout of the house, it is always kept simple with a core of two to three rooms and a kitchen. Ancillary uses like the bathrooms and storage are built close to the main house only as and when it becomes necessary. This need-based construction not only facilitates flexibility of space but also brings down the cost of construction (Suresh 2010). These examples of architecture fit perfectly into several definitions of the term vernacular.

Apart from these tribal houses, there is another style of architecture very unique to Kerala. This formalised science of building construction in Kerala is known as *Thachu-Shastrai* and has its own set of manuals that rule over how a building is constructed. The buildings themselves are known as kettidam that literally translates to something that is tied or bound together. The *Thachu-Shastrai* is based on the ancient system called the *Manasara* that was imbibed from the ancient Vedic texts. Trying to understand in totality the workings of the *Manasara* is vastly out of scope for this research as, yet it is essential to have a basic knowledge of it in order to grasp the complexities of the style. The layout of the house is based on what is called the Mandala, which means circle in Sanskrit. It is a spiritual and ritualistic symbol that imitates the metaphysical structure of the cosmos. It represents the schema of the cosmos in a geometric grid of squares and is considered the link between human beings and the universe. Each grid square is representative of a particular aspect of the cosmos (Sachdev 2005). The central square represents the Supreme Being Brahma and therefore becomes the core area of the house layout. Therefore, it is at the centre of the Kerala house that the open courtyard is placed around which all the other rooms are arranged. The central squares are surrounded by other lesser gods and deities, followed by the ring of squares representing the human realm beyond which is the realm of the demons and goblins that have no direct contact with the world of the gods. This entire system is believed to be placed over the Asura (fallen god) that forms the foundation of the entire house. The position of this Asura is what dictates the position of the different functions.

Although the rules of these *Manasara* texts are followed commonly throughout the country, each region has its own interpretation of the rules and is manifested in a form unique to the region. Rules and methods of construction also vary according to the function housed by the structure. Varying degrees of reticence was adopted depending on whether a temple, palace or ordinary dwelling was being constructed. It is believed that what belongs to the gods cannot be present in the houses of ordinary people even if he was king (Kramrisch 2002). Therefore, carvings and other elaborate decorative features were well thought of and placed strategically in a temple. While in a dwelling where function was of a more immediate consequence, the decorative features were kept to a minimum.

37.4.4 Climate Responsiveness

Traditional methods of construction have had the aspect of thermal comfort, rainwater harvesting and several other energy-efficient systems in-built in their design. Several examples can be cited from various traditional techniques. The case of the pit dwellings, for example, that uses the fact that the structures underground for better temperature control. In the north-eastern desert state of India, houses have wind catchers that are connected via water cisterns to cool rooms. They also have Jalis that are windows with small perforations to cool the hot air by the sudden expansion of air through the perforations. In the case of Kerala, the existence of the large central courtyard allows steady air movement that is essential in a hot and humid climate that the region experiences. A study conducted by Dili et al. (2010), reinforces this theory. It states that the warm outdoor air enters the house interacting with the cool air that has settled near the floor as the windows in traditional houses are lower (around 1 foot from the ground). This interaction in turn brings the temperature down and establishes a thermally comfortable internal space. Apart from this internal courtyard and lower windows, the chuttu verandah provides sufficient overhang and provides good protection for the walls. The sloping roofs allow for the rainwater to flow down easily and also the shape provides for a buffer space between the rooms and the first floor and the outside, thus cooling them down significantly. Specific to our dwelling of study, as it acted as an ancillary unit to the main ancestral house, there is an absence of the courtyard but the chuttu verandah was present.

37.4.5 Structural Systems

Earliest traditional buildings made of wood are usually timber frame, whereas later structures using stone are load bearing. The wooden joinery uses wooden pegs and nails. The perfect prefabrication system could be followed in traditional construction due to the perfection achieved by the craftsmen (Thanpuran 2010). This tradition of using wood still continues today.

37.4.6 Materials and Construction Style

The materials used for traditional construction are very few and as many traditional construction systems go, are locally available and are usually naturally sourced. The earliest buildings were made primarily of wood and thatch. As time progressed, mud, stone and laterite stone were used as structural elements and wood and tiles were used as roof material (Thanpuran 2010). As mentioned previously, temples always held a more important place in terms of buildings. Therefore, stone, wood, brick and copper shingles as roofing were materials reserved mainly for temple construction, while the houses were built of mud, light woods and thatch or clay tiles as a roofing material (Kramrisch 1953). Plasters are usually lime based and repaired annually usually over the summer months, although this practice of using a lime was plaster for the walls are fast disappearing and can be seen more commonly in rural areas.

37.5 Research Methodology and Project Background

The Anavangot Tharavadu is a century-old ancestral home of the Anavangot Family. The present generation has now migrated out of the village and has left dwelling in disuse for over ten years after the demise of the previous generation. Before the house was dismantled, it was documented for capturing the features depicted in the following sections.

37.5.1 Detailed Physical Documentation

As an outcome of a hand survey, the following drawings were prepared that show the layout of the dwelling. From the plan in Fig. 37.2, one can see the simplicity of the layout of the structure. This draws strongly from the tribal layouts mentioned earlier, with structures like kitchen added later. Being an agrarian society, you can see the presence of large verandah (Refer Fig. 37.4) in order to enable meet and greet of the farm workers without having them to enter the core of the household. Another interesting feature being the grain store located in the northern section of the plan void of any windows (Refer Fig. 37.4).

Apart from the plan, key sections (Refer Fig. 37.3) were also captured in drawings. In this, what was noticed was that the core area is buffered from direct sunlight from both the top as well as the sides with an attic space verandah and store spaces along the periphery (Fig. 37.4).

37.5.2 Process of Discovery of Earth Elements

The various components of mud in the building were discovered during its dismantling. Mud was used in three contexts:

- Walls Material—Adobe bricks were used in the walls that ranged from 23 to 53 cm as seen in Fig. 37.5 in thickness in different locations in the house. Fired clay bricks were also noticed although it was hard to detect why they had been incorporated. The size of the brick was found to be ranging from 22–24 cm × 10–12 cm × 5–7 cm (LXBXH).
- Mortar—Mud mortar was used between the bricks

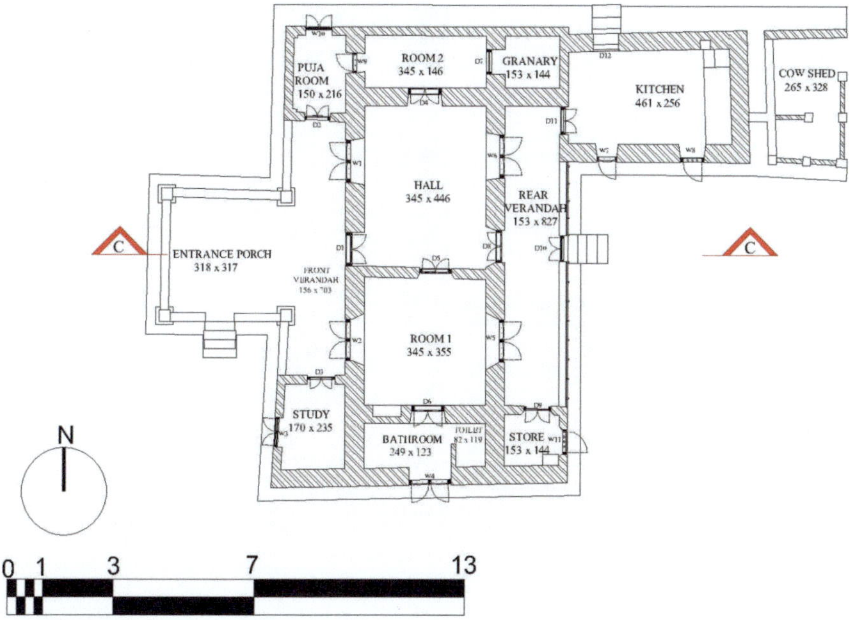

Fig. 37.2 Plan of the dwelling

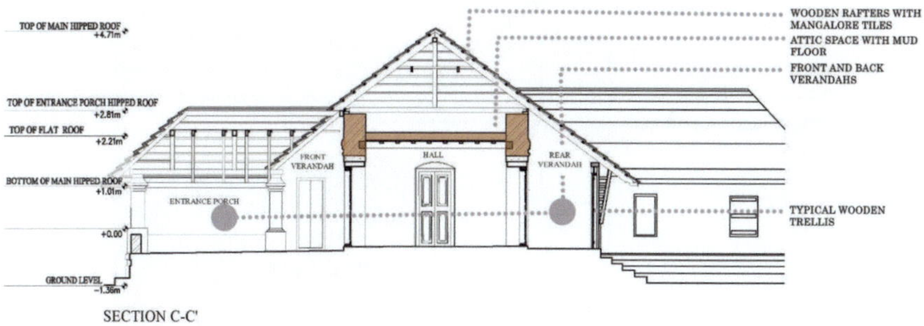

Fig. 37.3 Sections of the dwelling with mud components indicated

Fig. 37.4 Large verandah and granary

Fig. 37.5 Cross section of adobe wall

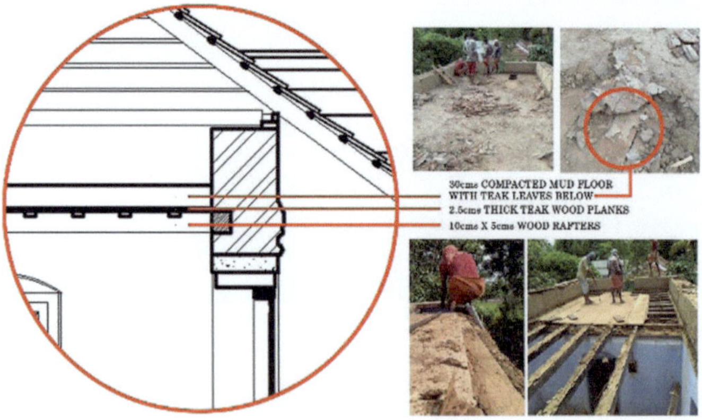

Fig. 37.6 Details of mud roof

- Roof—A 300 mm layer of mud was observed as the flat roof that created the attic space above. The detail of this technique is illustrated in Fig. 37.6. This is a special kind of roofing system seen extensively all over the country in different forms. In this case, there are teak wood beams spanning the room on top of which are teak wood planks. On top of these planks, during dismantling, teak leaves were found probably used as a layer to retain the mud above or as a buffer layer between the mud and the wooden members.

Fig. 37.7 Termite attack on walls, damages in roof that causes issues in the wall structures (L-R)

Fig. 37.8 Plaster cracks, plaster detachment, improper treatment (cement plugging) of termite infested wood (L-R)

37.5.3 Condition Assessment and Understanding Pathologies

The issues related to the structure were not very different from any heritage home that has been abandoned. The main pathologies observed are in Figs. 37.7 and 37.8.

37.6 Conclusion

There were several observations and learning that came about during the entire process of documentation and dismantling of the dwelling. The component of mud was not only limited to the walls but was also used as a major component of the roof in a different context. Although the versatility of the material is clear, due to the infestation of termites, the walls were weakened and the wooden components in the building fell prey to attacks.

This study had its limitations in the ability to study the mixes used as well as the study durability of the bricks that survived. These components can be looked at in future research. Also, there are several other similar buildings in the region. Through the study of this living heritage, other factors like thermal performance, light and ventilation and other socio-cultural aspects can also be looked at.

References

Dili A, Naseer M, Varghese T (2010) Thermal comfort study of Kerala traditional residential buildings based on questionnaire survey among occupants of traditional and modern buildings. Energy Build 42(11):2139–2150

Fitch J, Branch D (1960) Primitive architecture and climate. Sci Am 203(6):134–144

Kramrisch S (1953) Dravida and Kerala: In the art of Travancore. Artibus Asiae. Supplementum (II):5–51

Kramrisch S (2002) The Hindu temple. Motilal Banarsidass, Delhi

Rapoport A (1991) House form and culture. Prentice-Hall, Englewood Cliffs

Sachdev V (2005) A *Vastu Text in the Modern Age*: Vishvakarma Darpan, 1969. J R Asiat Soc 15(02):165–178

Suresh K (2010) Indigenous agricultural practices among Mavilan tribe in North Kerala. Stud Tribes Tribals 8(2):103–106

Thanpuran A (2010) Return of traditional systems in modern times. Sthapathi 8(3):86–90

Vellinga M, Oliver P, Bridge A (2007) Atlas of vernacular architecture of the world. Routledge, New York

Part VII
Codes and Design Guidelines

Chapter 38
Engineering Design of Rammed Earth in Canada

Timothy J. Krahn and Kristopher J. Dick

38.1 Introduction

Starting in 2005, the national model building code for Canada adopted an objective-based system for evaluating and adjudicating proposed designs. A comprehensive set of objectives and functional statements are key provisions of the code, defining the regulations goals in terms of safety, health, accessibility, fire and structural protection and the environment (Canadian Commision on Building and Fire Codes 2015). The challenges and opportunities raised by this code system, as well as the differences in plan examination scrutiny between urban and rural jurisdictions were discussed in an earlier conference paper (Dick and Krahn 2015). To illustrate, two case studies were examined with an emphasis on the regulatory system—including the building permit application process, through construction and ultimately the issuance of occupancy permits.

This paper presents the final engineering designs resulting from the regulatory constraints for each case study. Both assemblies examined are stabilized rammed earth (SRE) wall systems, reinforced with steel and/or high-density polyethylene (HDPE) geo-grid, configured as double-wythe assemblies with rigid insulation installed continuously between the inner and outer rammed earth elements. Apart from code compliance, structural design considerations included freeze/thaw durability, minimization of thermal bridging across assemblies, development length and capacity of rammed earth/reinforcement bond, inter-wythe connections, wall geometry, slen-

T. J. Krahn (✉)
Building Alternatives Inc., Codrington, Ontario, Canada
e-mail: built.alternatives@gmail.com

K. J. Dick
University of Manitoba, Winnipeg, Canada
e-mail: kristopher.dick@umanitoba.ca

B. V. V. Reddy et al. (eds.), *Earthen Dwellings and Structures*,
Springer Transactions in Civil and Environmental Engineering,
https://doi.org/10.1007/978-981-13-5883-8_38

derness ratio, lateral support for out of plane loading, in-plane shear for wind and seismic loading, axial capacity for uniformly distributed loads, point load capacity, and integration with other structural systems including steel, dimensional lumber and engineered wood products.

38.2 Engineering Design

38.2.1 Canadian Engineering Design Standards

Plain and reinforced concrete, regular and reinforced masonry, structural wood and steel all have national design standards in Canada. They are published by the Canadian Standards Association and are vetted by committees made up of practising professionals, academics and industry representatives (Canadian Standards Association 2005, 2007). The national model building code refers to these design standards, explicitly making them acceptable tools for meeting relevant objectives and functional statements. Materials that do not fall into these four categories pose a challenge for plan examiners working for authorities having jurisdiction. If they do not have personal experience with a given material or assembly, they must find another way to take the designer at their word—or demand proof of concept via a model or some other acceptable demonstration of competence.

38.2.2 Case Study 1: Allen Residence—Huntsville, Ontario

38.2.2.1 Structural Design Considerations

The Allen residence, located near the town of Huntsville in the Muskoka region of Ontario, was completed in 2012. The design and permit application process was completed under the 2006 Ontario Building Code (OBC), which was an adoption of the 2005 National Building Code of Canada. Engineering design followed some aspects of CSA A23.3-04 "Design of Concrete Structures", primarily for axial, in-plane and out of plane bending analysis of wall sections, beam design and guidance on reinforcing steel detailing. Design compressive strength value was based on pre-construction testing of a number of mix recipes. Preliminary mix recipes were based on David Easton and Bruce King's work with Nun's Canyon clay (King 1996). Local aggregate suppliers were engaged in the mix design process and were very helpful in providing, storing and delivering consistent mixes for both testing and construction phases of the project. Freeze-thaw testing was undertaken by the builder following protocol based on the CSA A23.2 concrete materials standard. 150 mm diameter × 300 mm height cylinders were used in pre-construction strength testing. Cylinders were damp cured and crushed at 14-, 28- and 56-day cure intervals. The final design

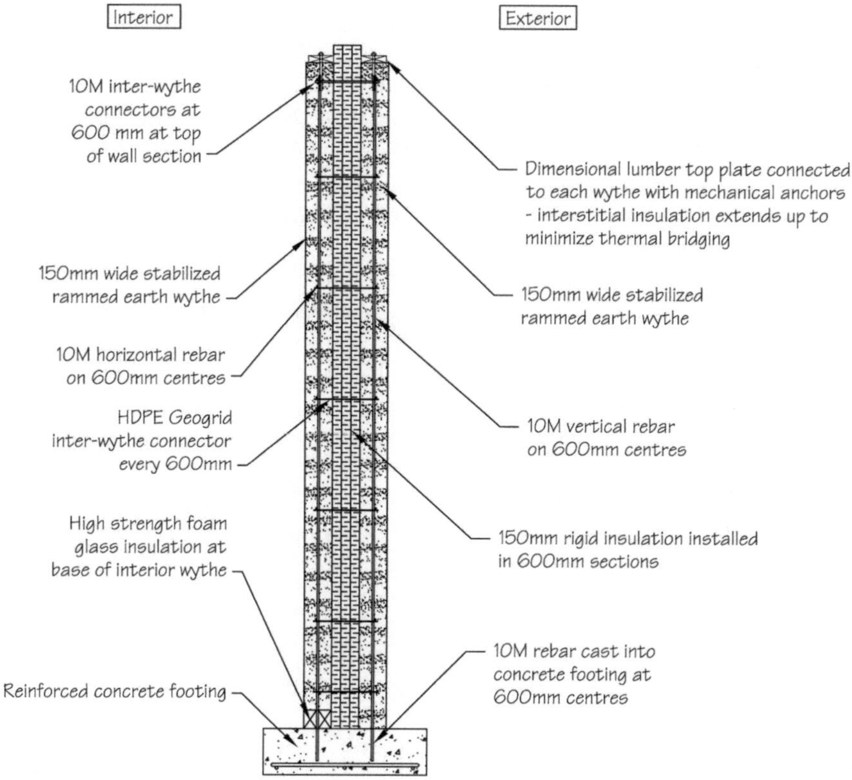

Fig. 38.1 Typical wall section—Allen residence

mix contained 8% ordinary Portland cement binder by weight. Based on the 28-day cure test data, a design compressive strength (f_c') of 15 MPa was selected. Using this design compressive strength, the reinforcement requirements for the common wall section—two 150 mm interconnected wythes of 5.5 m height—was 10 M vertical rebar at 600 mm centres, and 10 M horizontal rebar, also at 600 mm centres, installed in the centre of each wythe. Figure 38.1 shows a typical cross section of the reinforced SRE wall design for the Allen residence.

38.2.2.2 Thermal Performance Influences on Engineering Design

Thermal performance was a primary concern in this project's design, and reduction of thermal bridging across the inter-wythe insulation was an important objective. For this reason, the incorporation of HDPE geo-grid as a substitute for steel as inter-wythe connection material was explored. Geosynthetics have been used as soil reinforcement for decades, often incorporating compacted, engineered soil mixes in design. The New Zealand Earth Building Standard NZS 4297 lists geo-grid as an acceptable shear reinforcement material for standard (raw) rammed earth wall systems (Standards New Zealand 1998).

A discrete soil reinforcement model was developed based on the literature and manufacturer's product specification (Holtz and Lee 2002). It was determined that the relative tensile capacity of the HDPE geo-grid, embedded a minimum of 100 mm on either side of the insulation, continuously along the wall at 600 mm intervals was enough to consider the two 150 mm wythes laterally braced at each location, thereby reducing the slenderness ratio of the wall.

At the top of the wall, where roof loads come to bear, steel inter-wythe connectors were specified in order to avoid potential creep effects in the HDPE geo-grid at a critical location. This allowed the inter-wythe insulation to extend up past the top of the wall assembly and connect to the attic insulation, reducing thermal bridging and providing a continuous air seal. At the base of the wall, under the interior SRE wythe, two rows of 150 mm wide high strength expanded foam glass block were installed to provide thermal separation from the foundation.

The plan examination department in the town of Huntsville accepted the design as code conforming under part 4 of the OBC with the condition that the rammed earth construction is inspected by the design engineer.

38.2.3 Case Study 2: Smyth-Allcott Residence—Kemptville, Ontario

38.2.3.1 Structural Design Considerations

The Smyth-Allcott residence, located just north of Kemptville, Ontario, is within the city of Ottawa's capital region jurisdiction. As discussed in the 2015 conference paper, the plan examination department in Ottawa employs professional engineers who conduct thorough reviews of all non-prescriptive design proposals. Pre-permit application compressive strength testing had been carried out on cylinders, similar to those used for the Allen residence, and once again, a design compressive strength of 15 MPa was proposed. However, due to the presence of clay in the SRE mix, and a proposed design compressive strength of 15 MPa, the plan examiner would not accept the use of the CSA A23.3 concrete design standard, which does not allow particles below 80 μm in diameter and requires a minimum compressive strength of 25 MPa for reinforced concrete in bearing locations. Therefore, the CSA A304.1 masonry design standard was used in a redesign submitted for permit purposes.

The move away from a concrete-based design standard to a masonry-based design allowed for lower minimum compressive strengths for reinforced assemblies and put more emphasis on overall wall geometry rather than focussing primarily on reinforcement. Additionally, rectangular prisms were required for compressive strength testing, rather than cylindrical samples. The second pre-construction compressive strength testing programme resulted in a design compressive strength (f'_m)

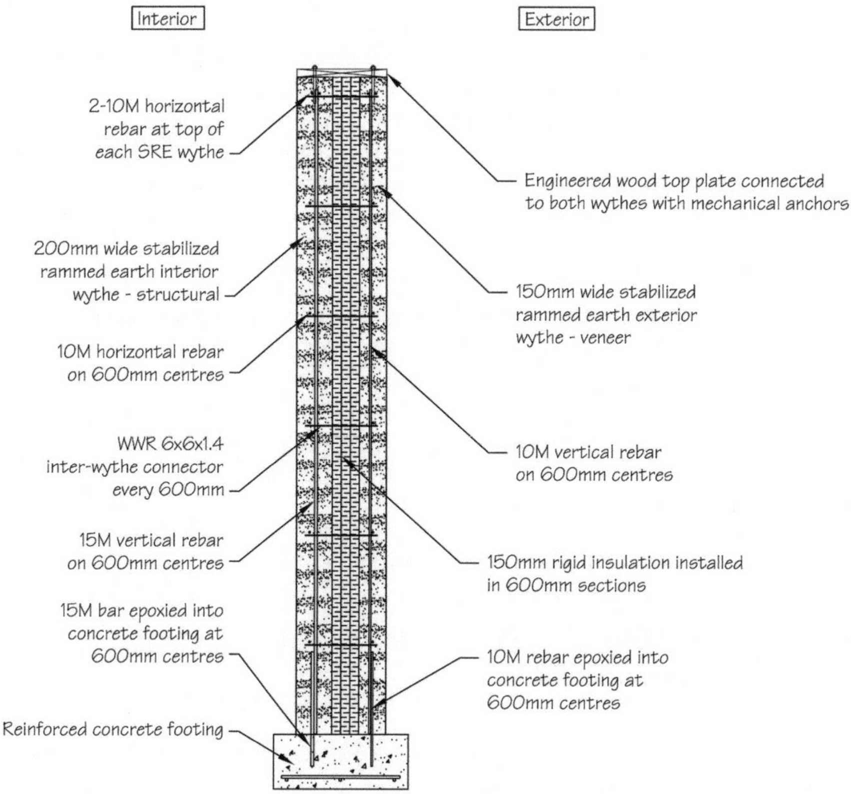

Fig. 38.2 Typical wall section—Smyth-Allcott residence

of 10.5 MPa. Another consequence of the scrutiny by the city of Ottawa's plan examiners was the elimination of innovative details and novel materials in the wall assembly. HDPE geo-grid was not allowed as an inter-wythe connecting material, nor was high-strength insulation allowed in the compressive load path. Welded wire mesh (WWR) reinforcing steel was used as inter-wythe connecting material and thermal bridging at the base of the wall was not avoided.

Because there is no published standard for engineering design of rammed earth in Canada, and no corresponding material standard, the city of Ottawa plan examiners would not allow the exterior SRE wythe to be considered a structural component of the assembly. Instead, it was considered a non-structural veneer—citing the allowance of low strength, reclaimed brick for exterior finishes in residential construction. This meant that the interior wythe had to be sufficient to handle all the structural loading. Figure 38.2 shows a typical wall cross section from the Smyth-Allcott residence. For the height of wall being proposed, 3.66 m, the minimum width of masonry is 200 mm. In order to resist the in-plane shear loads for the location, 15 M rebar was required vertically on 600 mm centres. Horizontal reinforcement was 10 M rebar, also at 600 mm spacing. At the top of the wall, a continuous connection was required, and an engineered wood product was employed to span the full 500 mm width of the wall. An extra run of horizontal rebar was requested at the top of the wall to reinforce the SRE near point and line load locations.

The maximum factored axial load resistance (P_r) for a reinforced masonry wall allowed in CSA S304.1 is given in clause 10.4.1 and reads as follows:

$$P_{r(\max)} = 0.80(0.85\varnothing_m f'_m A_e)$$

where $\varnothing_m = 0.60$, and A_e is the effective cross-sectional area being loaded.

For the Smyth-Allcott residence, this meant a maximum compressive resistance per metre of wall length was calculated to be:

$$P_{r(\max)} = 0.80(0.85 * 0.60 * 10.5\ \mathrm{N/mm^2} * 200\ \mathrm{mm} * 1000\ \mathrm{mm}) = 856.8\ \mathrm{kN}$$

38.3 Discussion and Recommendations

38.3.1 Appropriate Use of Concrete and Masonry Design Standards for SRE

Stabilized rammed earth, built with more than 5% cementitious binder by weight, exhibits similar material characteristics to concrete (Windstorm and Schmidt 2013). However, it is not defensibly equivalent for engineering design purposes in the absence of Canadian research and code referenced design resources. The use of a masonry design standard as the basis for engineering design is not without advantages, but masonry's characteristics are also not directly comparable to those exhibited by SRE. There is considerable international experience and research on the subject (ASTM International 2010), but it is not a simple matter for a Canadian engineer to adopt the results of international work into their designs when being asked to defend their proposal within current building codes.

At this time, an alternative solutions proposal may be requested to justify the use of the masonry design standard for engineering design of SRE walls. If the alternative solutions proposal is accepted, the engineer now has a defensible design path. However, elements of that design may be limited by masonry's material characteristics and the design standard. There are potential situations where design along conventional concrete engineering methods may be more accurate for the actual performance of an SRE wall. It may be possible to apply for separate alternative solutions for different design features, for instance; compliance via masonry design standards for global axial capacity and minimum reinforcement requirements versus compliance via concrete design standards for local bending and moment capacity.

38.3.2 Recommendations for Future Research

Research into ductility (Rd) and over-strength (Ro) values for SRE to use in seismic design is recommended to take advantage of potential ductility advantages that SRE may have over both masonry and concrete structures.

It has been noted that SRE wall systems are not well represented by cylinder type compression test methods (Krayenhoff 2015). The ability to use the more convenient and more commonly practised cylinder test method in a manner that does not unfairly penalize SRE wall systems would be very useful for economical construction with this material in the Canadian context. Recent work by Venkatarama Reddy at the Indian Institute of Science does speak to the equivalence of test results done on cylindrical samples to prismatic samples of similar material (Reddy et al. 2017).

References

ASTM International (2010) Standard guide for design of earthen wall building systems, ASTM E2392/E2392M-10. West Conshohocken, PA, USA

Canadian Commission on Building and Fire Codes (2015) National Building Code of Canada. National Research Council of Canada, Ottawa

Canadian Standards Association (2005) CAN/CSA S304.1-04—design of masonry structures. Canadian Standards Association, Mississauga

Canadian Standards Association (2007) CAN/CSA A23.3-04—design of concrete structures. Canadian Standards Association, Mississauga

Dick KJ, Krahn TJ (2015) Preparing regulatory challenges and opportunities for small to medium scale stabilized rammed earth buildings in Canada. In: Ciancio, Beckett (eds) Rammed earth construction. CRC Press, Taylor & Francis Group, London, pp 93–96

Easton D (2007) The rammed Earth house. Chelsea Green Publishing Company, White River Junction

Holtz RD, Lee WF (2002) Internal stability analyses of geosynthetic reinforced retaining walls. Washington State Department of Transportation, Olympia

King B (1996) Buildings of Earth and Straw. Green Building Press, San Rafael

Krayenhoff M (2015) Rammed Earth in a concrete world. In: Ciancio, Beckett (eds) Rammed Earth construction. CRC Press, Taylor & Francis Group, London, pp 111–114

OBC (2012) Ontario building code. Ministry of Municipal Affairs and Housing, Markham

Reddy BVV, Suresh V, Nanjunda Rao KS (2017) Characteristic compressive strength of cement-stabilized rammed Earth. J Mater Civil Eng 29(2):04016203 (American Society for Civil Engineering)

Standards New Zealand (1998) Earth Building Standard NZS4297-1998, Wellington

Windstorm B, Schmidt A (2013) A report of contemporary rammed Earth construction and research in North America. Sustainability 5(2):400–416. www.mdpi.com/journal/sustainability. ISSN-1050

Chapter 39
Hygrothermal and Hydromechanical Behaviours of Unstabilized Compacted Earth

Antonin Fabbri, Longfei Xu, Henry Wong and Fionn McGregor

39.1 Introduction

Crude earth is composed of clays, silts, sands and possibly gravels and fibres. One particularity of this material is the presence of micro- to nano-connected pores. Thanks to these several levels of porosity, as well as their strong affinity water molecules, which is due to their clayey content (Murad et al. 2012), earthen-based materials are classically classified as good to excellent hygroscopic materials. It denotes their ability to buffer moisture and improve indoor air quality while keeping the internal temperature relatively stable. This affinity with water molecules also significantly impacts their mechanical behaviour. For example (Bui et al. 2014), underlines that an increase of the water content from 2 to 12% leads to divide the compressive strength and the stiffness by, at least, 4 for soils of several compositions compacted according to the proctor procedure (energy of compaction equal to 0.6 kJ/dm^3). Similar results are notably reported notably in Jaquin et al. (2009), Fabbri and Morel (2016) or Gerard et al. (2015). It follows that any abnormal increase of water content would induce too low strength values to allow the wall stability. This collapse induced by an increase of water content is a well-known phenomenon is soil mechanics, and it is taken into account in most of unsaturated elasto-plastic laws (Alonso et al. 1990; Lai et al. 2016, etc.).

A. Fabbri (✉) · L. Xu · H. Wong · F. McGregor
LTDS, UMR5513 CNRS, ENTPE, Université de Lyon, 69100 Vaulx-en-Velin, France
e-mail: antonin.fabbri@entpe.fr

L. Xu
e-mail: longfei.xu@entpe.fr

H. Wong
e-mail: kwaikwan.wong@entpe.fr

F. McGregor
e-mail: fionn.mcgregor@entpe.fr

B. V. V. Reddy et al. (eds.), *Earthen Dwellings and Structures*,
Springer Transactions in Civil and Environmental Engineering,
https://doi.org/10.1007/978-981-13-5883-8_39

In addition to this main disease, during their lifetime, earthen walls have to face important variations of indoor and outdoor relative humidity. These variations may lead to quite important variation of their strength and even induce some shrinkage and swelling processes (Champiré et al. 2016; Bruno 2016). In consequence, the changes in hydromechanical behaviour and/or dimensional variations which may be caused by the natural cycles of water sorption–desorption and/or wetting–drying would be an important concern. To date, there is no, to clear consensus of their impact on the material behaviour in the particular cases of earthen constructions materials. Anyway, it seems quite reasonable to think that the impact of these cycles on the material performance would depend on the nature of the raw earth (mineralogy, grain size distribution) and of the construction technic (CEB, rammed earth, adobe, cob, etc.).

At the macroscopic scale, the importance of these cycles has been studied experimentally by Bui et al. (2009) on different types of unprotected rammed earth walls exposed during 20 years to external climatic solicitations in a wet continental climatic environment (Grenoble, France). This work has eventually demonstrated an extrapolated lifetime which is over than 60 years (mean erosion of 6.4 mm or 1.6% of the thickness of the wall after 20 years of exposure) without any mineral stabilization of the earth.

It follows that a proper assessment of the global performance of earthen materials should take into account these thermomechanical–hydromechanical couplings. In that context, this paper aims at assessing the couplings between the stress, strain, humidity and temperature fields in the partially saturated compacted earthen material through a poromechanical approach. To reach this goal, the impact of hygrothermal processes on the mechanical behaviour (hardening/softening and swelling/shrinkage phenomena) is estimated through experimental investigation on centrimetric samples in a temperature and hygrometry controlled triaxial cell. The obtained relations are implemented in a fully coupled poromechanical approach.

39.2 Hygrothermal Couplings

The first step to understand correctly, the hydromechanical behaviour of earthen material is the precise quantification of the humidity and temperature field. The main equations of hygrothermal couplings which are used to reach this goal are nowadays quite well known by the scientific community. One of the most detailed and comprehensive descriptions on the hygrothermal couplings are provided in Janssen et al. (2007). Today, one of the main goals within this topic concerns the development of accurate and user-friendly software that can predict, with sufficient accuracy, the coupling between mass and heat transfers and their impact on the overall performance of a building. This research activity gives rise to commercial software such as WUFI (Fraunhofer, s.d. IBP/WUFI. [online] Available at: http://www.wufi.de), which can provide reliable results on a wide range of materials and climatic loads but cannot always accurately reproduce the hygrothermal behaviour of unconventional materials such as earth when they are submitted to substantial hygrometry and temperature variations. To overcome some of the restrictions of the WUFI code (e.g. but not limited to, the lack of sorption-desorption hysteresis and dependence on temperature, the quite simple form imposed for the variation of the transport parameters with water content), a fairly high number of codes and procedures have been developed by the scientific community. One of the more visible studies in this field was conducted within the HAMSTAD project (see, e.g. Hagentoft et al. 2004).

At equilibrium, the mass conservation equations that drive the hygrothermal couplings can be restricted to a system of three partial differential equations, two of them based on mass balance equations, and one on the heat balance. There are several ways to write these equations, and one of them is:

$$\frac{\partial m_w}{\partial t} = \underline{\nabla} \cdot \left[D_L \left(\underline{\nabla} P_L - \rho_L \underline{g} \right) + \frac{p_v}{P_G} \frac{M_L}{M_G} D_G \underline{\nabla} P_G + P_G \delta_p \underline{\nabla} \left(\frac{p_v}{P_G} \right) \right] \tag{39.1}$$

$$\frac{\partial m_G}{\partial t} = \underline{\nabla} \cdot [D_G \underline{\nabla} P_G] + \dot{m}_{\rightarrow v} \tag{39.2}$$

$$\rho_d c \frac{\partial T}{\partial t} = \underline{\nabla} \cdot (\langle \lambda \rangle \underline{\nabla} T) - \left(\sum_{i=v,a,L} c_i \underline{\omega}_i \right) \cdot \underline{\nabla} T - \Delta h_v \dot{m}_{\rightarrow v} \tag{39.3}$$

In these equations, m_w is the mass of water per unit of total initial volume (both liquid and vapour), m_G is the mass of gas per unit of total volume (both dry air and vapour), ρ_L is the density of water, ρ_d is the dry density of the material, M_L and M_G are, respectively, the molar mass of water and gas, D_L is the effective liquid water permeability coefficient, D_G is the effective gas permeability coefficient, δ_p is the vapour diffusion coefficient, $\langle \lambda \rangle$ is apparent thermal conductivity of the material, c is the apparent specific heat capacity of the material, $\left(\sum_{i=v,a,L} c_i \underline{\omega}_i \right) \cdot \underline{\nabla} T$ the heat convectively transported by the fluid phases, $\dot{m}_{\rightarrow v}$ is the mass of water which evaporate per unit of total initial volume while $-\Delta h_v \dot{m}_{\rightarrow v}$ denotes the heat consumption or supply due to the water phase change process. Since $\Delta h_v > 0$, heat consumption occurs during evaporation (when $\dot{m}_{\rightarrow v} > 0$) while condensation (when $\dot{m}_{\rightarrow v} < 0$) leads to heat supply. Finally, P_L and P_G are, respectively, the liquid and wet air total pressures (all the pressures are expressed in Pa), T is the temperature (all the temperatures are expressed in Kelvin) and p_v is the partial pressure of vapour. This latter is linked to the relative humidity, denoted by (RH), through the relation:

$$p_v = (RH) p_v^{\text{sat}} \tag{39.4}$$

where p_v^{sat} is the partial pressure at saturation, whose values are quite well tabulated and which is, at first order, only function of the temperature. The gas, liquid and vapour pressures are linked together by the Kelvin's law, which traduces the chemical equilibrium between the liquid water and its vapour:

$$P_G - P_L = -\frac{\rho_L RT}{M_L}\ln\left(\frac{p_v}{p_v^{\text{sat}}}\right) \tag{39.5}$$

Assuming that the heat convectively transported by the in-pore fluid phases is ignored, that the mass variation of vapour is negligible when compared to the mass variation of liquid water and that the impact of evaporation/condensation processes on the overall gas pressure is negligible (the consequence of the dilute mixture assumption of the gas phase), these previous equations combine to provide:

$$\phi_0\rho_d\frac{\partial w_L}{\partial RH}\frac{\partial RH}{\partial t} = \nabla\cdot\left(A_p\nabla P_G + A_T\nabla T + A_{RH}\nabla RH\right) \tag{39.6}$$

$$\rho_G\left[\left(\phi_0 - \frac{\rho_d}{\rho_L}w_L\right)\left(\frac{1}{P_G}\frac{\partial P_G}{\partial t} - \frac{1}{T}\frac{\partial T}{\partial t}\right) + \frac{\partial RH}{\partial t} - \frac{\rho_d}{\rho_L}\frac{\partial w_L}{\partial RH}\frac{\partial RH}{\partial t}\right] = \nabla\cdot\left(D_G\nabla P_G\right) \tag{39.7}$$

$$\rho_d c\frac{\partial T}{\partial t} - \rho_d\Delta h_v\frac{\partial w_L}{\partial RH}\frac{\partial RH}{\partial t} = \nabla\cdot\left(\langle\lambda\rangle\nabla T\right) - \left[\rho_L D_L\left(\frac{RT}{M_L}\left(\frac{\nabla RH}{RH} + \ln(RH)\frac{\nabla T}{T}\right)\right)\right] \tag{39.8}$$

with

$$A_p = D_L + \frac{p_v}{P_G}\left(\frac{M_L}{M_G}D_G - \delta_p\right)$$

$$A_T = \frac{\rho_L R\ln RH}{M_L}D_L + RH\frac{dp_v^{sat}}{dT}\delta_p$$

$$A_T = \frac{\rho_L R\ln RH}{M_L}D_L + RH\frac{dp_v^{sat}}{dT}\delta_p$$

where is w_L is the gravimetric water content, which is assumed to be only a function of the relative humidity (cf. Fabbri et al. 2017), and ϕ_0 is the initial porosity. The variation of w_L with RH is obtained through the sorption–desorption curves. The curves obtained for the tested material using the DVS method are reported in Fig. 39.1. In the range of relative humidity studied in this paper (that is between 35 and 97%), $\partial w_L/\partial HR$ was found to be almost the same either in sorption or in desorption. In consequence, a single bijective function $\partial w_L/\partial HR = f(HR)$, estimated from the sorption curve, was considered.

39.3 Hydromechanical Behaviour

39.3.1 Materials and Methods

The earthen material considered in this study came from an existing construction located in "Rhône-Alpes" region in the south-east of France. Decimetric blocks were sampled from rammed earth walls of this construction during operations of new doors

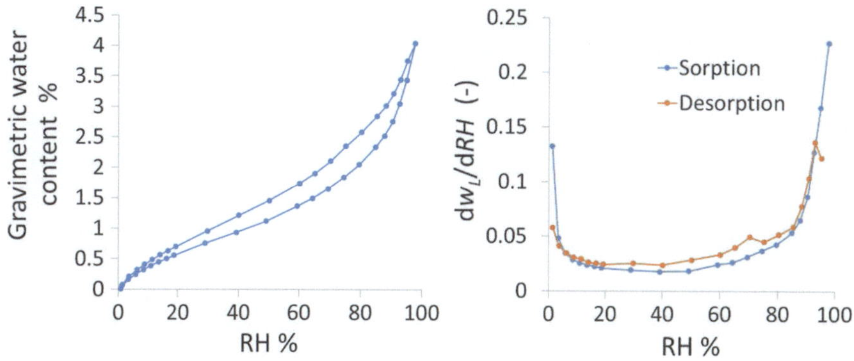

Fig. 39.1 Sorption–desorption curves of the tested compacted earth sample obtained through the DVS method (left) and calculation of $\partial w_L/\partial HR$ (right)

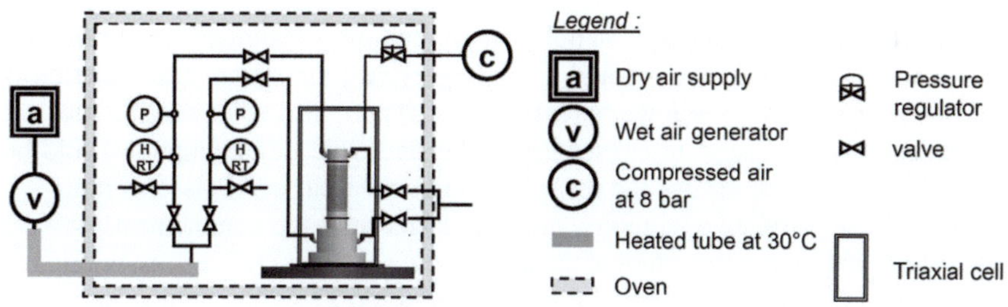

Fig. 39.2 Schematic diagram of the experimental set-up. Depending on opening or closing of the valves, drained, undrained, simple flow or double flows triaxial tests can be realized

and windows opening. This choice ensures that the studied material is suitable for building sustainable earth constructions. The blocks were manually crushed, first with a pick and then with a soft hammer. Triaxial compression tests were performed with the electro-mechanical press (Z020TN, Zwick Roell, Ulm, Germany). The accuracy of the press sensor was about 20 N for the axial load and 2 μm for its displacement. Four non-contact sensors (accuracy of 0.4 μm) were installed on 1/3 and 2/3 of specimen's height, thereby measuring the axial deformation; meanwhile, three non-contact sensors were placed on the middle of the specimen with the included angle of 120° to measure the radial deformation. A schematic representation of the experimental device is reported in Fig. 39.2.

Three tests conditions (unconfined, 1 bar of confinement pressure and 6 bars of confinement pressures) are considered at two temperatures (23 and 30 °C) and 3 values of RH (35, 75 and 97%). $RH = 35\%$ and $RH = 97\%$ were chosen because they correspond, as for us, to upper and lower bounds of the in-pore air relative humidity in normal conditions. $RH = 75\%$ was chosen because it is the relative humidity at which the slope of the sorption curve begins to significantly increase (cf. Fig. 39.1, right).

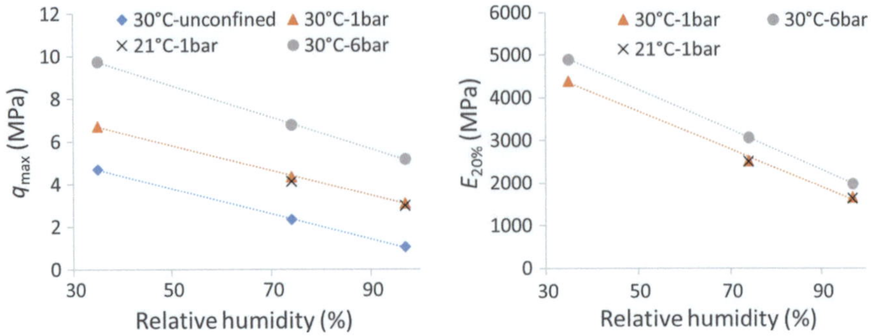

Fig. 39.3 Evolution of the deviatoric strength at failure, q_{max} (left), and of the Young's modulus (right) with relative humidity, temperature and confinement pressure

Whatever the test condition, samples firstly exhibit a contraction, followed by a dilatancy. Under both loading and unloading cycles, the stress–strain relation is almost linear especially at lower stress levels (20 and 40% of deviatoric stress at failure, denoted by q_{max}). In consequence, the linear-elastic assumption could be adopted in order to determine the secants Young's modulus and Poisson's ratio.

An increase of relative humidity and/or a decrease of the confinement pressure induce both an obvious reduction of deviatoric stress at failure (cf. results reported in Fig. 39.3). On its side, temperature difference between 23 and 30 °C does not bring any significant variation. Similarly, the Young's modulus estimated from the loading and unloading cycle at 20% of q_{max} significantly decrease with an increasing relative humidity.

39.3.2 Swelling Behaviour

The swelling behaviour was investigated through two experimental protocols. The first consists in an unjacketed test at increasing imposed relative humidity. This test is called "homogeneous swelling test", since, at each step, the relative humidity within the sample is homogeneous. The second one consists in circulating a moist air of increasing relative humidity through a jacketed sample. In this configuration, called the "continuous swelling test", the sample is under a gradient of relative humidity. To establish the relation between the volumetric strain and the relative humidity, this latter is calculated at the displacement measurement points using the hygrothermal model presented in the previous section. These two experiments lead to the results which are reported in Fig. 39.4.

A good consistency is observed between all the swelling tests. This result gives some confidence on the accuracy of the hygrothermal model which is used to estimate the relative humidity within the sample for the "continuous swelling test".

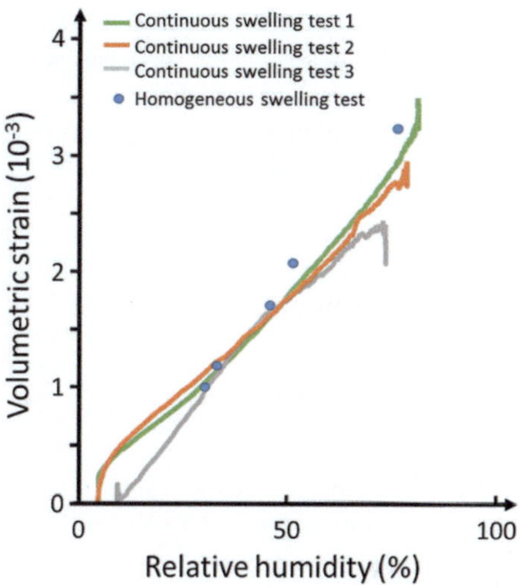

Fig. 39.4 Volumetric strain induced by the increase of relative humidity within earthen samples

These tests underline that the swelling process in the hygroscopic regime of saturation is not negligible. Indeed, this experience leads to swelling strains in the range of 1 mm/m for a variation of relative humidity from 30 to 50%. This phenomenon should thus be taken into account for a proper design of earthen constructions. In addition, while the internal and external sides of an earthen wall are subjected to daily variations of relative humidity, in the middle of the wall this latter is much more stable and it is rather submitted to seasonal variations. An earthen wall is thus submitted to strong gradients of relative humidity during its lifetime, which may lead to non-negligible stresses due to the swelling/shrinkage processes.

Finally, the shape of the relation between the volumetric strain and the relative humidity appears to be almost linear. If the local equilibrium between the adsorbed water, the in-pore capillary water and the water vapour are assumed, this tendency can be reproducing using a Gibbs potential of adsorbed water similar to the one developed by Lei et al. (2014). It leads to the following expression for the swelling strain, denoted by ϵ_v:

$$\mathrm{d}\epsilon_v = \frac{\mathrm{d}m_{\mathrm{ads}}}{\rho_L(1-\phi_0)} \tag{39.9}$$

where ρ_L is the liquid density and ϕ_0 is the initial porosity and m_{ads} is the mass of the interlayer water per unit of material initial volume. It is linked to the relative humidity (denoted by RH) through the relation:

$$m_{\mathrm{ads}} \approx a\,RH \frac{\rho_d^0}{1-a\,RH} \tag{39.10}$$

where ρ_d^0 is the initial dry density and a is a positive dimensionless coefficient between m_{ads}/m_s and 1, whose value increases with the swelling potential of the material. A good correlation is observed between the experimental results of swelling tests and of the theoretical curve obtained from the relations (39.9 and 39.10) for a swelling coefficient $a = 0.0017$.

39.4 Poromechanical Model

39.4.1 Law of Behaviour

The experiments at several values of relative humidity have underlined that the Young's modulus (denoted by E) varies quite significantly with water content, while the Poisson's ratio (denoted by ν) seems to remain almost constant. On the other side, the tests at 23 and 30 °C do not exhibit strong differences in mechanical behaviour. In consequence, using the general framework of poromechanics (Coussy 2010), the following linear law may be proposed for the stress tensor (ΔX stands for $X - X_0$, X_0 being the initial value of X):

$$\Delta\sigma = \frac{E}{\nu + 1}\left(\epsilon + \frac{\nu}{1 - 2\nu}\epsilon_v\delta\right) - \left(\Delta P_\phi + \frac{E}{3(1 - 2\nu)}\left(3\alpha\Delta T + \frac{\Delta m_{ads}}{\rho_L(1 - \phi_0)}\right)\right)\delta \tag{39.11}$$

where $\boldsymbol{\sigma}$ and $\boldsymbol{\epsilon}$ are the stress and strain tensors, P_ϕ is the equivalent pore pressure, α is the thermal dilatation coefficient and $\boldsymbol{\delta}$ is the second order identity tensor.

39.4.2 Comparison with Experimental Data

The model, composed by the conservation equations, is solved with COMSOL Multiphysics (coupling between PDE and linear elasticity modules). Its predictions are compared to swelling experiments at constant temperature under hydric gradients.

The cylindrical sample is initially under a confinement pressure of $p_{\mathrm{conf}} = 1.5$ bar above the atmospheric pressure and equilibrated at a relative humidity of 3% thanks to a dry air flow of $Q = 50$ mL/min through the sample (from the bottom side to the top side). At $t = 0$, the relative humidity of the air injected into the sample increases from 3 to 90% while the flow rate is kept constant. Evolutions of axial and radial displacements are monitored, as well as gas pressures and relative humidity at top and bottom surfaces. The experimental results, in comparison with the numerical application, are reported in Fig. 39.5. All the material parameters used for the simulation have been measured independently.

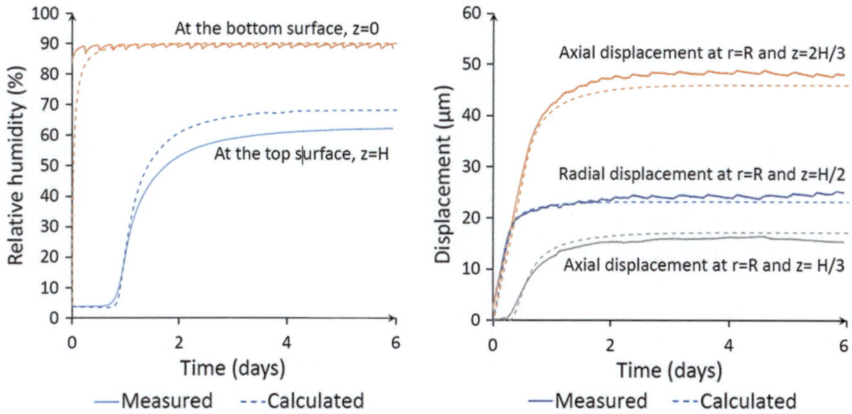

Fig. 39.5 Comparison of the experimental results with model predictions. r and z are, respectively, the radial and the axial coordinates, R is the radius of the sample and H is its length

The poromechanics model developed here leads to consistent results. In particular, the quite good correlation between both relative humidity and deformation evolutions gives some confidence on the ability of the simplistic form given for the interlayer activity to reproduce the good tendencies.

39.5 Conclusion

Hydromechanical triaxial behaviour of compacted earth at different temperatures (23 and 30 °C) was investigated. This study shows that mechanical characteristics of earthen materials have a strong dependence on the relative humidity at which the samples are stored and in parallel yields some preliminary observations on their thermal-mechanical behaviour. In particular, dimensional variations of the sample under variation of air relative humidity have been experimentally investigated, and they were found to be not negligible.

These experimental results were used to develop a thermomechanical–hydromechanical model for unstabilized compacted earth. This former model was made on the framework of poroelasticity and considers the swelling-shrinkage phenomena through a semi-empirical approach. The comparison with experimental data underlines its ability to correctly reproduce the water mass transfer through the porous network and its consequence on the dimensional variations of the material.

Acknowledgements The present work has been supported by the French Research National Agency (ANR) through the "Villes et Bâtiments Durables" program (Project Primaterre no. ANR-12-VBDU-0001). The second author is supported by the China Scholarship Council (CSC) with a Ph.D. Scholarship for his research work.

References

Alonso E, Gens A, Josa A (1990) A constitutive model for partially saturated soils. Géotechnique 40:405–430

Bui QB, Morel JC, Venkatarama Reddy BV, Ghayad W (2009) Durability of Rammed earth walls exposed for 20 years to natural weathering. Build Environ 44:912–919

Bui Q-B, Morel J-C, Hans S, Walker P (2014) Effect of moisture content on the mechanical characteristics of rammed earth. Constr Build Mater 54:163–169

Bruno AW (2016) Hygro-mechanical characterisation of hypercompacted earth for sustainable construction. Ph.D. thesis, Université de Pau et des Pays de l'Adour

Champiré F, Fabbri A, Morel JC, Wong H, McGregor F (2016) Impact of relative humidity on the mechanical behavior of compacted earth as a building material. Constr Build Mater 110:70–78

Coussy O (2010) Mechanics and physics of porous solids. Wiley, New York

Fabbri A, Morel J-C (2016) Earthen materials and constructions (Chap. 10). In: Harries KA, Sharma B (eds) Nonconventional and vernacular construction materials. Woodhead Publishing

Fabbri A, McGregor F, Costa I, Faria P (2017) Effect of temperature on the sorption curves of earthen materials. Mater Struct 50:253

Gerard P, Mahdad M, McCormack AR, François B (2015) A unified failure criterion for unstabilized rammed earth materials upon varying relative humidity conditions. Constr Build Mater 95:437–447

Hagentoft C-EE et al (2004) Assessment method for numerical prediction models for combined heat, air and moisture transfer in building components: benchmarks for one-dimensional cases. J Therm Envel Build Sci 27(4):327–351. https://doi.org/10.1177/1097196304042436

Jaquin PA, Augarde CE, Gallipoli D, Toll DG (2009) The strength of unstabilised rammed earth materials. Géotechnique 59:487–490

Janssen H, Blocken B, Carmeliet J (2007) Conservative modelling of the moisture and heat transfer in building components under atmospheric excitation. Int J Heat Mass Transf 50:1128–1140

Lai B, Wong H, Fabbri A, Branque D (2016) A new constitutive model of unsaturated soils using bounding surface plasticity and a non-associative flow rule. Innov Infrastruct Solut 1:3

Lei X, Wong H, Fabbri A, Limam A, Chang YM (2014) A thermos-chemo-electro-mechanical framework of unsaturated expansive clays. Comput Geotech 62:175–192

Murad M, Bennethum L, Cushman J (2012) A multi-scale theory of swelling porous media: I. Application to one-dimensional consolidation. Transp Porous Media 19:93–122

Author Index

B. V. V. Reddy et al. (eds.), *Earthen Dwellings and Structures*,
Springer Transactions in Civil and Environmental Engineering,
https://doi.org/10.1007/978-981-13-5883-8

MIX
Papier aus verantwortungsvollen Quellen
Paper from responsible sources
FSC® C105338

If you have any concerns about our products,
you can contact us on
ProductSafety@springernature.com

In case Publisher is established outside the EU,
the EU authorized representative is:
Springer Nature Customer Service Center GmbH
Europaplatz 3, 69115 Heidelberg, Germany

Printed by Libri Plureos GmbH
in Hamburg, Germany